Praise for *The Power of the Sea*

"Anyone who appreciates the fact that the sea remains something we cannot control will love this book from Bruce Parker. You will come away with a better understanding of why the sea will leave us in awe till the end of time."

—Jim Cantore, The Weather Channel

"From the ancient tides to ingenious ways of predicting tsunamis and weighing the impact of climate change today, Bruce Parker's *The Power of the Sea* is an engaging and essential history of science. It's also a terrific account of survival on our wild blue planet."

—David Helvarg, author of *Saved by the Sea: A Love Story with Fish*

"Bruce Parker is not only a brilliant scientist but a natural storyteller as well. *The Power of the Sea* presents the destructive nature of ocean waves in human terms. For me, the power of *The Power of the Sea* lies in the compelling personal stories that make the book immensely readable. From Napoleon's near death encounter with a raging Red Sea tide, to the vital importance of predicting tide and swell before the D-Day landings, to the individual acts of heroism during the tragic 2004 Indian Ocean tsunami, to the epic storm surges that continue to flood Bangladesh and Myanmar today, Parker never loses sight of the uneasy alliance between man and sea. All who play or live near the sea should read this book. Be warned, you may never look at the ocean the same way again."

—John Kretschmer, author of *At the Mercy of the Sea* and columnist for *Sailing* magazine

"In a changing world driven by the sea, it is important for everyone to understand how it works. Bruce Parker has blended history and science into a book that clearly and often dramatically explains how and why the sea will affect their lives—now and in the future. This is a must-read for anyone who has ever been awed by the ocean."

—Dan Basta, Director, NOAA Office of National Marine Sanctuaries

"Rarely does a book written by a practicing scientist grab you like this one. Intelligent, accurate, and accessible, *The Power of the Sea* reads like a *Believe It or Not* of aquatic destruction. The largest wave in history? What did the tides have to do with the Normandy invasion? What should we have done about Katrina? For the answers to these and other questions you never thought to ask, read Bruce Parker's wonderful book."

—Richard Ellis, author of *The Empty Ocean* and *Tuna: Love, Death, and Mercury*

"A vivid portrayal of sea disasters and the important role that the ocean has played in so many historic events. An illuminating scientific look at how we have learned to predict such disasters and what still needs to be done to safeguard us from future global calamities."

—Curtis Ebbesmeyer, author of *Flotsametrics and the Floating World*

"The Power of the Sea is the best book I have ever read about tsunamis, storm surges, or rogue waves. It dramatically demonstrates the need to better understand the awesome power of the sea if we are to save lives and property."

—Jerry Schubel, President, Aquarium of the Pacific

"Whether you love history, science, or just want to know how the world shapes our lives, this is both an informative and enjoyable read."

—Margaret Davidson, Director, NOAA Coastal Services Center

"This richly researched, eloquent volume gives the reader a front-row seat where the action plays out. Dr. Parker's thorough knowledge of the subject is abundantly evident and his comfortable and informative style makes this a must-read for anyone interested in the environment. The power of the sea is palpable in Dr. Parker's treatment of a fascinating diversity of historically significant events."

—Richard Spinrad, Director, NOAA's Office of Oceanic and Atmospheric Research and former Technical Director to the Oceanographer of the Navy

"An appealing overview of sea movements. Former National Ocean Service Chief Scientist Bruce Parker . . . mixes hair-raising descriptions of disasters with efforts to understand them, followed by advances, mostly since 1800, in predicting sea movements, a complex process that today involves satellites, supercomputers, and worldwide warning networks. Focusing on water alone—leaving marine life to Rachel Carson and others—the author provides a lucid, original contribution to popular-science writing."

—*Kirkus Reviews*

THE POWER OF THE

THE POWER OF THE
SEA

Tsunamis, Storm Surges, Rogue Waves, and Our Quest to Predict Disasters

BRUCE PARKER

THE POWER OF THE SEA

First published in 2010 by
PALGRAVE MACMILLAN®
in the United States—a division of St. Martin's Press LLC,
175 Fifth Avenue, New York, NY 10010.

Where this book is distributed in the UK, Europe, and the rest of the world, this is by Palgrave Macmillan, a division of Macmillan Publishers Limited, registered in England, company number 785998, of Houndmills, Basingstoke, Hampshire RG21 6XS.

Palgrave Macmillan is the global academic imprint of the above companies and has companies and representatives throughout the world.

Palgrave® and Macmillan® are registered trademarks in the United States, the United Kingdom, Europe, and other countries.

ISBN: 978-0-230-61637-0

Library of Congress Cataloging-in-Publication Data

Parker, Bruce B.
The power of the sea : tsunamis, storm surges, rogue waves, and our quest to predict disasters / Bruce Parker.
p. cm.—(MacSci)
Includes index.
ISBN 978-0-230-61637-0
1. Ocean. 2. Ocean and civilization. 3. Ocean—Environmental aspects.
I. Title.

GC28.P37 2010
551.46—dc22 2010013626

A catalogue record of the book is available from the British Library.

Design by Newgen Imaging Systems (P) Ltd., Chennai, India.

First edition: November 2010

10 9 8 7 6 5 4 3 2 1

Printed in the United States of America.

To Diane

Contents

Figures

INTRODUCTION

When the Sea Turns against Us

Escaping the Sea's Fury through Prediction

When the sea turns its enormous power against us, our best defense is to get out of its way. But to do that we must first be able to *predict* when and where the sea will strike.

If we could have predicted that on December 26, 2004, a tsunami would strike the coasts of the Indian Ocean, 300,000 lives would not have been lost in twelve nations. Even a thirty-minute warning could have saved people from the hundred-foot wall of water that bulldozed entire towns out of existence on the northwest coast of Sumatra. The sudden loss of life on such a terrible scale is difficult to fully comprehend, yet it has happened many times over the centuries. In 1970 a tropical cyclone in the Bay of Bengal generated a storm surge, twenty feet high and a hundred miles wide, that flooded the coast of Bangladesh and drowned more than 300,000 people. As recently as 2008 more than 140,000 were killed by an eighteen-foot storm surge that violently washed over the coast of Burma.[1] If the people had been warned, they could have escaped inland and thousands of lives would have been saved. On the high seas, ninety-foot rogue waves have capsized passenger liners and broken oil tankers in half. Not only were those terrifying waves not predicted, but they were not even believed to be real by most scientists—until recently rogue wave reports were dismissed as exaggerated sea stories. If the seventy-foot rogue wave that struck the *Norwegian Dawn* off Virginia in April 2005 could have been predicted, that luxury liner would have taken a different route and escaped the damage that forced her into port. Even the tide, the most predictable of all ocean phenomena (because of its well-understood generation by the gravitational effects of the moon and sun), can still bring death when tide predictions are ignored—when, for example, a boat gets caught in a violently swirling tidal whirlpool or when a thirty-foot tide comes rushing back over fishermen who stayed too long on mudflats harvesting oysters at low tide.

The power of the sea is even more immense on a global scale, forcing weather patterns around the world and changing our climate. The climatic fluctuation of an El Niño that begins with the warming of the waters off Peru causes heavy rains and floods in some regions of the world and droughts in others. At the end of the nineteenth century two large El Niños caused catastrophic droughts in Asia that led to millions dying in India and China from the resulting famine and disease. More recently the strong 1998 El Niño had major impacts around the world, including torrential rains and mudslides in California, where homes sliding into the sea became front-page news. Our primary defense against the harmful global effects of an El Niño is to be able to predict it and prepare for it before its effects begin to be noticed.

The sea also plays a critical role in long-term climate change—in determining whether our planet will move into an ice age or into a period of extreme warming. During an ice age, mile-thick ice sheets cover huge portions of land and sea level drops hundreds of feet. During a warm period, ice sheets melt and a rising sea level floods coastal lands around the world. Any defense we might have against climate change begins with accurate prediction. If we can predict exactly how future climate will be altered in different parts of the world, then we can prepare for that future. Or perhaps we can even change that future—if indeed, as so many believe, it is our own actions that have been pushing Earth's climate in a dangerous direction. Whether the cause of the recent enhanced global warming is increased carbon dioxide from our use of fossil fuels, or centuries of global deforestation, or the Earth's chaotic response to particular astronomical cycles, or all three, the sea plays a critical role in the climatic changes that are occurring.

It has not been an easy endeavor to develop reliable methods to predict the violent motions of the sea, and throughout history humankind has sought to understand the forces that cause the sea to move. This book is about the scientific journey from our earliest strange ideas about the sea to modern marine prediction using computer models fed with gigabytes of real-time oceanographic data collected from satellites and thousands of land- and sea-based instruments. It relates stories of scientific discoveries interwoven with stories of unpredicted natural disasters. Over the centuries, while scientists and mariners have been trying to learn how to predict the motions of the sea, the sea has killed millions, destroyed untold billions of dollars in property, and changed history. This book includes stories of historical events whose outcomes were determined by the sea's power. The leading characters in these tales include some of the most famous names from history—including Napoleon, Moses, Alexander the Great, Julius Caesar, Columbus, and the U.S.

Marines on Tarawa in World War II. And those who sought ways to predict the sea's movements include some of the most famous names in science—Aristotle, Newton, Laplace, Kelvin, Galileo, Leonardo da Vinci, Benjamin Franklin.

From early humans' first dealings with the sea, whether living by its shores or sailing on its surface, they became all too aware of the sea's fickleness—it was a source of food and the path to exploration, trade, and prosperity, but it also had the power to end life and destroy property at any time without any warning. To live with the sea, people needed to be able to predict when it would rise up against them. But how did one go about predicting what the sea would do? It began with those who lived by the sea recognizing a connection between the movement of the moon and the rise and fall of the ocean's surface. They also realized that the tidal range changed throughout the month with the changing phases of the moon. Early humans learned to predict the tide long before they understood what caused it. Ship captains could thus avoid running their ships aground when they returned to port because they knew when it would be low tide, and Chinese junks could avoid being smashed by a tidal bore barreling up the Qiantang River because they knew when it would come. Mariners also noticed that large waves occurred when the winds blew, so they kept their ships in port when the winds became too strong. But sometimes large waves crashed over their ships even when there was no wind, greatly discouraging those who hoped to predict large waves based on the local wind. When mariners ventured out onto the vast surface of the ocean, they found that at particular locations its waters moved rapidly, pulling their ships along with it. They learned to use these ocean currents to explore a whole new world. Whenever possible, people tried making predictions based on correlations they noticed between episodes of natural destruction and changes in the world that preceded them. And occasionally, when one of those predictions came true, lives were saved.

Prediction is the very essence of science. We do not believe scientific theories unless they can predict specific phenomena. But from a more practical point of view, we use science, and the prediction capability it gives us, to have some control over our lives and to protect us from the environment around us. To accurately predict the sea's actions we had to first learn how the sea works—why it moves like it does in so many varied and complex ways, what drives those movements, and where its power comes from. We learned that ultimately the power of the sea comes from three sources—the sun, the moon, and the Earth. Although the sea stores some of the energy it receives, thus keeping our planet's environment stable and protecting its living creatures from extremes that would kill them, the sea also transmits energy, sometimes concentrating it and amplifying it and at times causing incredible destruction.

Each of the sea's destructive phenomena that we will talk about in this book can be traced back to one of these three energy sources.[2] Storm surges and wind waves originate from the energy of the sun. The winds that generate these surges and waves are produced by the uneven distribution of heat from the equator to the poles, that heat having been derived from sunlight hitting the Earth. In extreme cases winds push a huge volume of water onto the coast in the form of a large storm surge, as when Hurricane Katrina flooded New Orleans. High winds also generate large waves, such as those that threatened the success of the Allied amphibious landing at Normandy in World War II. The tides derive their power from the moon and (to a lesser degree) from the sun through their gravitational effects on the oceans. Twice a day the tides move water into and out of all the bays and estuaries of the world, and what is merely a foot or two of vertical movement in the open ocean is amplified to produce tidal ranges as great as fifty feet in the Bay of Fundy, on Canada's southeast coast. Tsunamis derive their power from energy inside the Earth, the heat from the Earth's formation along with some additional heat from radioactive decay. That heat produces very slow convection in the Earth's mantle, which moves huge crustal plates, their collisions resulting in underwater earthquakes and volcanic eruptions. Both can generate tsunamis, which are very long waves until they reach the shallow water near a coast, at which point they rise up to overwhelm everything in their path, as happened after the great earthquake near Lisbon in 1755 and after the volcanic eruption of Krakatoa in 1883.

That the sea greatly affects the entire world should not come as much of a surprise. The sea covers 70 percent of the Earth's surface. It contains approximately 97 percent of the Earth's water, that amount varying somewhat as sea level drops during glacial periods (when water from the sea goes into snow that accumulates as mile-high glaciers over vast areas of land) and rises during warmer interglacial periods (when water that had been locked up as ice on the land returns to the sea). The sea is the greatest solar collector on Earth and stores heat four thousand times more efficiently than the atmosphere. The sea also stores five hundred times as much carbon as the atmosphere, and it absorbs up to half the carbon dioxide produced by burning fossil fuels. In addition to direct warming by the sun, the oceans of the world receive solar energy from the winds, gravitational energy from the moon, and tectonic energy from the Earth. How these sources of energy produce the movements of the sea is described by the branch of ocean physics known as *hydrodynamics,* which provides the basis for all our prediction techniques. Over the years scientists formulated theories to explain how the sea moves, and they derived mathematical representations of these theories. They also invented instruments to measure the ocean—its temperature, its salinity, the changing height of its surface, the speed and direction of its flows, and the oscillating motion of its waves. In this book we will see an evolution from our earliest attempts at making oceanographic measurements to today's worldwide effort to install and integrate a vast global network of oceanographic sensors—on buoys, on ships, on islands, along coasts, and on satellites, all connected in real time to supercomputers that run prediction models. Ultimately, the size and scope of this task demanded international cooperation and the implementation of the Global Ocean Observing System (GOOS).

Ocean measurements slowly began to show how really complex the sea's motions were. Time and time again scientists had to modify their theories when those theories failed to explain observations made with their instruments. With the exception of tides and ocean currents, scientists would for centuries have very little success in predicting how the sea moved under average conditions, much less during devastating events. The theories eventually formulated to match these measurements would show that the motion of the sea is erratic and chaotic, the regularity of the tide being the only exception. The mathematical equations derived to describe the sea's motions were so complex that they could not be fully solved until computers were invented in the twentieth century. And only then could such computer models be fed with massive amounts of real-time marine observations, these data coming from instruments whose technology was developed only late in that century. As we will see, GOOS is now beginning to provide the real-time data needed by an international array of sophisticated hydrodynamic computer models to make the marine predictions we need. Although this has led to many successes, significant problems still remain in predicting tsunamis and rogue waves and critical aspects of El Niño and climate change.

The sea is a single large, complex geophysical system, but we learned about it in pieces, each new acquired bit of knowledge motivated by our instinct for survival as we searched for ways to predict when and how a particular marine phenomenon would assail us. But these marine phenomena are all interconnected, and they are best studied using an integrated observing system such as GOOS and its U.S. component, the Integrated Ocean Observing System. When fully implemented, GOOS and the ocean models that it supports will be the culmination of centuries of marine scientific research and will finally provide the marine predictions needed around the world. Today more than half the world's population lives near the sea, but even those who live far from its shores are affected by the sea—by the millions of products that come to their country on enormous container ships, by the fish they eat, by the weather that originates over the sea, by the changing climate that is controlled by the sea. Being able to predict what the sea will do, tomorrow or a hundred years from now, affects every person on the planet.

Humankind's earliest success at marine prediction was the ability to roughly predict the tide by watching the moon. And although the tide's astronomical forcing, namely, the periodically varying effects of the moon and sun, makes it unique among oceanographic phenomena, that early success would lay a foundation for all later oceanographic observations and predictions. In centuries past, the consequences of being surprised by the tide could be extremely serious, as a very famous French general once found out, as we will see in the next chapter.

CHAPTER 1

The Earliest Predictions for the Sea

The Tide

On July 25, 1798, Napoleon Bonaparte entered Cairo as the master of all Egypt after defeating the Egyptian Mamluks in the Battle of the Pyramids.[1] Such victories are not always long lasting, however, and Napoleon realized that another battle was imminent, this time against the British and the Turks. While he prepared for that battle, Napoleon immersed himself in the local culture, exploring his newly conquered territory and establishing an institute of arts and sciences at Cairo, staffing it with the 167 scientists and artists he had brought with him from France.[2] In December Napoleon visited Suez to examine the remnants of a canal built many centuries earlier by Egyptian Pharaohs to connect the Nile River with the Red Sea, back when sea level had been higher. He also inspected the site of a planned new canal that was to connect the Mediterranean Sea to the Red Sea. On the morning of December 28 Napoleon intended to take a small band of soldiers to visit the Wells of Moses, on the other side of the Gulf of Suez at the northern end of the Red Sea. The Gulf of Suez extended about three miles farther north of the port of Suez in 1798 than it does today. The point where Napoleon expected to cross the Gulf of Suez was about one mile wide and always dry, or at least fordable, at low tide. Caravans from Tor and Mount Sinai regularly crossed at this spot.

Predicting the time of low tide at that location on the morning of December 28 would have been no problem for Napoleon's scientists. By 1798 tide prediction had finally become a scientific endeavor.[3] But even before then the French, the English, and other peoples living by the sea had developed their own approximate methods for predicting the twice-daily rise and fall of the sea. Tidal ranges were large along all the coasts of France except the

Mediterranean coast. The largest were along the coasts of the English Channel, where the rise from low tide to high tide was greater than twenty feet and at a few locations even as great as forty-five feet. In such areas, beaches that were little more than a few yards wide at high tide could become a mile wide at low tide, six and a quarter hours later. Thus, it had been out of necessity that methods of tide prediction were developed, for such large tides affected the lives of those who lived along the coast, as did strong *tidal currents*, the oscillating horizontal water flow that accompanies the rise and fall of the tide. A fisherman who went to dig up oysters or scallops from mudflats that were uncovered at low tide had to know when the water would come rushing back, or he would drown. A sea captain leaving a harbor had to know the time of high tide, so his ship would not run aground and sink. Or his ship might not be able to buck a strong tidal current flowing into the harbor (referred to as the *flood current*). He would thus have to know the time when the current would reverse and flow outward (the *ebb current*). Monks working in a tide mill, a mill that used the power of the tide to grind wheat, planned their entire work schedule according to the two times of low tide each day, when the tidal power would be maximum.

The older tide-prediction techniques were developed when mariners noticed that the oscillating tide correlated with the phases and movement of the moon. They observed the surface of the sea rising to a highest level and then six and a quarter hours later falling to a lowest level, and then after another six and a quarter hours the sea returned to the highest level again. Then the whole cycle repeated, there being roughly two such cycles in a day.[4] Then they noticed that the height of the high waters changed throughout the month, from cycle to cycle. The highest high waters (as well as the lowest low waters) occurred near the time when the moon was full and round and at its brightest. The *tidal range,* the vertical distance from low water to high water, was largest near this time of full moon, now referred to as *spring tide.* But two weeks later, when the moon was totally dark, referred to as *new moon,* the tide again had its largest tidal range, another spring tide. Those trying to figure out how the moon caused the tides were confused by a dark new moon causing tidal ranges as large as a bright full moon. Surely, they thought, a bigger, brighter moon should be able to pull more water with it than a dark moon, but it didn't work that way. The smallest tidal range, called a *neap tide,* occurred halfway between the times of new moon and full moon, when half the moon shone. Mariners had also observed at night, when the moon rose in the east and moved across the sky, that the sea rose, reaching high water around the time when the moon was highest in the night sky. Then, as the moon fell toward the western horizon, the waters along the shore fell back. Coastal dwellers found that the tide changed with a pattern they could rely on, day after day, month after month, year after year—a pattern that allowed them to roughly predict the changing height of the sea's surface.

For those living on the coast, the timetable of their lives was usually determined by the tide, rather than by the solar timetable of days and nights. The tide has a lunar timetable. The two high waters that occur each day occur a little later than they did the day before, and the next day they are later still. And, of course, the same is true for the low waters, so when oyster fishermen journeyed out to the mudflats at low water, it was a little later each day. On a particular day the first low water might be at noon, but the next day it would be almost an hour later, and the day after that almost another hour later. After six or seven days, low water would be in the evening and the fishermen would be digging up oysters in the dark, or they would shift to the next tidal cycle and dig up oysters early the next morning.

Most French people, including Napoleon, had heard the stories about the dangerous tides near the cone-shaped rocky isle of Mont-Saint-Michel at the southeastern end of the Gulf of St. Malo, which is connected to the English Channel. The tide at Mont-Saint-Michel rises forty-five feet from low water to high water in roughly six and a quarter hours, which means that the water rises more than seven feet every hour. That doesn't leave a fisherman much time to dig oysters on mudflats revealed at low tide and safely leave before the sea returns. But it's not just how fast the water rises that is of concern; it's also the way the sea returns to cover the mudflats. Throughout the centuries the locals living along this French coast have described the incoming tide as arriving like galloping horses, swiftly encircling the Benedictine abbey that has been on Mont-Saint-Michel since AD 708. Fishermen caught on this expanse of mudflats at the wrong time found themselves suddenly surrounded by water rushing in at them on all sides. Many drowned.

The tidal range at the northern end of the Red Sea was not nearly as large as the tidal range near Mont-Saint-Michel, but even an eight-foot range could cause problems for one caught on a mud or a sand flat when the tide comes in. So an accurate tide prediction was important for Napoleon. When he and his men reached the shore of the Gulf of Suez at the scheduled time, he found a mile-long expanse of sea bottom exposed at low tide as his scientists had predicted. His small band of soldiers on horseback easily crossed the dry flats to the other coast before the tide began to rise.

Napoleon reached the Wells of Moses in good time. He stayed throughout the afternoon, meeting with Cenobites from the convent of Mount Sinai and with some Arabian chiefs from Tor. Then late in the afternoon he and his men left and began their return trip to Suez. The sun had set by the time Napoleon and his soldiers reached the seashore. The tide seemed to be out far enough for them to begin crossing the exposed sea bottom. But the sea bottom did not stay exposed for long. Suddenly the tide began rushing in at them, seemingly from all directions. Surrounded by rapidly rising water, and with darkness adding to their confusion, they were thrown into disorder and panic. They could not see a shoreline in any direction. As the tide rose, the water quickly became deeper and threatened to engulf them. Their only chance was to find a shoal where the water might still be shallow enough to walk on. Napoleon calmed his men and ordered them to form concentric circles around him, each horseman facing outward as part of several straight lines pointing in different directions, like the spokes of a wheel. He then ordered each line of horsemen to advance outward. When the lead horse of a line reached deeper water and had to begin desperately swimming, that column drew back and followed one of the columns still walking on the sea bottom. Eventually, each of the columns lost their footing until only one remained, which everyone followed to an ultimate escape from the Red Sea. In spite of the rising water surface on which wind waves rode and crashed against them in the darkness, only one of Napoleon's contingent came close to being lost, a general with a wooden leg who had trouble sitting firmly on his horse with water up to his waist.[5]

Back on shore Napoleon is said to have remarked, "Had I perished in that manner, like Pharaoh, it would have furnished all the preachers of Christendom with a magnificent text against me."[6] He was referring to the famous Exodus from Egypt by the Children of Israel led by Moses. Napoleon had been told that this route was purportedly the same route across the Red Sea used by the Israelites.

Three thousand years earlier the Children of Israel had been camped on the shore of the Gulf of Suez at the northern end of the Red Sea, or the Sea of Reeds, as many believe it was called back then, because of the reedy marshes growing in brackish water at its northern end.[7] Their location was probably farther north than where Napoleon almost met his demise, because in the time of Moses sea level was higher than it was in Napoleon's time. The Israelites had left Egypt behind and, they hoped, their lives as slaves under the rule of Pharaoh. But Moses knew that Pharaoh would send his army after them. When the dust clouds raised by their chariots were finally seen, still miles in the distance, they filled the Israelites with fear, for now they were trapped between Pharaoh's army and the Sea of Reeds.

The dust clouds from Pharaoh's chariots were, however, probably an important part of Moses' plan, for they would have allowed him to calculate how soon Pharaoh's army would arrive at the seacoast. For Moses surely had a plan. One does not attempt such a massive undertaking as leading an entire people into freedom without a plan, even if hoping for help from on high. And Moses must have known that his key to success was timing—exact timing. In his earlier years Moses had lived in the wilderness.[8] He knew the area by the Sea of Reeds. He knew the night sky. And he must have known where caravans crossed the Sea of Reeds at low tide. Pharaoh, on the other hand, living along the Nile River, which was connected to the almost tideless Mediterranean Sea, probably had little experience with the tide in the Sea of

Reeds. It would not have occurred to him that the tide would soon wash over what had been the Israelites' path to freedom.

Knowing when low tide would occur, how long the sea bottom would remain dry, and when the waters would rush back in, Moses could plan an escape across the Sea of Reeds that took advantage of the tide. Choosing a full moon for their escape (perhaps merely to light their way on a nighttime crossing) would have given them a spring tide and a larger tidal range. That would have meant a lower low water, and thus a longer-lasting dry sea bottom and more time for the Israelites to cross, followed by a higher high water to better engulf Pharaoh's pursuing army. Timing was crucial. The last of the Israelites had to cross the dry sea bottom just before the tide returned, enticing Pharaoh's army of chariots onto the exposed sea bottom where they would drown in the returning tidal waters.[9]

The Bible mentions a strong east wind that blew all night and pushed back the waters. Ocean physics tells us that wind blowing over a shallow waterway pushes back more water than a wind blowing over a deep waterway.[10] Thus, if a wind did by chance fortuitously blow as the Israelites were crossing the Red Sea, it would have had more effect at low water than at any other time, uncovering even more sea bottom.[11] Such a wind would surely have been assigned to divine intervention, and as the story of the Exodus was retold, that aspect would have overshadowed Moses' planning their escape to take advantage of a predicted low tide. But Moses could not have predicted a suddenly beneficial wind, and therefore he could not have based his plan on such a wind. He could have predicted only the tide. And thus the successful Exodus of the Israelites would have depended on timing based on that tide prediction.

Today at the northern end of the Gulf of Suez the spring tidal range averages just under five feet and reaches six feet at certain times of the year, not counting wind effects. Tide measurements did not begin in the Red Sea until the 1890s, but a hundred years earlier, at the time of Napoleon's escape from the onrushing tide, his private secretary and the author of his memoirs, Louis de Bourrienne, said that at high tide the water rose five or six feet and as much as nine or ten feet with the wind blowing in the right direction.[12] We know from astronomical calculations that the tide that engulfed Napoleon and his men was only a neap tide. The etching in figure 1.1 is correct; the moon was in its third quarter. Napoleon might not have escaped the Red Sea if there had been a nearly full or new moon, which brings the larger spring tide. Napoleon's sudden encounter with the onrushing tide was shocking enough, and such a relatively small tide during the Exodus might have been enough to defeat Pharaoh's army, especially with chariot wheels getting stuck in wet sand. If that is what happened, the Exodus story would have certainly become more and more dramatic with each retelling, until it included walls of water drowning Pharaoh's army of chariots.

Figure 1.1 Napoleon's escape from the Red Sea on December 28, 1798. The moon pictured in this etching was correct, meaning that it was a neap tide. (*Harpers New Monthly Magazine, 1852*)

However, there is a very good chance that the tidal range at the northernmost end of the Red Sea was larger in Moses' time. Three thousand years before Napoleon, at the time the Exodus is believed to have taken place, there is evidence that sea level was higher than it is now and that the Gulf of Suez extended farther north.[13] A lengthened Gulf of Suez, or a smaller basin connected to the Gulf of Suez, would most likely have increased the amplification of the tide, producing tidal ranges at the time of the Exodus that were larger than those seen today or in Napoleon's time. In that case the Red Sea event wouldn't have needed that much exaggeration as it was passed down from generation to generation and then finally written into the Bible (with different pieces of the story contributed by at least three different authors, we are told by biblical scholars).[14] If the tide was indeed involved in Moses' "parting" of the Red Sea, one might rightly say that this was the most dramatic tide prediction in history.

But as it turns out, my suggestion that Moses might have planned to cross the Sea of Reeds at the predicted time of a low tide turns out not to be a new suggestion. In *Praeparatio Evangelica* (*Preparation for the Gospel*), Eusebius of Caesarea (ca. AD 263–339) quotes from a book by the Hellenistic historian Artapanus (80–40 BC). In that quote Artapanus gives two versions of the crossing of the Red Sea. One is a story similar to what appears in the Bible, which he attributes to the people of Heliopolis. The second version he summarizes thus: "Now the people of Memphis say, that Moses being acquainted with the country waited for the ebb, and took the people across the sea when dry."[15]

As one might expect, the earliest written references to the tides are from geographic areas where large tidal ranges coincided with early civilizations. Along the west coast of India in the Gulf of Cambay north of what is today Mumbai, formerly Bombay, the tidal range can reach over thirty feet. Farther north in the Gulf of Kutch the range is still over twenty feet. The ancient Harappan civilization that developed in the Indus Valley by 2300 BC reached south to include these two areas. It is not surprising then that the first known written reference to the tides was in the *Sāmaveda,* a series of ancient Indian hymns probably written around 1100 BC.[16] Not only are the tides described, but their main cause is correctly attributed to the moon—and with this understanding they were most likely able to crudely predict the tide. There is, however, evidence that the Harappans were well aware of the tides a thousand years before the *Sāmaveda* was written. The Archaeological Survey of India carried out excavations at Lothal, at the northern end of the Gulf of Cambay, that revealed a dockyard for berthing ships that was built around 2300 BC. The amazing thing about this dockyard was its lock system. A ship would enter the channel at high tide, at which time a wooden door was inserted in grooves on both sides of the channel and lowered to close off the channel, keeping the water in the basin as the tide fell and thus allowing the ship to stay afloat. The Harappans who lived farther north, in the Indus Valley, were also familiar with a dramatic manifestation of the tide, the tidal bore. Twice a day a tumultuous wall of water charged up the western branch of the river—a distorted, perpetually breaking wave, which, as we shall see shortly, eight hundred years later would come as an almost fatal surprise to a famous visitor from the Mediterranean Sea.

It may at first seem surprising that the early philosophers in the civilizations around the Mediterranean Sea did not know about the tide. The Mediterranean Sea, however, has such a small tide that it is easily hidden by the random water level fluctuations caused by the winds.[17] The Greeks did not learn about the tide until 450 BC, when Herodotus traveled to the Red Sea and observed the tides there.[18] It was a century later that Pytheas of Massilia corroborated the existence of tides when he traveled to the Atlantic Ocean and witnessed the huge tidal ranges along the coasts of the British Isles.[19] Still another century later, Seleucus, a Hellenistic mathematician from Babylonia, observed the tides in the Persian Gulf and might have been the first person to make regular oceanographic measurements. He apparently tabulated the times and heights of high and low waters over many days and even months, noting the changes in these high and low waters from tidal cycle to tidal cycle. From these data he was the first to recognize that the two high tides on any given day could be quite different in height and that the greatest difference occurred when the moon was farthest north or south of the equator.[20]

But recognizing that the tide exists, and in some cases even seeing a connection with the moon, which allowed a crude prediction capability, does not mean that anyone knew how the tides were produced. There were many conjectures, some of which seem comical today, even though suggested by some of the most famous Greek and Roman thinkers.[21] Plato believed that the Earth was a large animal and that the tide resulted from the oscillations of fluid within this Earth animal, an idea still being promoted four hundred years later by Apollonius of Tyana, although he said the tide was due to either the Earth animal's breathing or to its drinking and spitting out water. Aristotle, a student of Plato, rejected this Earth-animal theory and proposed instead that the tide was caused by winds produced by the sun or the moon striking the water. Around this same time Timaeus suggested that the tides were caused by rivers discharging their waters into the sea. Still other Greek and Roman philosophers were less imaginative and simply said the tide was caused by the gods. Even Seleucus, who understood the pattern of the tides and its connection to the moon better than any ancient thinker, did not know how the moon caused the tides. His best guess seems to have been that the moon compressed the atmosphere, which pushed down on the sea, a theory that would be proposed again seventeen centuries later.

The tide was probably unknown to Alexander the Great in July 325 BC as he stood on the banks of the Indus River in western ancient India as the most powerful man in the world.[22] Eight years and three thousand miles earlier, Alexander's army had left his home in Macedonia near the shores of the Mediterranean Sea and fought its way through Asia. His military conquests had just culminated in the defeat of the army of King Porus and his terrifying elephants. Standing there in the city of Pattala, in what is now Pakistan, looking at the waters of the Indus, Alexander could never have guessed what the tide had in store for him. Even the most powerful man in the world can still be at the mercy of nature, or as he might have regarded it, at the mercy of the gods to whom the unknown aspects of nature were usually attributed. Even if Alexander had acquired from his teacher Aristotle some feeling for a physical world not simply controlled by gods on Mount Olympus, he would still have had no clue about what he was about to face in the Indus River. Nor would he have dreamed that the tide would almost accomplish what all the armies of Egypt, Persia, Tyre, and India had been unable to do.

After eight years of war Alexander had finally decided to go home, because his army was exhausted. Though victorious once again, they had been especially shaken by their last battle against the Indian elephants. They were, in fact, on the verge of mutiny. But before Alexander began his return journey to Macedonia, he wanted to do some final exploration, for Alexander had acquired from Aristotle and his other teachers a curiosity about the world and about nature and a desire to explore and learn. When he had left Macedonia on this journey of conquest, he brought with him a host of surveyors, geographers, botanists, zoologists, and other scientists. Once he returned, the information they had acquired would become the basis for several scientific works published by Greek and Roman philosophers. Of course, these geographers and scientists were also useful for other more immediate and practical purposes, such as being sent ahead on scouting missions into the new lands that Alexander was entering. Their maps and other information were very useful for his battle plans.

At Pattala the Indus River split into two main branches, and Alexander decided to journey down the western branch. Leaving some of his army behind to construct a harbor and dockyard, he took the rest downriver on his fastest ships. But he lacked a local pilot to guide them, all his previous local guides having managed to escape, and this time he had not sent his scientists ahead to scout the river, since he was not expecting another battle. Alexander thus headed down this unknown river with an uncharacteristic lack of preparation, because of an "obsessive desire to see the Ocean and reach the ends of the earth."[23] After traveling four hundred *stades* (about forty-six miles), he stopped to capture some locals and ask them how far the sea was. The locals replied that they "had not even heard of the sea but that after three days one would reach bitter-tasting water which spoiled the fresh water."[24] Alexander knew that the salt water of which they spoke meant he was heading in the right direction. They rowed on for three more days before Alexander finally ordered them to moor their vessels at an island in the middle of the river and separate into parties to forage.

Suddenly, to everyone's shock, the river began to flow backward. The water that now was furiously flowing up the river smelled of the sea. But then an even bigger and truly frightening surprise occurred—a turbulent and thunderous wall of water came barreling up the river at them. A single steep wave crest, covering the entire width of the river and sounding like a herd of the Indian elephants that they had just fought, crashed into their ships and flooded the land. Having never seen anything like it, the Macedonians "thought they were witnessing prodigies and signs of heaven's displeasure."[25] The ships were lifted by the tide, scattering the entire fleet. Men who had been put ashore desperately tried to get back to their vessels, but many vessels had already left their moorings when their officers were stricken with panic. Other vessels crashed into each other. The rising waters quickly inundated all the surrounding shores, leaving only the tops of knolls like little islands to which the remaining men desperately swam. But then just as suddenly the waters began to withdraw. The receding waters now violently drained the previously flooded areas of land, leaving shipwrecks on dry land

and the fields strewn with baggage, arms, pieces of oars, loose planks, and some types of sea life that to the Macedonians looked like terrifying monsters.

As night approached, Alexander sent horsemen downriver to provide a warning in case another wall of water attacked them. His men began repairing the ships, all the while listening for the roar of the wall of water. After a while, about twelve and a half hours after the first wave had come, they heard the hoofbeats of racing horses. The horsemen were riding as fast as they could, trying to stay ahead of another approaching wave. Luckily this time the moving wall of water was not as large, and the "shore and banks rang with cheers from the soldiers and sailors as they welcomed their unexpected rescue with exuberant joy."[26]

Alexander and his army repaired their ships and eventually reached the sea, now called the Arabian Sea, part of the Indian Ocean. Here Alexander sacrificed to Neptune and the local deities and then began his long trek home to Macedonia. We do not know whether Alexander, once he reached the Arabian Sea, noticed the large tidal range along its shores, which can reach twelve feet today. Such a tidal range was something Alexander would never have witnessed in the Mediterranean Sea, and he would have had no way of knowing that the turbulent wall of water that raced up the Indus River from the sea was a tidal bore.

To understand how a tidal bore is produced in a river like the Indus, one must first understand the dynamics of the tide in the world's oceans. In the ocean the tide is actually a very long wave, which, as we will see later in this chapter, is generated by the gravitational forces of the moon and the sun.[27] When the crest of this long wave reaches the coast, we have high tide, and when the trough reaches the coast, we have low tide. An observer on shore does not easily recognize that the tide is a wave, because its *wavelength*—the distance from one wave crest to the next—is so long, hundreds or even thousands of miles. It also takes a long time to go from high water to low water and back to high water, approximately twelve hours and twenty-five minutes, referred to as the *tidal period*. This very long tide wave travels along the coast and up rivers with a speed that depends on the depth of the waterway, traveling faster in deep water and slower in shallow water. Shallow water not only slows the tide wave but also shortens the tidal wavelength. Shallow water also normally wears down the height of the tide wave, because energy is lost due to the friction of the moving water of the tidal current rubbing along the sea bottom.[28] But a river whose width narrows as the tide wave moves up it has the opposite effect. The tide wave's height increases because more water is being forced through a smaller area as the wave moves into the narrower parts of the river. If the river narrows quickly over a short distance, the amplification of the tide can be large enough to counteract the effect of friction trying to wear the wave down. Another effect of shallow water explains the shape of the tidal bore. In shallow water the depth under the crest of the wave (high tide) is greater than the depth under the trough of the wave (low tide). This means that in shallow water the wave crest travels faster than the wave trough. Thus, as the wave travels, each wave crest gets closer to the wave trough in front of it, while that wave trough falls behind the crest in front of it. This leads to water piling up at each crest, increasing its height, and making it very steep. After traveling far enough in shallow water, the shape of the wave thus becomes distorted compared to the simple sinusoidal shape it had in the deep ocean. Each crest (high water) eventually looks like a steep, perpetually breaking wave, in fact, like a wall of turbulent water moving up the river. If there is a large tidal range at the mouth of the river, then a very large tidal bore will be produced upriver. This is what happened in the Indus River in 325 BC when Alexander was hit by the tidal bore. Twice a day, tidal bores arrive like clockwork, and each bore's arrival at particular points along the river is very predictable, like all tidal phenomena. But, of course, such predictability is possible only if one has already seen the bore enough times to have noticed that its arrival correlates with the movement of the moon and its height correlates with the phases of the moon, that is, the bore is larger at new and full moons. Tidal bores have occurred in many other rivers of the world, and shortly we will look at a famous one in China.

As we have mentioned, the Greeks did not have much experience with the tide. Alexander's teacher, Aristotle, was apparently so frustrated by his complete lack of understanding of the tide that it might have provoked his death. However, there are at least three versions of the story dealing with Aristotle's supposed anxiety over not being able to determine the cause of the reversing currents in the Euripus, a narrow shallow strait between the east coast of Greece and the island of Euboia. We now know that these reversing currents are tidal currents, even though the tide at both ends of the strait, in the Aegean Sea, is small enough to be unnoticeable. The small differences in the tidal ranges and times of high water at both ends of the Euripus are enough to drive tidal currents within the strait, which become fairly fast—six miles per hour, a significant speed to swim or row against—where the strait is very narrow and shallow. But this was not obvious to Aristotle or to any of the other Greek philosophers, and so they could not predict which direction the waters would flow at any given time.

The first version of Aristotle's story was given by Procopius in *History of the Wars*. He wrote that Aristotle "observed the strait which they call Euripus in an effort to discover the physical reason why sometimes the current flows from the west, but at other times from the east," and that Aristotle said that it was not connected with changes in the wind. He goes on to say that Aristotle "observed and pondered a long time, until he worried himself to death with anxious thought and so reached the term of his life."[29] In the second version the Roman Justin Martyr made the story more dramatic. In his hortatory address to the Greeks in AD 150 he said that Aristotle "departed this life because he was overwhelmed with the infamy and disgrace of being unable to discover even the nature of the Euripus." And finally, in the third version Elias the Cretan made the story theatrical. He said that when Aristotle failed to understand the Euripus currents, he threw himself into the water and drowned, saying, "Comprehend me because I did not comprehend you."[30] All these stories might have started because late in his life Aristotle sought refuge and lived in the city of Halkida (Chalkida) near Euripus, where he eventually died. The confusion over the changing currents of Euripus became infamous enough to enter the vernacular of the ancient Greeks. Those of unsteady character and changing mood were often called "Euripus."

The tide presented other dangers to ancient peoples besides tidal bores and the relatively fast tidal currents in Euripus. The speeds of currents in Euripus were considerably slower than those in other narrow straits. Some have speeds comparable to a raging river, such as the twenty-mile-per-hour currents in Saltstraumen Strait, on the western coast of Norway, and in Seymour Narrows, between Vancouver Island and the mainland of British Columbia, Canada, or the twelve-mile-per-hour tidal currents in Kanmon Strait, Japan. Such current speeds can swamp a boat unlucky enough to have its anchor line tied to its aft instead of its bow or can carry swimmers helplessly downstream, or upstream as the case may be. Only during *slack water,* the momentary still water during the current reversal from flood current to ebb current, or vice versa, would there be a chance of escape for a swimmer. (Offshore, in the ocean and in wide bays, tidal currents do not have this reversing flow. They instead have a flow that slowly changes direction, rotating around the compass over one tidal cycle. These *rotary* tidal currents do not have a slack water.[31])

But the most dangerous manifestation of fast tidal currents is the *tidal whirlpool*, also called a *maelstrom,* a violently rotating funnel-shaped hole of water that can suck ships underwater. The most famous tidal whirlpool in history is in the Strait of Messina, a narrow and shallow waterway between Sicily and the southern tip of the Italian mainland.[32] The whirlpool occurs a short time after the tidal current reverses direction and is caused by the sudden width change at the northern end of the strait near Cape Peloro.[33] It was only natural that myths and legends would surround a violent whirlpool that appeared to eat ships. This tidal whirlpool was made famous by Homer in his *Odyssey* and was so famous that it also appeared in Virgil's *Aeneid* and in Apollonius's *Argonautica.* Homer represented the whirlpool as the second of the two monsters Ulysses faced, Scylla and Charybdis. Charybdis alternately sucks down black water and belches it back out. (The water actually does have a dark color due to decayed nutrients and dead creatures from the deep waters of the Ionian Sea brought to the surface by the upwelling caused by the tidal currents hitting the shallow sill of the strait.)[34] In Homer's time the idea that the reversing flows and churning dark waters were caused by the intake and outflow of water from a monster's mouth was as good an explanation as anything else, and it certainly had a dramatic flare. In the *Odyssey* the sorceress Circe tells Ulysses that to avoid being sucked into Charybdis he must have his men row as fast as they can along its edge, advice that also made sense from a navigational point of view. In this story, however, she also advised this method because they would be rowing past Scylla, the six-headed monster, who ended up grabbing and eating some of Ulysses' men. Outside of the myths and legends, however, experienced navigators, even centuries ago, learned how to predict when the Charybdis whirlpool would occur, it being a tidal phenomenon whose occurrence could be correlated to the movement of the moon and whose strength could be correlated with the phases of the moon.

Once the Greeks and especially the Romans ventured to other parts of the known world beyond the Mediterranean, predicting tidal heights became important to them. Julius Caesar had to cope with the twenty- to forty-foot tidal ranges of the English Channel during his conquests of Gaul and England. In his *Commentaries about the Gallic War,* Caesar wrote that in 56 BC the Veneti, a powerful seafaring tribe who lived in what is now Brittany, took advantage of the high tides in defending their coastal towns. Built on the ends of headlands, they were approachable from land only at low tide and were difficult to approach by sea because of the strong tidal currents.[35] In 55 BC Caesar crossed the English Channel with warships and troop transports powered by hundreds of rowers, his intent being to conquer England. This was probably the first military amphibious landing in history where the tide would play an important role. Caesar's naval experience had been in the tideless Mediterranean Sea, where he could run his ships onto the beach and not worry about losing them (unless they were captured by another military force with the necessary rowers and maritime knowledge). He quickly learned, however, that along the English coast if he did not beach his ships at high tide, the rising sea would float them, allowing them to crash into each other if there was any kind of wind, or they might even float away. To avoid the necessity of leaving many men on board at all times, he made sure to beach his ships at high tide. But Caesar had not learned that the height of high tide changes throughout the month, being highest near times of full moon or new moon (spring tide) and lowest when the moon is in the first or third quarter (neap tide). Unfortunately Caesar had unknowingly selected a neap high tide to beach his warships and to anchor his troop transports in shallow water nearby. About seven days later, when there was a full moon and thus a spring high tide, a storm arose at night, which because of the higher high tide was able to waterlog some of the beached ships and to cause others to refloat or break their short anchor lines, these ships crashing into each other and sinking. Caesar's men were eventually able to repair only enough ships to take them back across the English Channel before winter set in.[36] He returned the following year and did not make the same mistake, his tide-prediction skills now including knowledge of the higher tides at full and new moons.

There would be no further advancements in tide prediction over the next thousand years throughout western Europe. During the Dark Ages only a few books were written on tides, primarily by religious writers who simply repeated information found in ancient works such as Strabo's *Geography* and Pliny the Elder's *Natural History*.[37] Of course, the twenty- to forty-foot tidal ranges along most British coasts continued to affect everyday lives, especially those who made their living from the sea, but the tides sometimes played a prominent role in more dramatic events. During the Battle of Maldon in 991, it was a high tide that temporarily protected Anglo-Saxons on Northey Island from being attacked by Vikings on the mainland and almost saved the day.[38] In 1020 King Canute the Great, the Viking king who ruled Britain, was told by flatterers that his power was so great that he could command the mighty tides of the sea to go back. But Canute, a newly devout Christian, had his throne carried to the seashore and unsuccessfully commanded the tide to stop to make his point that the power of kings was nothing compared with the power of God (who apparently was the only one who could stop the powerful tide). In September 1066 it would be the Normans' turn to conquer England, and in preparation for the Battle of Hastings, William the Conqueror studied Caesar's invasions of 55 and 54 BC. Learning from Caesar's mistakes, he beached his ships as far up on the shore as he could get them during high tide to avoid the loss of ships that Caesar had suffered during the higher high tides near full moon. Nine years earlier William had used the tide to his advantage in defeating King Henry I of France and Count Geoffrey at the crossing of the River Dives. The rapidly incoming tide cut Henry's army in two, allowing William to easily slaughter the separated halves.[39] In October 1216, the tide might have indirectly killed another king, John, the unwilling signer of the Magna Carta. His army was crossing the Wash, a body of water connected to the North Sea, when it was surprised by a high tide that washed away all his baggage, including all his treasure. Historians of the time said that because of this event King John became very upset, contracted dysentery, and died a week later in Newark, to be succeeded by nine-year-old Henry III.

In retrospect, it is somewhat surprising that King John would be caught unaware by the tide. By that time England was a maritime nation and the English had devised various techniques to predict the tides, some of them prized family secrets passed down from generation to generation. In fact, the first printed tide-prediction table in England (and in Europe), for London

Bridge, was published a few years before King John's problem with the tide. That English tide table, however, was not the world's first tide table. That honor goes to one produced at least two hundred years earlier which was used to predict the arrival of the tidal bore in the Qiantang River near Hangzhou, China.[40]

Those who lived next to or on the Qiantang River were well aware of the danger of the tidal bore, which came up the river twice a day, although some days it was much larger and more dangerous. This huge wall of turbulent water, a two-mile-wide white wave crest, moved swiftly up the river and capsized Chinese junks not in the safety of special shelters built as part of the stone seawall along the shoreline.[41] The thunderous roar that preceded the bore gave some warning, but the tide-prediction table provided more advanced warning and gave people more time to get out of the bore's way.[42] This was especially needed at spring tides, when the bore rose to twenty-five feet and traveled fifteen miles per hour up the river. Its large height was due to the amplification caused by the great reduction in river width, from fifty-five miles at its mouth down to only a few miles near Hangzhou.

The tide table was originally carved in stone on the Zhejiang Ting pavilion on the bank of the Qiantang at Yanguan, where the bore could be quite large. In AD 1056 a printed version appeared.[43] There is some evidence that Chinese tide-prediction tables might have been in use by AD 850, but these tables have never been found and the tide table from 1056 is the earliest unearthed. By the time it was produced, Chinese knowledge had progressed significantly from earlier ideas about the tide. One popular view had been that the tide was the result of a whale or sea serpent going into and coming out of its cave. Another was that the tides were due to the overflowing of the Milky Way as it went from the heavens above to under the sea.[44] The Chinese had probably known by the second century BC that the bore was largest at full moon, and by the first century Wang Chung had written that the tide was connected with the moon and the sun and that he believed that the extreme nature of the tidal bore was in some way due to the narrowness and shallowness of the river into which the tide flowed.[45] By the time the 1056 tide table was produced, Chinese scientists had recognized that the tides are different when the sun is north of the equator (summer in the Northern Hemisphere), when it is south of the equator (winter), or when it is over the equator (spring or autumn).[46] The tide table for the bore thus consisted of three tables, one to be used in the summer, one for winter, and one for spring or fall.[47] Each table gave the times, to the nearest two hours, for two daily occurrences of the tidal bore in terms of the "twelve earthly branches," which were twelve two-hour time intervals starting at 11:00 P.M., each branch named for an animal—rat, ox, tiger, hare, dragon, snake, horse, sheep, monkey, cock, dog, and pig. The heights of the tidal bore, from lowest to highest, were represented by seven characters that translate as lowest, very low, low, fairly high, high, very high, and highest.

Although Chinese scientists knew that the bore was caused by the tide and could even predict it, the Chinese populace preferred to look to legends to explain the bore. The most popular legend in China about the bore was that centuries earlier, around 484 BC, a virtuous minister named Wu Tzu-Hsu was unjustly forced to commit suicide because of supposed disloyalty to the king of Wu, who threw the minister's body into the river. As revenge, the spirit of Wu Tzu-Hsu roused a destructive wave twice a day.[48] Religious significance was given to the times when the bore was largest and went the farthest up the river. At the midautumn full moon, "the 18th day of the 8th month," when the bore attained its greatest height of the year, people from hundreds of miles away would gather on the shore and watch junks and fishing boats plunge against the wave while others dove into the water and swam with flags in their hands to meet Tzu-Hsu. Since many people drowned during this festival, governors as early as the eleventh century tried to ban the activity, but to no avail, and the festival continued into the nineteenth century. Even today, each arrival of the tidal bore brings out a crowd of onlookers and attracts tourists from all over China to the most popular viewing location, in Yanguan next to the Zhan'ao Pagoda. Not far from this spot they also visit the Temple of the Sea God, where larger-than-life figures of Wu Tzu-Hsu and the king stand.

Surprisingly, those tide tables produced in 1056 do a fairly good job of predicting the arrival of the tidal bore at Yanguan today.[49] Since the river's bathymetry and shoreline have probably changed many times since the eleventh century, and since such changes will affect the size and timing of the bore, one can only guess that the river is back to a similar condition as when those tide tables were written.

At certain times of year the Qiantang River tidal bore can still reach twenty-five feet high, sometimes flooding over the riverbanks and catching tourists by surprise. This bore and one of comparable size in the Amazon River are the only large tidal bores left in the world. In other rivers around the world, dams or dredging or other changed river conditions have eliminated or reduced the bores that used to exist. There is no longer a bore in the Seine River in France or in the Colorado River in the United States. Today, tidal bores rarely occur in the Indus River, unless the wind helps a little. We don't know whether the small tidal bore that still occurs regularly in Turnagain Arm was larger back in 1778 when Captain James Cook first explored the northern end of Cook Inlet near the future city of Anchorage, Alaska. But whether huge or not, it is disconcerting to see a miles-wide wave crest, covering the entire width of a waterway, coming at you. More than once, seeing that the tide was too strong for the survey

Figure 1.2 Tidal bore in the Qiantang River near the Zhan'ao Pagoda in Yanguan, China. (*The Century Magazine, 1900*)

crew he had sent out, Captain Cook had to signal to them to turn around, and he later named the area River Turnagain.[50] Today surfers admire the tidal bore in Turnagain Arm for its seemingly endless ride, which can go on for miles.

It took two hundred years for western Europe to catch up to the Chinese in tide prediction. But for centuries after that there was no real advancement, although numerous rather artistic graphical techniques were developed to help seafarers use the most popular, though still crude, tide-prediction method of the time. Whether it was a tide table, a tidal almanac, a tide chart, or a tide clock, the same method of tide prediction was used. For a particular harbor, on a day when there was a new moon or full moon, an observation was made of the time difference between the moon's transit overhead and the following high water, that time difference being called the *establishment of the port.*[51] Then to predict the time of high tide on any day with a new or full moon, that time interval would be added to the time of the moon's transit overhead, which astronomers could predict quite accurately. For each day after new moon, forty-eight minutes were added to produce the high-water prediction on that day, although, as we shall see, adding fifty minutes would have been more accurate. The London Bridge tide table mentioned earlier, which came from the Abbey of St. Albans, was a simple one-page table also based on the time of the moon's transit.[52] Later tide-prediction tables were more elaborate and artistic, with special versions for the king of the nation where they were produced. A circular table displaying tide information for many ports was included in the Catalan Atlas produced for King Charles VI of France in 1375.[53] It provided the values of the establishment of the port for fourteen locations in France and England along the English Channel. In the 1540s the Brouscon tidal almanac, with thirty ports, was used extensively by seafarers because it was pocket size and on vellum.[54] Even the fancy and finely crafted tide clocks of the seventeenth and eighteenth centuries, including the large ones on church towers, used the same crude method of tide prediction based on the use of a tide interval relative to the moon's transit.[55]

But as clever as philosophers, scientists, and sailors had been in developing methods to predict the tide, such predictions were not going to become reliable until someone finally figured out how the moon and the sun *cause* the tides. By the seventeenth century, almost two thousand years after the work of Seleucus in Babylonia, there was still no answer to this mystery, in spite of its importance to the many seagoing nations of the world. Midway into the seventeenth century there were three primary schools of thought on how the tides were produced.[56] First, there were those who believed the theory of the French mathematician and philosopher René Descartes, who had proposed that space was not empty but instead was filled with an invisible ether. As part of his theory, Descartes thought the tides were caused by the moon moving through the ether and compressing it, which pushed down the sea surface. This would have produced a low tide when the moon was overhead, instead of a high water as observed, but in spite of this problem there were many adherents to this theory, primarily in France.

Second, others embraced the theory put forth by the Italian scientist Galileo Galilei, who surprisingly did not think the moon was involved at all. Galileo started with a correct idea, that the waters in an ocean slosh back and forth with a natural period that depends on the ocean's length and depth. The longer a basin, the slower the oscillation of water.[57] For example, water in a bathtub sloshes back and forth more slowly than water in a coffee cup but much faster than water in a lake. It's analogous to the natural period of a pendulum increasing with its length. But in his desire to use the tides to prove that the Earth moved around the sun, Galileo proposed that such oscillations in the ocean were initiated by changes in acceleration due to a combination of the Earth's rotation and its movement around the sun. He believed that the tides had a twelve-and-a-half-hour period, because he thought that the natural period of the Atlantic Ocean was twelve and a half hours (the natural period of the Atlantic Ocean is nineteen hours). Galileo meant his tidal theory to be a proof of the Copernican view that the Earth was not the center of the universe, but it conflicted with the Vatican's view and almost cost him his life. He managed to avoid being burned at the stake as a heretic by the Vatican, but he was imprisoned in his home for the rest of his life and forbidden to meet anyone with whom he might discuss scientific issues, including the tides.[58]

The third theory of the tides was due primarily to the German scientist Johannes Kepler and involved gravitation, which itself was an idea based on earlier work on magnetism. The English scientist William Gilbert had discovered that the Earth acted like a magnet. He proposed, in a book not published until after his death, that the tides were a consequence of the magnetic attraction between the Earth and the moon. Before Gilbert, the Flemish mathematician Simon Stevin had theorized that the tides were due to an attractive force from the moon. But Kepler, who is most famous for his mathematical laws describing planetary orbits around the sun, went further than Stevin or Gilbert and developed a mathematical treatment of the tides based on gravitational attraction. Kepler was correct about the gravitational attraction of the moon pulling on the sea, but he was unable to explain the most fundamental aspect of the tides, namely, why there are two high tides per day.

The problem was, all three schools of thought were wrong. Galileo and Kepler each understood a piece of the puzzle, but neither had gotten it right. And Descartes wasn't even close. It would take the English scientist and mathematician Sir Isaac Newton to finally put these pieces together and explain how the tides are generated. In 1687 in *Philosophiae Naturalis Principia Mathematica,* Newton showed that the generation of tides depends on both the gravitational attraction of the moon (as suggested by Kepler) and the centrifugal force of the moon-Earth orbit (involving accelerations not unlike the ones that Galileo had based his theory on). Gravitational attraction pulls the Earth and the moon toward each other, but at the same time centrifugal force pushes them apart because they both are revolving around a common point. (Because the Earth is eighty-two times more massive than the moon, this common point is much closer to the center of the Earth and is, in fact, inside the Earth, which is why it looks like the moon is revolving around the Earth.) Gravitational attraction and centrifugal force exactly balance at the center of the Earth. However, on the surface of the Earth on the side closest to the moon, the gravitational force is greater than the centrifugal force, and on the opposite side of the Earth, the centrifugal force is greater than the gravitational force. Thus, on an Earth covered with water there will be two bulges, one bulge toward the moon on the side of the Earth closest to the moon, and one bulge away from the moon on the side of the Earth farthest from the moon. On an Earth rotating under this ocean of water with its two bulges, there will be two high tides a day.

The period from one high tide to the next high tide (from one bulge to the next) will be approximately half a day. One might expect this to mean that the primary tidal period would be twelve hours, but Newton showed why the tidal period is twelve hours and twenty-five minutes. The Earth completes one complete rotation relative to the sun in twenty-four hours, or a *solar day.* But the moon is revolving around the Earth in the same direction that the Earth is rotating. So by the time the Earth has completed one complete rotation *relative to the sun,* the moon has moved farther around and the Earth must travel a little longer to make one complete rotation *relative to the moon.* A *lunar day* is twenty-four hours and fifty minutes, and thus the time to get from one lunar tidal bulge to the next is twelve hours and twenty-five minutes. And so not quite two full tidal cycles occur per solar day. The 1.93 cycles that are completed per solar day are called the *primary lunar frequency.* In the terminology of the tides, the *tidal frequency* is used more often than the *tidal period.* The *period* is the time required to complete one whole cycle, in this case, from one high tide to the next high tide in twelve hours and twenty-five minutes. *Frequency* is how many cycles are completed in a particular time interval, for example, in a solar day (twenty-four hours); in this case, about 1.93 cycles per day. We will see the term *frequency* again later.

The sun also produces tides, and although the sun is much more massive than the moon, it is ninety-three million miles away. Because of that great distance, the sun has less than half the tide-producing effect of the moon.[59] Newton worked out a complete mathematical treatment taking into consideration both the moon and the sun. He showed why the tidal range was greater at new moon and full moon and smaller when the moon was in first or third quarter, which had been noticed for thousands of years earlier but never understood. Newton showed that the larger spring tides at new or full moons occur because at those times the moon and the sun are lined up with the Earth and their tidal bulges add together. The smaller neap tides, when the moon is in first or third quarter, occur because at those times the moon and the sun are working against each other.

Newton also showed that there is another variation in the tide which causes its height to vary over a month, because the moon-Earth orbit is elliptical, not circular, and the distance between the moon and the Earth varies over the month. When the moon is closest to the Earth (called *perigee*), the tidal range is greater. When the moon is farthest from the Earth (called *apogee*), the tidal range is smaller. The distance between the moon and the Earth changes over a 27.6-day period as it goes from perigee to apogee and back to perigee. Newton also explained why in some places every other high tide can be larger than the previous high tide. The difference between two successive high tides is known as the *diurnal inequality* and was first noticed by Seleucus two thousand years earlier. Newton showed that this is because the axis of the Earth tilts with respect to the moon. The diurnal inequality changes over a 13.7-day period because the amount of this tilt changes over that period. When there is no axis tilt and the moon is seen directly over the equator (called *equatorial declination*), the two tidal bulges on the Earth produced by the moon are symmetrical and there will be no difference between consecutive high waters or between consecutive low waters. But when the axis is tilted and the moon is seen either north or south of the equator (called *northern lunar declination* or *southern lunar declination,* respectively), then the two tidal bulges on the Earth are asymmetrical and there will be differences in the heights of consecutive high waters, or consecutive low waters, or both.[60]

And yet with all this worked out, it took decades for Newton's correct tidal theory to become widely accepted. This was due partly to the *Principia* not being published in English until 1729 and partly to the difficulty many people had grasping these new ideas. To try to promote Newton's work, Edmund Halley, discoverer of the comet that bears his name and also a tidal researcher, published a summary of Newton's theory of tides that he had originally written for King James II.[61] This apparently did not convince the Académie Royale des Sciences in Paris, where there stubbornly remained strong support for Descartes' ether-based theory. Nine years after Halley's publication, the Paris Académie, claiming to be unhappy with progress in tide prediction, sponsored a prize for the best essay on "the flood and ebb of the sea." Two years later that prize was shared by four scientists from four different countries, for the tides had become an international science. One of these was a French scientist named Antoine Cavalleri, who rejected Newton's gravitational theory and tried to further develop and promote Descartes' theory. The other three winners—the Swiss mathematician Leonhard Euler, the Scottish mathematician Colin Maclaurin, and the Dutch mathematician Daniel Bernoulli—however, fully supported and elaborated on Newton's theory. Euler probably made the most important contribution, making a correction to Newton's theory by pointing out that the vertical component of the difference between gravitational and centrifugal force is too small to directly lift the water. More correctly, it is the horizontal component of this force that pushes the water sideways toward the equator, causing the bulges on both sides of the Earth.[62]

Unfortunately, even with this improvement in the scientific understanding of how tides are caused, based on Newton's work, there was no immediate benefit for the art of tide prediction. Part of the problem with the new scientific theories of Newton, Euler, and others was that they did not take into account how the oceans and adjoining bays responded hydrodynamically to the tide-producing forces of the moon and the sun. The moon and the sun cause tidal oscillations in the oceans, but it is the size of the basin that determines how large those oscillations will be. In other words, it is astronomy (the periodic movements of the moon, sun, and Earth) that determines the twelve-hour-and-twenty-five-minute tidal period as well as the other periods involved, but it is the *hydrodynamics* (the physics of the movement of the sea) that determines the times and heights of high and low waters at specific locations. The water surface of a basin, when disturbed, oscillates with a natural period that depends on the basin's length and depth. If that natural period of oscillation happens to be fairly close to the tidal period of twelve hours and twenty-five minutes, the result will be a much larger tidal range, an effect called *resonance*.[63] It is also hydrodynamics that explains why the speed and the wavelength of the long tide wave depend on the ocean depth and how the tide wave is affected by bottom friction, by the Earth's rotation, and by the changing width of a waterway. In some places in the world there is only one high water a day and one low water a day, such as along the southern coast of China. This is another hydrodynamic effect in which the size of the basin causes the *diurnal* (once-daily) tidal oscillations to grow larger than the *semidiurnal* (twice-daily) oscillations. In this case, the diurnal inequality caused by the tilt of the Earth's axis, which is typically small, has been made much larger by the hydrodynamics.[64]

Galileo had been the first one to understand the hydrodynamic effects on the tides. He understood the concept of an ocean basin having a natural period of oscillation determined by its dimensions. Unfortunately, he misused this concept in his erroneous tidal theory. In his prize-winning essay, Colin Maclaurin, one of the four winners who shared the Académie's prize, explained yet another hydrodynamic effect. He showed that the rotation of the Earth pushes tidal currents to the right in the Northern Hemisphere (and to the left in the Southern Hemisphere), this effect making the tidal range higher on one side of a bay than on the other.[65] But it wasn't until 1776 that a theory was developed that fully described the hydrodynamics of tidal motions. The French scientist Pierre-Simon, Marquis de Laplace was the first to mathematically describe how the oceans dynamically respond to the tide-producing forces caused by the changing gravitational effects of the moon and the sun.[66] In 1776, using the newly created calculus, which had been invented by Newton and also separately by Gottfried Leibniz, he derived three equations for the global ocean.[67] The first equation was based on the principle of conservation of mass, and the other two were based on the conservation of momentum in two horizontal directions.[68] Laplace included the effect of the Earth's rotation, which would later be called the Coriolis force—although Gaspard Gustave Coriolis "discovered" it fifty-nine years after Laplace did and ninety-five years after Maclaurin first suggested it.[69]

The Laplace tidal equations marked the beginning of the hydrodynamic modeling of the oceans—the first real application of physics to the sea. Variations of these equations, however, would end up being used for far more than tide prediction. They marked the beginning of a field that would come to be known as *geophysical fluid dynamics.* Similar equations would be used to model and predict storm surges, ocean currents, waves, and tsunamis. With added thermodynamic equations, that is, equations dealing with the transfer of heat, they would be applied to the atmosphere, forming the basis of modern weather-prediction models. Ocean models and atmosphere models would later be combined to predict El Niño and climate change. The Laplace tidal equations were very complex and could not be solved except for the most simplified situations in which some terms in the equations were neglected and the waterway was assumed to have a very simple shape. Solutions to the full equations for real-world oceans and bays would have to wait for the arrival of supercomputers in the late twentieth century that could handle numerical methods. Such solutions are at the heart of all modern prediction models for the sea, and for the atmosphere.

Although the Laplace tidal equations are complex, for the astronomical tide there was an immediate benefit that did not require a complete solution of these complicated mathematical equations. Laplace demonstrated that the tide is unique among all phenomena in the oceans and the atmosphere in that all its energy will be found only at specific frequencies. For example, most of the energy will be found at 1.93 cycles per day, which is due to the effect of the moon. Additional energy will be found at 2.00 cycles per day, due to the effect of the sun, and at 1.90 cycles per day due to the effect of the elliptical orbit of the moon (the changing distance between the moon and the Earth).[70] There are also other frequencies produced by other motions associated with the rotation of the Earth, the revolution of the Earth-moon system, and the orbit of the Earth around the sun. This contrasts dramatically with nontidal energy in the ocean from wind- and weather-induced changes in water level, in which the energy is spread across all frequencies in a random and always changing manner. Laplace proposed that if one could calculate the energy at each of the most important tidal frequencies, then one could accurately predict the tide. This insight would be at the heart of all future tide-prediction methods and ultimately would lead to one of the most elegant computing machines ever invented.

But that elegant and accurate tide-predicting machine would not be built for another ninety years. In the meantime, the tide predictions for use by ship captains and fishermen would continue to be calculated using methods not much different from what had been used for centuries, based on correlations observed between the tides and the movement and phases of the moon. A few entrepreneurs came up with more complex formulas in hopes of producing more accurate tide predictions and selling more tide tables than their competitors. But most tide predictions still relied on techniques using the establishment of the port, described earlier in the chapter as the time difference between the new moon's transit overhead and the following high water. Often they did not work very well for captains bringing their ships into port, and there were consequences. In 1777 William Hutchinson, the dockmaster at Liverpool, England, described a few of those consequences in his book *A Treatise on Practical Seamanship.*[71] In one case he was in a West Indian ship trying to sail over a bar in front of a harbor in Ireland, and the captain was using a tide rule for that harbor that turned out to be very poor. The ship ran aground on the bar, tearing off its rudder and a great deal of the stern and keel, the hold filling with seven feet of water. They survived only by being able to run the ship ashore to prevent it from sinking. At Liverpool, he had seen many ships coming in at neap tide but following a tide-prediction rule that did not give proper differentiation between tidal heights at springs versus at neaps. So without enough water depth these ships struck the bottom, often sinking and often with many lives lost.

In the next chapter we will see how such poor tide predictions were finally replaced by much more accurate predictions produced by large "brass brains," and how those accurate tide predictions became critical during World War II for a totally different purpose.

CHAPTER 2

The Moon, the Sun, and the Sea

The Tide Predictions for D-Day

What with the heightened emotions at Boston's Old South Meeting House on the evening of December 16, 1773, it is doubtful the American colonists would have changed their plans had they known that their intended assault on three British East India Company ships would take place during a perigean spring low tide. The Boston Tea Party certainly would have happened no matter what, but because of the extremely low tide, there was one aspect of that famous event that did not go quite as planned. At perigean spring tide, not only were the moon, sun, and Earth lined up so that the moon's and the sun's tidal forces added together, but the moon was also at its closest to the Earth, increasing the moon's tidal forces.[1] This combination of effects typically occurs only a couple of times a year and produces the highest high tides and the lowest low tides. In this case it was the extremely low tide, occurring at 7:23 P.M.,[2] that caused the problem when colonists dressed as Indians took over the three ships docked at Griffin's Wharf in Boston Harbor. The *Dartmouth*, the *Beaver*, and the *Eleanour* were in only two feet of water. Although tied up to the wharf, they were, in fact, aground.[3]

That night the American colonists broke open 342 casks of tea, totaling forty-five tons, and threw it over the sides of the ships. But with a water depth of only two feet and no strong tidal current to wash it away since it was near slack water, the tea merely sat on the bottom. The tea began to pile up, quickly rising above the waterline, so that soon immense stacks of tea were around all three ships. Some of the piles rose above the sides of the ships, and tea fell back into the ships. Colonists had to spread out tea on the tidal flats, mixing it in what little water there was and trampling it into the mud. Later, when the tide started to rise, the piles of tea rose with the water surface, and the increasing tidal current moved the floated piles of tea toward the shore. One firsthand account said that early the next morning a long windrow of tea, "about as big as you ever saw hay," extended from the wharves down to Castle Island. A party of colonists had to go out in boats and stir the tea into the waters of Boston Harbor. For weeks tea continued to wash up on the shores around Boston, some of it still floating above the water, so that colonists had to beat it with oars to make it unusable.

The Boston Tea Party was a rare event with respect to the tide, not because it was a perigean spring tide, which occurs a couple of times a year, but because in the heat of the moment the colonists forgot about the tide. Americans rarely forgot about the tide. Tide predictions were too important to the way of life of most colonists. Trade with Europe and among the colonies depended on shipping, and tide predictions were critical for bringing ships safely into port. Naval defense of the colonies depended on safe navigation and knowledge of tides and tidal currents. Clam and oyster fishermen could make their catch during only a narrow window of time, determined by tide predictions. That is still true today, but in Colonial America other businesses, businesses we do not have today, depended on the tide. The tides were used for power. Hundreds of sawmills and gristmills were powered by the tide. Two in Glen Cove on Long Island were typical. One cut lumber for the building boom in New York City in the late 1600s. The other ground wheat and corn for the biscuits that were the staple of generations of sailors.[4] In the Carolinas, rice farming, the dominant crop before cotton, depended critically on tidal irrigation.[5] The business cycle for tide mills and tidal irrigation was based on tide predictions.

Most Americans got their tide predictions from one of the many almanacs published throughout the colonies. Almanacs like Benjamin Franklin's *Poor Richard's Almanack* were the best sellers of their day.[6] In creating the tide tables found in these almanacs, the authors used the establishment of the port for each harbor (explained in the last chapter), namely, the time difference between a new moon's transit overhead and the following high water at that harbor. Then forty-eight minutes were added for every day after new moon. Typical was the tide table found in *Benjamin Banneker's Pennsylvania, Delaware, Maryland, and Virginia Almanack*. That tide table had rows for all the days of each month and five columns.[7] The first column showed the *age of the moon*, that is, the number of days after new moon. The last four columns gave the times of high water for four locations around Chesapeake Bay: Cape Charles, Point Lookout, Annapolis, and Baltimore. To the right of the table Banneker gave these simple instructions: "To find the time of High-Water by this Table—Look in the column of the moon's age, against which day you will have the time of High-Water, at the places named at the head of the table."

The importance of tide predictions to the American colonists became even more obvious during the Revolutionary War. There were several events during the war whose outcomes depended on tide predictions—which were used by the colonists but apparently were not used by the British. The American colonial attack on the English naval schooner *Gaspee* in June 1772 was the first use of organized force against the English, predating the celebrated shot at Lexington and Concord by three years. The *Gaspee,* with its eight cannons, had been sailing around Narragansett Bay enforcing an unpopular British trade regulation. Knowing the predicted time of high water, the American packet boat *Hannah* enticed the *Gaspee* to chase her into shallower water. As the water depth decreased with the ebbing tide, the larger-draft *Gaspee* ran aground and became stuck. A short while later a colonist walked down the main street of Providence, Rhode Island, beating a drum and informing the inhabitants that the *Gaspee* was aground on Namquid Point and that according to the tide prediction it would not float until three o'clock the next morning. He invited "persons who felt a disposition to go and destroy that troublesome vessel to repair in that event to Mr. James Sabin's house."[8] That evening eight longboats of men rowed down the river to the helpless *Gaspee,* which by then had rolled over on her side as the tide reached low water. The colonists captured the British and set the *Gaspee* on fire.[9]

Two years later another British ship, the HMS *Cancaeux*, was heading for Portsmouth, New Hampshire, to reinforce Fort William and Mary, because the fort had a large store of ammunition and only six English soldiers defending it. An American pilot enticed the *Cancaeux* behind a shoal at high tide and kept her there long enough for the tide to go down and trap her. But in this case the *Cancaeux* was trapped for more than a few hours. It was a spring high tide, the highest tide for half a month, and when the following high tide came, it was not high enough for the *Cancaeux* to get past the shoal.[10] She remained trapped there for days, giving Paul Revere plenty of time to ride to Portsmouth and warn the colonists. They stormed the fort on December 13, 1774. The muskets, cannons, and one hundred barrels of gunpowder taken from the fort eventually found their way to Americans who fought in the Battle of Bunker Hill.[11]

Four months later Paul Revere made his more famous midnight ride to warn American colonists that the British were coming—to capture John Hancock and John Adams in Lexington and seize weapon stores in Concord. The tide played a role here also. On the night of April 18, 1775, the British army crossed the Charles River in Boston and headed toward Lexington (this was the second choice of the famous "one if by land, two if by sea"). Revere began his horseback ride to Lexington by also crossing the Charles River. Revere planned his route to diagonally cross the Charles in an upstream direction so that he moved with the tidal current. The British troops crossed diagonally heading downstream, moving against the tidal current. The tidal current retarded the British troops while helping Revere, who got to Lexington before the British.[12]

The simple tide-prediction methods found in almanacs may have been useful, but only because they were applied by mariners with acquired local knowledge of the bays and waterways, people who could adjust these approximate predictions on the basis of that local knowledge. In general, tide predictions were crude. In 1776 Laplace provided the correct theory for producing much more accurate tide predictions, but it was not until 1867 that someone used his ideas to develop a practical method that took advantage of the fact that tidal energy is found only at certain known frequencies (see the last chapter). The English scientist Sir William Thomson developed the *harmonic method* to analyze a time series of tide measurements to determine how much energy each tidal frequency had.[13] The amount of energy at each tidal frequency is different at every location because of hydrodynamics, that is, because of the way bays and waterways affect the oscillation of the tide. But the brilliance of the harmonic method was that no understanding of hydrodynamics was required. One simply needed to analyze a long-enough data record at each location to calculate the effects of hydrodynamics. William Ferrel of the U.S. Coast Survey also developed a technique for harmonic analysis and prediction of tides, independent of the work of Thomson but also inspired by the work of Laplace.[14] George Darwin (son of Charles) further refined the harmonic prediction method, as did several other scientists at the U.S. Coast Survey.[15]

The portion of the tide due to the energy at a particular tidal frequency can be plotted as a cosine curve that has a particular amplitude and a particular time for its high water (relative to some reference time, such as the time when the moon is overhead). One such cosine curve will represent the part of the tide due to the effect of the moon. A second cosine curve will represent the part of the tide due to the effect of the sun. Other cosine curves will represent the effect of the moon's elliptical orbit (the moon moving from perigee to apogee and back again), the effect of the moon's changing position relative to the Earth's equator, and the effect of the sun's changing position relative to the Equator, as we saw in Chapter 1. When these cosine curves are added together, the result is a curve that closely matches a measured tide curve. *Harmonic analysis* is the method used to determine what the amplitude and timing of each of these cosine curves must be in order to have the predicted tide curve match the measured tide curve as closely as possible. These specific tidal frequencies are referred to as *harmonic constituents.* The calculated amplitude and phase difference (which determines the timing) for the harmonic constituents are called the *harmonic constants.* They are obtained by analyzing a series of hourly water level observations from a tide gauge.

Dozens of other pairs of harmonic constants for other harmonic constituents, representing the effects of other aspects of the Earth-moon orbit and the Earth's orbit around the sun, can be calculated. Most of these have a much smaller effect on the prediction, but some of them are usually included to achieve maximum accuracy.[16] Later it would be discovered that in shallow water nonlinear hydrodynamic effects transfer tidal energy to still other frequencies, which can also be easily included in the harmonic prediction method. These other harmonic constituents include higher harmonics of the basic astronomical frequencies, namely, constituents with two or three times the frequency (which is equivalent to one-half or one-third, respectively, of the period). They are called *overtides* (analogous to musical overtones). Overtides are what make a tide curve for a shallow-water area looked distorted compared with the simple cosine curve for the tide that one sees for deep-water areas. In shallow-water areas the tide curve shows a faster rise to high water and a slower fall to low water. The extreme case is the tidal bore that we saw examples of in Chapter 1.[17]

To use Thomson's or Ferrel's harmonic method required a lot of data. Water level measurements had to be made frequently (usually hourly) for at least a couple of weeks. But the longer the data series, the more harmonic constituents that could be obtained from the analysis, thus representing more of the tidal energy and making the tide prediction more accurate. For centuries, tide measurements were made using a *tide staff,* a long stick with precise markings like those on a ruler or yardstick, permanently mounted on the side of a pier.[18] But obtaining hourly tide data by having an observer look at a tide staff every hour for fifteen days was very labor intensive and thus rarely done. Luckily, by this time automated data acquisition had become possible with the invention of the *self-registering tide gauge.* The first such gauge had been built by Henry Palmer in 1831. It consisted of a float inside a long well, called a *stilling well* because it "stilled" the rapid oscillations due to wind waves, which otherwise would adversely affect the gauge's accuracy.[19] The stilling well was attached to a pier. Inside, a float was attached by a wire running over pulleys to a pen that moved up and down as the tide rose and fell, drawing a tide curve on a rotating drum of paper. The gauge had a clock mechanism to provide accurate timing.

In 1851 the U.S. Coast Survey had built its own self-registering tide gauges, designed by Joseph Saxton.[20] Three years later the Saxton gauge was installed at San Francisco, beginning the longest-running water level data record in the United States. As we will see in Chapter 7, in 1854 this San Francisco tide gauge and another gauge installed in San Diego became the first instruments ever to detect a tsunami, it having crossed the Pacific Ocean from Japan where an earthquake had generated it. The tsunami showed up as very small oscillations on the San Francisco and San Diego tide curves. These gauges were also the first instruments to remotely detect the eruption of Krakatoa in

1883 by measuring its tsunami signal.[21] Saxton self-registering gauges were installed at many other coastal locations along the Atlantic, Pacific, and Gulf coasts of the United States, beginning the first oceanographic observation system in the United States. In other nations as well, tide-gauge networks were installed, primarily to support the navigation community, but as we shall see, they would also provide data important for predicting other oceanographic phenomena besides the tide.

In 1893 two of the U.S. tide gauges, at Reedy Island in the Delaware River and at Fort Hamilton in New York Harbor, were enhanced so that their measurements could be seen immediately by mariners. This was perhaps the earliest example of real-time oceanographic data. It was accomplished by attaching a large tidal indicator, which mechanically took the latest measurement from the tide gauge and displayed it on a thirty-foot-high semicircular dial (see figure 2.1). An arrow pointed to the height of the tide, and a second

Figure 2.1 A tidal indicator built on Reedy Island in the Delaware River in 1893. (*U.S. Coast and Geodetic Survey, 1897*)

arrow indicated whether the tide was rising or falling.[22] Such a display could be seen many miles away by a ship captain with a long glass, who could then determine whether the water depth would be great enough for his ship to pass through that part of the waterway without running aground. The tidal indicator had the advantage of showing the actual height of the water level, which included not only the astronomical tide but also any water level changes due to the wind (discussed in the next chapter). The delivery of oceanographic data in real time essentially means data delivered quickly enough to make decisions based on the conditions represented by those data. As we will see, real-time data will play an important role in the prediction of all the oceanographic phenomena talked about throughout this book.

Using a self-registering tide gauge, a long time series of tide data could be acquired and analyzed to calculate the size and timing of the most important tidal harmonic constituents. But how could tide predictions be made using these harmonic constants without going through a long and arduous mathematical process? Thomson, having developed the harmonic method, now came up with an ingenious invention that allowed him to automate the prediction process using the harmonic constants calculated using his analysis method. In that precomputer era he designed a *mechanical analog tide-predicting machine*. It was made up of dozens of gears and pulleys over which ran a wire, which was connected to a pen that touched a moving roll of paper.[23] Each tidal constituent was represented by one gear rotating with a speed that was specific to the frequency of that particular constituent. (For example, the gear of a semidiurnal constituent rotated about twice as fast as the gear for a diurnal constituent.) A pin and yoke arrangement transformed the rotating motion of each gear into an up-and-down motion that pulled on the wire, thus providing that tidal constituent's contribution to the tide curve being drawn on the paper. Each pin and yoke pair was adjusted to provide the correct height and timing of each tidal constituent as determined by each pair of harmonic constants from the analysis of tide data.

Thomson's first tide-predicting machine was built in London in 1872 by the Légé Engineering Company and summed the contributions of the ten most important tidal constituents. It was made out of brass and finely crafted, and it would later be known as Kelvin's tide machine after Thomson became Lord Kelvin in 1892. In the United States, Ferrel designed a nineteen-constituent machine built in Washington in 1882 by Fauth and Company.[24] In 1906, Edward Roberts, who had worked out the gear ratios for Kelvin's original machine, designed a forty-constituent tide-predicting machine built by Légé and acquired by the Liverpool Tidal Institute. A second U.S. machine, with thirty-seven tidal constituents, was designed by Rollin Harris in 1894 and completed in 1912 in the workshops of the U.S. Coast and Geodetic Survey (see figure 2.2).[25] Using these machines, tide predictions were made for an entire year for all important ports and harbors. From these predictions the

Figure 2.2 The U.S. Coast and Geodetic Survey tide-predicting machine No. 2 was completed in 1910 and handled up to thirty seven tidal constituents. It was used during World War II to make the tide predictions for amphibious landings in North Africa and the Pacific. (*U.S. Coast and Geodetic Survey, 1915*)

heights and times of all high and low waters were published in annual tide tables. "Old Brass Brains," as the Harris machine came to be known at the U.S. Coast and Geodetic Survey, was used to make tide predictions until the 1960s, when electronic computers took over.

By the time World War II broke out in 1939, many maritime nations had one of these huge brass tide-predicting machines, and accurate tide predictions for all the major ports of the world had become fairly commonplace. Although the prediction process was fairly efficient, a considerable manual effort was still required to analyze water level data in order to acquire the harmonic constants that were put on the tide-predicting machines. To speed up this analysis the hourly water level observations were written on special forms and overlain by a series of keys, which were sheets of paper with holes revealing only particular observations. Even using these keys, many tedious summations of thousands of data values had to be made by human "computers," and it took weeks to harmonically analyze a year's worth of data, something that on today's electronic computers can be done in a couple of seconds.

Mariners and those living by the coast got their tide predictions from an official Tide Table, which was published annually and consisted of two types of tables. The first gave daily predictions of the times and heights of high and low waters for a number of *reference stations,* usually in harbors, and thus called *standard ports* in some countries. These daily predictions were produced on a tide-predicting machine using harmonic constants from analyzed tide data. The second type of table listed thousands of *subordinate* (or *secondary) stations,* with one station per row of the table. In each row, values were given for the difference between the times of high water at the subordinate station and a nearby reference station, and likewise differences in heights. Similar values were also given for low waters. Tide predictions for a subordinate station were calculated by applying these differences to the daily tide predictions at the designated reference station. Comparable Tidal Current Tables provided predictions of tidal currents but needed additional columns, because four prediction points were given (maximum flood, slack before ebb, maximum ebb, and slack before flood) instead of the two used for tides (high water and low water).

Tide tables were always important for predicting when high waters would occur so that large ships could be navigated into or out of shallow harbors without running aground, but with the outbreak of World War II, tide predictions became important for another purpose, a truly life-or-death purpose—amphibious landings. Capturing Japanese-held islands in the Pacific often meant knowing whether enough water would be covering treacherous coral reefs so that U.S. Marines could safely cross them and reach the beaches. And in the European theater the Allies typically wanted to land troops near the time of high tide in order to reduce the length of beach they would have to cross under heavy German fire.

The best day for landing troops on an enemy-held beach was usually selected based on when the high tide was predicted to occur and how high it was predicted to be. But there were other considerations as well, such as the times of sunrise and sunset and how dark the night would be, namely whether it would be a bright full moon or a dark new moon or somewhere in between. A special prediction product called the *Tide and Light Diagram* was produced by the U.S. Coast and Geodetic Survey for most islands in the Pacific.[26] This diagram depicted graphically for a one-month period the predicted daily times of high and low water, moonrise, moonset, sunrise, sunset, and various degrees of twilight. These top-secret diagrams were designed by Walter Zerbe, Chief of the Section on Tide Investigations, assisted by Bernard Zetler. During World War II the section produced almost a thousand tide and light diagrams for the Pacific and North Africa. The section grew to twenty-five people and eventually came under the Joint Army-Navy Intelligence Service. Many of the diagrams were created for coastal areas that the United States had no intention of attacking, in case Japanese spies got a copy. More routine tide predictions, such as for the annual tide tables, were prepared four years in advance in case the tide-predicting machine in Washington, DC, broke down or was sabotaged.

The U.S. Coast and Geodetic Survey had harmonic constants for many islands in the Pacific.[27] However, a few islands had no harmonic constants and no tide data to analyze. Most infamous of these was the Tarawa atoll in the Gilbert Islands in the central Pacific, where the terrible loss of life during the Marines' amphibious landing on November 20, 1943, was blamed on "the tide that failed."[28] The United States needed the airfield on Betio Island at the southwest corner of the Tarawa atoll to attack their main objective, Kwajalein in the Marshall Islands, 540 miles to the northwest. Betio was so heavily fortified by the Japanese that their commander, Rear Admiral Keiji Shibasaki, is said to have boasted that a million men could not take Tarawa in a hundred years. With the enemy having aimed its most powerful defenses at the ocean side of Betio, the U.S. Marines decided to attack on the lagoon side. But that meant crossing a coral reef in their landing boats. It was therefore critically important to know when there would be enough water over the reef.

A fully loaded Higgins boat drew three and a half to four feet of water. Very accurate prediction of the high tides over a period of a week or two was needed to determine the best time to make the landing. They needed a high tide that occurred early in the morning, because landing in the afternoon would mean that reinforcements on the next tidal cycle would have to land at night. This high tide also had to occur long enough after sunrise that there would be sufficient daylight for accurate bombardment before the landing. Most important, this high tide had to be high enough for the Higgins boats to be able to pass over the reef without getting stuck. If the boats got hung up on the coral, the Marines would be sitting ducks for the Japanese guns, and they would have to leave their boats and wade hundreds of yards to shore through a stream of bullets.

November 20, 1943, was the day the planners chose to land on Tarawa. The problem was that November 20 came only one day after an apogean neap tide,[29] one of the two smallest high tides of the whole year, when the moon was farthest from the Earth and the Earth and the sun were working against each other. But with no tide data to analyze, no one knew whether this apogean neap high tide might still be high enough to allow the Higgins boats to cross the reef. The only known tide data for Tarawa had been measured in 1901. The Coast and Geodetic Survey had used these data to calculate the time of high water along with an estimated tidal range. In the 1943 U.S. Tide Tables, Tarawa was a subordinate tide station, showing a mean tidal range of four feet and a mean spring range of five and a half feet—raising expectations of an apogean neap tidal range much less than four feet.[30] For the purposes of making a tide prediction, however, the time and height differences for high and low waters were referenced to the Apia tide station in the Samoa Islands, more than a thousand miles away. For a tide prediction at a subordinate station to be accurate it must have very similar tidal characteristics to its reference station, which is not likely if the two stations are separated by a great distance.[31] Thus, no one knew how accurate tide predictions using values from the 1943 tide tables would be. Navy intelligence found some Australians and New Zealanders (given the nickname the "Foreign Legion") who were familiar with the Gilbert Islands. They expressed concerns about the tide predictions, yet somehow they decided that on November 20 there would probably be about five feet of water over the reef at 10:01 A.M. but there was a chance that it might reach only four feet under unspecified special circumstances. One of the Foreign Legion disagreed with the others and said there would be only three feet of water.

All agreed that putting the assault off for a couple of days would mean higher tides, but the commander of the Fifth Amphibious Force, Rear Admiral Richard Turner, decided to go on the twentieth, because he feared an increased chance of west winds and high waves. Fearing the worst, Lieutenant Colonel David Shoup, who was responsible for the assault plan, decided to add armor to seventy-five thin-skinned amphibious tractors (also called LVTs), which normally carried supplies, and use them to carry men over the reef.[32] A typical LVT had a shallower draft than a Higgins boat as well as tracks that could travel on coral, but it lacked adequate armor to protect itself from gunfire.

On the morning of November 20 three waves of LVTs crawled across the reef and delivered fifteen hundred U.S. Marines to the beach at 9:13 A.M. Most of the LVTs were shot to pieces by the Japanese, so the rest of the assault depended on a high enough tide to let the Higgins boats pass over the reef. When the first Higgins boat reached the reef, the horrible high-pitched sound of coral scraping on metal made it frighteningly clear that the tide was not high enough. With a sudden lurch each Higgins boat came to a screeching halt. The Marines left their boats and tried desperately to reach the beach eight hundred yards away, but Japanese fire began cutting them down. The few remaining LVTs took men off a few of the stranded Higgins boats and brought them in over the reef, but not enough were still operational to be of much help. When the tide reached its peak a little after noon, it was still not high enough to free the Higgins boats, and then the tide began getting lower. It would not reach its second high water until midnight, and even then it would not be high enough. After seventy-six hours of desperate, bloody fighting the Marines prevailed, but at a cost of more than a thousand lives and more than two thousand wounded, most between the reef and the beach. The Japanese, who had refused to surrender and fought to almost the last man, lost more than four thousand men.

It was the costliest victory in Marine history up to that point—and for such a tiny island. A furor erupted in the United States. On November 30 the *New York Times* quoted Secretary of the Navy Frank Knox as saying the Marines' difficulties were caused by "the sudden shifting of the wind," which unexpectedly lowered the waters around Tarawa.[33] But that was not the case.

A naval inquiry board was convened at Pearl Harbor to investigate the battle, including the tide predictions. Starting on November 30 the Navy began making tide measurements at Tarawa, sending the data to the Coast and Geodetic Survey for analysis. On January 26 a letter, signed by J. H. Hawley, Acting Director of the Coast and Geodetic Survey, was sent to the Navy stating that the Survey's analysis showed that the time of high tide was about fifty minutes later than listed in the tide tables, but that the predicted tidal heights were correct.[34] The naval tide data were collected until May 3. Harmonic analysis of these data (and also of British data taken in 1948–1949) showed that the predicted tidal heights for the assault were not in great error and confirmed that on November 20 a high tide had occurred a half hour after noon, but with a height too low to allow a Higgins boat to cross the reef. Waiting two days to attack would have provided an extra foot of water over the reef at high tide and would have given the Navy four hours of water level higher than the highest tide on November 20. Of course, even if this had been recognized, the assault on Tarawa might not have been delayed, for it had already been postponed five days, and it was believed that any further delay would compromise the important Marshall Islands campaign to follow.

Tarawa was the first U.S. amphibious landing against such a heavily fortified enemy, and the first where tide predictions were so important. But there would be many more. And seven months later a much larger tidal range would play a critical role in the greatest amphibious landing ever attempted, against an even stronger enemy.

The twenty-foot tidal range along the beaches of Normandy assured that tide predictions would be crucial in the Allies' plans for the D-Day landings that would begin freeing Europe from Nazi rule. Tide predictions were a pivotal consideration in choosing the date and time of the landings in Operation Overlord.[35] The twenty-foot tidal range exposed large sections of beach at low tide, and German Field Marshal Rommel had gone to extraordinary lengths to take full advantage of this large tidal range to improve German coastal defenses. Rommel was convinced that Germany could avoid defeat only by stopping the Allies at the beaches.

Rommel was sure that the Allies would land at high tide.[36] Four years earlier, when it looked like the Germans would invade England, the English had also believed the Germans would land at high tide to minimize the length of beach they would have to cross under fire. The new Prime Minister, Winston Churchill, was a former First Lord of the Admiralty and knowledgeable of the tides. On June 26, 1940, Churchill had asked the Admiralty for "a chart of the tides and moons to cover the next six weeks."[37] He wanted the naval staff to use these charts to tell him "on which days conditions will be most favorable for a seaborne landing."[38] The Admiralty responded that the most likely time would be "when high water occurs near dawn, with no moon," and they selected dates in July and August that met these criteria for beaches near eight English ports.[39] On November 14 of that year the Admiralty banned the publishing of tide-prediction tables because they would be useful to the enemy.[40]

When Rommel took over responsibility for strengthening the Atlantic Wall (the Germans' name for their defenses along the coasts of France and Belgium), he covered the beaches between the low-water line and the high-water line with millions of steel, cement, and wooden obstacles. They were positioned to be covered by water by midtide so they could rip out the bottoms of Allied landing craft. The Allies first observed Rommel's obstacles from the air in mid-February 1944, and "thereafter they seemed to grow like mushrooms,...until by May there was an obstacle on every two or three yards of front."[41] They came in all shapes and sizes. There were rows of hedgehogs, made of three six-foot iron bars crossed at right angles so that they looked liked giant children's jacks. There were Belgian gates, seven-by-ten-foot steel frames planted upright. There was row after row of sharpened, half-buried logs pointed upward at an angle, many with Teller mines on them.

Throughout his improvements to the Atlantic Wall, Rommel remained convinced that the Allies would land at high water. The Allies would indeed have liked to have landed at high tide, so the troops would have less beach to cross under fire, but Rommel's underwater obstacles changed that. Now the initial landings would have to take place near low tide so that demolition

Figure 2.3 General Field Marshal Rommel at low tide inspecting some of the thousands of underwater obstacles he had built between the low-water line and the high-water line next to the beaches of Normandy. (*Dwight D. Eisenhower Library*)

teams could blow up enough obstacles to open up corridors through which the landing craft could navigate to the beach.[42] The tide also had to be rising, so that the landing craft could unload and return without danger of being stranded by a receding tide, which would drop by three feet an hour. But other conditions also had to be met. For secrecy, Allied forces had to cross the English Channel in darkness, but the Navy needed thirty to ninety minutes of daylight to bombard the coast before the landings and for accurate pilotage to the beaches. Thus low tide had to coincide with first light, with the landing to occur an hour after. An airborne assault would occur the night before the landing, because paratroopers needed to land in darkness but they also needed moonlight to see their targets. Thus a late-rising moon was required. Only three days in June 1944 met all these requirements—June 5, June 6, and June 7. The second-choice period was two weeks later, when tidal conditions would be similar but there would be no full moon.[43]

A twenty-foot tidal range meant that water would rise at a rate of at least three feet an hour, perhaps much faster at certain times because of shallow-water effects. If the times of low water and the hourly rise were not known exactly, there might not be enough time for the demolition teams to blow up a sufficient number of Rommel's obstacles. In addition, times of low and high waters differed at each of the five landing beaches (from west to east, Utah, Omaha, Gold, Juno, and Sword), with almost one and a quarter hour between Utah and Sword. H-Hour on these beaches would have to be staggered according to the tide predictions, and they had to be known precisely.[44] Tidal currents (or tidal streams, as the British called them) were another important consideration, since strong tidal currents could easily push amphibious landing craft down the beach, away from their intended landing spots. Tidal currents were even more difficult to predict than the tides, and more difficult to measure.

All the British Admiralty tide and tide current predictions for the war effort were produced by the Liverpool Tidal Institute at Bidston Observatory, led by Dr. Arthur T. Doodson, at that time the world's leading authority on tide prediction.[45] He used two tide-predicting machines, the Kelvin machine (built in 1872 but overhauled in 1942) and the Roberts-designed machine (built in 1906). They were put in separate rooms in the Observatory to minimize the chance of a bomb destroying both machines.[46] Worry over this possibility heightened during one of the Nazi propaganda radio broadcasts during the Blitz, when the infamous Lord Haw Haw (the traitor William Joyce) promised "that by morning, Bidston Observatory will be no more."[47] Many bombs did fall near the Observatory, damaging doors and shattering hundreds of windows, but both tide-predicting machines survived and were kept running from early morning to late at night seven days a week.[48] To play it safe, predictions for all the ports in the British Admiralty Tide Tables were completed two years ahead of time. Additional predictions were produced for wherever around the world the Allies needed them.

On July 15, 1943, Lieutenant General Frederick Morgan sent a plan to the British Chiefs of Staff in London that strongly recommended the Normandy beaches as the best site for the Allied invasion of Nazi-held Europe, a recommendation that the Allied command accepted the following month.[49] It fell upon Commander William Ian Farquharson, the Superintendent of Tides for the British Admiralty at the British Hydrographic Office, to find a way to provide harmonic constants, or the data to calculate them, that could be used by Doodson to produce the most accurate tide predictions possible for the landing beaches. Because of extremely tight security, Farquharson could not tell Doodson the biggest secret of the war—that the Allied landings were to take place at Normandy. During the war years, Farquharson, in Bath in southern England, and Doodson, in Liverpool in the north, exchanged hundreds of letters and sometimes telegrams.[50] Farquharson frequently had to refer to particular locations by a code rather than by name. If a letter was intercepted by an enemy agent, the agent must not be able to figure out the D-Day landing location from the predictions that Farquharson was asking for from Doodson. In one letter Farquharson asked Doodson for hourly tidal heights for June and July 1944 for IHB no. 102, asking him to refer "to it by that number, omitting lat and long."[51] He was asking for tide predictions at Cherbourg, which was station number 102 in the *Special Publication of Harmonic Constants* by the International Hydrographic Bureau in Monaco. Germany had left that organization in 1933, but this reference still didn't amount to a very safe code.[52]

To produce accurate tide predictions for the Normandy beaches, Doodson ideally needed accurate harmonic constants calculated from water level data measured at or very near those enemy beaches. Unfortunately, such data did not exist, or if any did, they were in Nazi-controlled Vichy France. The British had accurate harmonic constants for the two nearest major French ports, Le Havre to the east and Cherbourg to the west, bracketing the Normandy beaches in the Bay of the Seine. But the tide on the Normandy beaches was not likely to be similar enough to the tide at Le Havre or at Cherbourg to be able to use their constituents. Interpolation also would not work, because the shallow-water conditions were very different. Shallow-water distortion of the tide leads to an even faster rise from low water to high water, giving demolition teams even less time, and so accurate shallow-water tidal constituents from the Normandy beaches had to be included in the prediction. Three locations in the 1943 Admiralty Tide Tables were nearer the Normandy beaches: Port-en-Bessin, a small fishing harbor between Omaha and Gold Beaches; Courseulles-sur-Mer, on the small river Seulles near Juno Beach; and Merville and Ouistreham, two towns on opposite sides of the River Orne, both just east of Sword Beach.[53] However, all these were secondary stations that used Le Havre as a reference station, and there were warnings in the British Tide Tables about the possible inaccuracy of predictions for these locations.[54]

Desperately needed were water level data from the Normandy beaches themselves. The British carried out several secret midnight reconnaissances of the enemy beaches in small boats and two-man submarines, during which they measured beach gradients, collected samples of the sand (to see if it was compact enough to take the weight of tanks), and obtained a few water level and current measurements.[55] These tide and current data records were much shorter than normally required for tidal analysis. But somehow Farquharson had to use these data combined with tidal information he had for Cherbourg, Le Havre, and the three secondary stations, to come up with harmonic constants for Doodson.

On October 9, 1943, Farquharson sent Doodson a three-page handwritten letter marked urgent that had eleven pairs of tidal harmonic constants for what he referred to as "Position Z," asking Doodson to produce four months of hourly height predictions commencing April 1, 1944. "The place is nameless and the constants inferred," he wrote. "There is in fact very little data for it. I am gambling on the inferred shallow water constants giving something like the right answer."[56] It is still not known, even from his many letters to Doodson, how Farquharson came up with these eleven pairs of harmonic constants. Probably he modified the Le Havre constants in such a way as to match the shape of the measured tide curve determined from the little bit of water level data collected at one of the Normandy beaches. He may have also taken into consideration the time and height differences and other information found in the Admiralty Tide Tables. By whatever means he produced them, his harmonic constants for the Normandy beaches compare favorably with those derived from relatively recent numerical hydrodynamic tide models of the English Channel.[57] Doodson then put these harmonic constants on his tide-predicting machine to produce the tide predictions for D-Day. Years after the war Doodson would admit that from the harmonic constants he guessed that the Normandy beaches were the intended D-Day landing site.[58] At the British Hydrographic Office, Doodson's tide predictions had to be modified for each of the landing beaches and provided in a variety of convenient ways, including *Tidal Illumination Diagrams* (similar to the Coast and Geodetic Survey's *Tide and Light Diagrams* mentioned earlier), that added information about moonrise, moonset, sunrise, sunset, and various degrees of twilight (see figure 2.4).[59] These predictions were distributed to the American forces by the Intelligence Division of the U.S. Army's Chief Engineer.

Although the tide had been the key factor in selecting June 5, 6, and 7 as the three days most suitable to be D-Day, it would be the weather that would decide on which of those days the invasion would take place. Bad weather with high winds and high waves and surf on the beaches would make an amphibious landing impossible (as we will see in Chapter 6). It would also mean no air support and very inaccurate naval gunfire, as well as low clouds and poor visibility for the airborne operations during the night before the

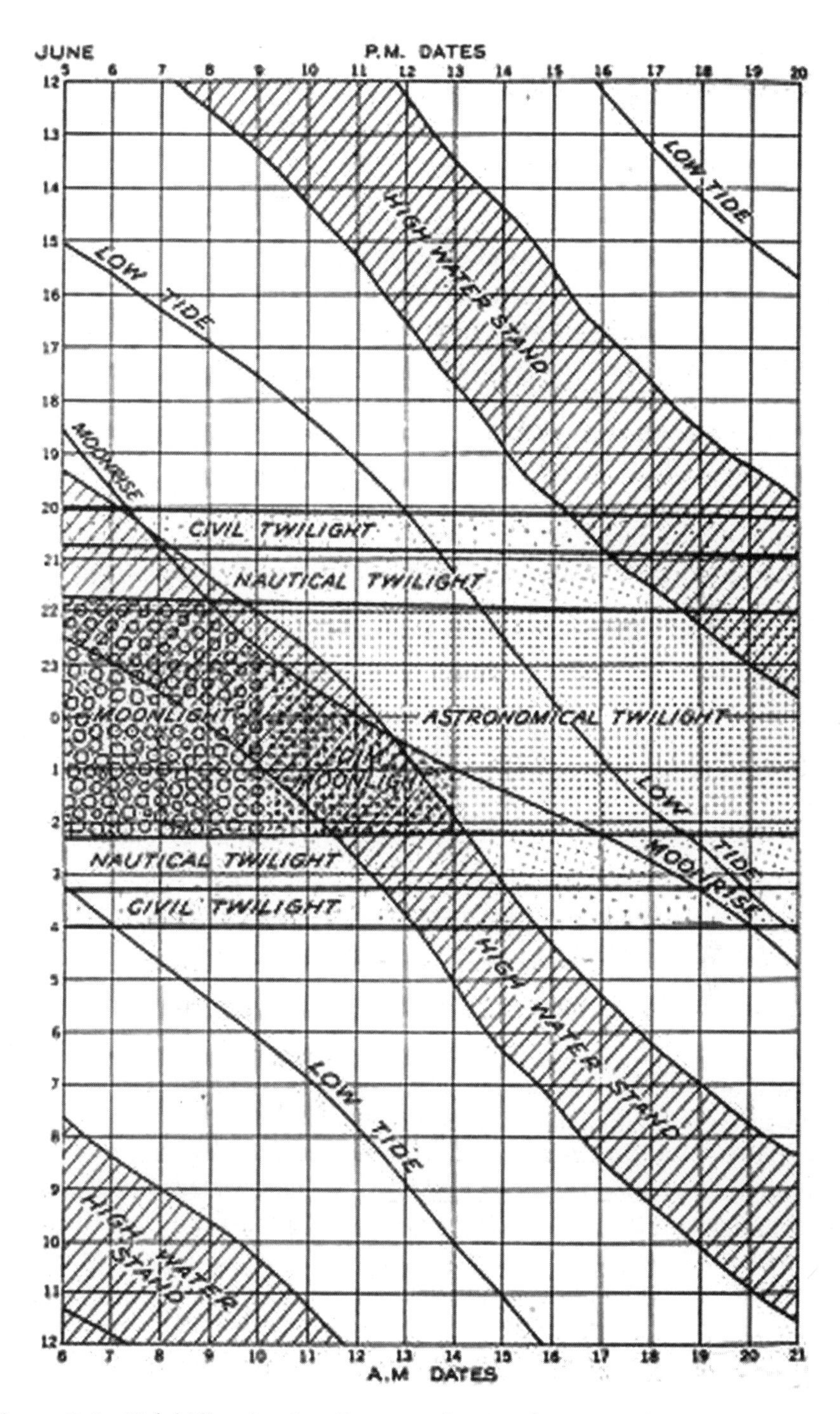

Figure 2.4 Tidal Illumination Diagram for Omaha Beach for June 5 through June 29, 1944. (*Society of American Military Engineers, 1947*)

landing. And although June weather was typically good, bad weather ended up delaying D-Day from the original June 5 date chosen by General Eisenhower and his staff. The decision by Eisenhower to go on June 6, based on the forecast of a thirty-six-hour break in the weather by his meteorologists, is credited with being a factor in surprising the Germans, who felt that the weather was too bad for the invasion to take place. Rommel had headed back to his home in Germany to spend June 6 with his wife on her birthday, after which he met with Hitler. However, another consideration for Rommel being willing to leave Normandy on June 6 was that he still believed, for unknown reasons, that the Allies would come at high tide, and thus he thought the tides were wrong for an invasion on June 6.[60]

Once the invasion began, even though the tide predictions turned out to be quite accurate, the tide still had a considerable effect.[61] Demolition teams landed at Omaha Beach one hour after low water with the objective of blasting sixteen channels, each eighty yards wide, through the beach obstacles. Because of the speed of the rapidly rising tide and the heavy German gunfire from the bluffs high above the beach, they were successful only in blasting six complete gaps and three partial gaps, with more than half the engineers killed in the process.[62] The strong tidal currents were also a factor, frequently moving the landing boats farther down the beach than intended. At Utah Beach the landing craft were swept by the tidal currents to less strongly defended sections of the beach, so in that case the tidal currents helped.[63]

After World War II the most frequent use of tide and tidal current predictions was once again for safe navigation, although in 1950 the Korean War briefly brought back their use in a military amphibious landing at Inchon, where the tidal range is thirty feet. But a whole host of other uses developed, many in support of the environmental sciences. Tidal currents, for example, play a critical role in flushing polluted waters from estuaries. They also can transport harmful blooms of algae, sometimes toward public beaches, causing health problems.

In 1970 the U.S. Coast and Geodetic Survey became the National Ocean Service (NOS), as part of the newly created National Oceanic and Atmospheric Administration (NOAA).[64] Over the years NOS has installed over two hundred permanent automated tide gauges around the United States. However, modern technology has changed the gauges. The floats connected by a wire to a pencil drawing on a moving drum of paper are gone. Today acoustic gauges are used, in which a sound wave is sent down the tide well, where it bounces off the water surface and returns to a receiver, the length of time for the down and up trip determining the water level height very exactly. Also, many such acoustic signals can be sent rapidly, producing many measurements, which can be easily averaged over a few minutes to take out the effects of the rapid up-and-down movements due to wind waves. These gauges also detect oscillations due to tsunamis. There are also other newer ways to measure the tide, even far from land—for example, using a global positioning system (GPS) receiver on a buoy floating on the ocean surface.[65] Satellite altimeters can measure the height of the water level over large sections of ocean, the distance from the satellite to the ocean surface being measured using radar (radio signals).[66]

Tide gauges are now usually referred to as *water level gauges,* because their data are important for much more than tide prediction. NOAA's network of two hundred stations is thus called the National Water Level Observation Network and was the United States' first contribution to the Global Ocean Observing System, its data being used for a variety of purposes. Over the years thousands of additional temporary water level stations have been installed during NOS hydrographic (or bathymetric) and shoreline surveys, those surveys being carried out to produce accurate nautical charts. Today the tide and the tidal current are easily and accurately predicted anywhere water level or current data have been collected and analyzed. In addition, numerical hydrodynamic models, using equations similar to the ones first derived by Laplace, can be used to predict tides at locations for which there are no data.[67] Tide predictions for many thousands of locations around the world are routinely provided by NOS and other national agencies and private companies around the world, via hardcopy books, on the World Wide Web, and as software programs provided on CDs. The tides, of course, are no longer the mystery they once were. Because of their astronomical basis, such tide predictions can be made years, decades, or even centuries into the future—or even into the past (as long as the bathymetry and shoreline have not changed significantly).[68] And thus we can tell what the tide was during a historical marine event, just as we can tell what the phase of the moon was, as we have done with some of the stories in this chapter.

As we have seen, tide predictions are used by ship captains and pilots to determine the safest time to navigate large vessels into and out of harbors. However, in recent decades the drafts of commercial vessels such as oil tankers and container ships have grown so large that even the deepest U.S. harbors cannot handle them except near times of high tide. For economic and safety purposes, astronomical tide predictions are no longer enough, and ship captains must be aware of the much less predictable wind's effects on water level. The water level data from stations in the National Water Level Observation Network are therefore delivered in real time via the internet. Real-time water level data, as well as real-time current data, are also provided by NOS/NOAA in many ports in the United States using the Physical Oceanographic Real-Time System (PORTS). Forecast numerical hydrodynamic models for PORTS provide twenty-four-hour forecasts of water levels and currents that include the effects of forecast winds, those models being forced by coastal ocean models, which themselves are driven by weather forecast models from NOAA's National Weather Service (NWS). These water levels, delivered in real time, and the forecasts based on these data, can make all the difference to the safety of a large commercial vessel whose keel is barely above the bottom of a harbor.[69]

This real-time water level observation capability has also become critical for assessing the size of storm surges produced by the high winds of a hurricane or an extratropical storm, so that flooding forecasts can be fine-tuned and evacuation areas can be more accurately determined (see Chapter 4). When earthquakes occur in or near the ocean, these same real-time gauges are automatically put on a high sampling rate to detect possible tsunamis (see Chapter 7). Water level records from gauges in the Pacific were instrumental in developing an understanding of El Niño (see Chapter 10). These same gauges, many of them in operation for more than a hundred years, also provide critical data on how fast sea level is rising, a major issue in the discussion of climate change and the effects of global warming (see Chapter 10). So already we begin to see how interrelated oceanographic phenomena are and how ocean observing networks are multipurpose, important to many types of marine prediction. Harmonic tidal prediction was humankind's first step toward accurate marine prediction. And the installation of water level observation networks by nations around the world was our first step toward creating the Global Ocean Observing System (GOOS), which we will hear more about in later chapters.

Because accurate tide predictions have been commonplace for at least a century, many people forget how dangerous the tide can be—tidal whirlpools, tidal bores, raging tidal currents, and large tides rushing in over mudflats still bring death to those who are in the wrong place at the wrong time. Tide predictions are important for providing warnings of dangerous conditions. But even in today's age of far-reaching electronic communication, there are some who are unaware of such predictions or who ignore them, sometimes with tragic consequences. In Cook Inlet, near Anchorage, Alaska, ignoring the dangers of that region's forty-foot tidal range can be deadly. The miles and miles of mudflats uncovered at low tide at Turnagain Arm have areas that are like quicksand. A person who sinks up to their knees in the glacial silt becomes completely stuck, and when the water begins rising at almost seven feet an hour, there is little time to get out before the water is over their head. There has been more than one story about a trapped hunter covered by the rapidly rising tide trying to stay alive by breathing through the barrel of his shotgun while he desperately struggled to free himself from the suction of the mud.

Early in Chapter 1 we mentioned what was probably the oldest use for predicting the tide—fishermen digging shellfish on mudflats at low tide who needed to know when to leave so they would not drown when the sea returned. In some waterways the tide comes roaring back so fast that it cannot be outrun—as we saw in the Gulf of St. Malo and in the Red Sea at the times of Napoleon and Moses. Morecambe Bay on the English coast of the Irish Sea is one of those waterways with dangerous tides that can surround and overwhelm a person in a matter of minutes. It has a thirty-foot tidal range and the largest intertidal area in Britain, exposing 120 square miles of mudflats at low water. In some places during the ebb tide the water's edge falls back seven miles from the land.

On the evening of February 5, 2004, twenty-three Chinese immigrant workers were gathering cockles on these mudflats when they were caught by the tide. They could not escape because of the speed with which the water surrounded them.[70] A member of the Lancashire police said that some of the immigrants made desperate calls with their cell phones as they struggled in the swirling deepening waters. He said that in the background he could hear "the harrowing noises of people as they are trying to save themselves." They drowned in freezing water in the darkness of the night a mile from the nearest shore. It was "a very frightening and horrible way to die."[71]

The dead Chinese were illegal immigrants and included two women. They had been sent out onto the mudflats by their Chinese gangmaster to bring back bags of cockles.[72] Cockles are a popular shellfish in Europe, twelve million dollars' worth being harvested from Morecambe Bay in a typical year. The dangers of the bay's fast tides are well known to local cockle fishermen, who go out on the mudflats in tractors and leave before the time that the tide tables tell them will be dangerous. The previous dry summer had reduced the cockle population, driving up their price, which made the dangerous job of cockle gathering even more lucrative. This brought the gangs of illegal immigrants, unknowledgeable of the area and trusting in their gangmasters, who either did not bother to check the tide predictions or did not care. In the previous five weeks, there had been two other incidents of Chinese cockle pickers caught on the mudflats by the tide, but rescue workers had managed to save them. This time they were not as lucky. Two years later the gangmaster of the dead Chinese pickers was convicted in a British court of manslaughter.[73]

Along the coasts of most waterways the dominant cause of changing water levels is the astronomical tide caused by the gravitational effects of the moon and the sun. During hurricanes, however, storm surges produced by high winds often overwhelm the tide. In some areas these storm surges can become thirty or even forty feet high and violently flood large coastal regions. As we have seen, mariners and scientists developed methods to approximately predict the astronomical tide by the thirteenth century and quite accurately predict it by the end of the nineteenth century. But as we shall see in the next chapter, they would not learn how to predict deadly storm surges until late in the twentieth century. And millions would die before they finally developed that prediction capability.

CHAPTER 3

The Sea's Greatest Killer

Predicting Storm Surges

The morning of October 4, 1864, in Bengal was beautiful, sunny, and most important, dry. After nearly five months of torrential downpours brought on by the southwest monsoons, several days had now passed with little or no rain.[1] Although that five-month rainy season had been fairly typical, the people throughout India were still thankful it was over. The clear morning revealed a lush, green landscape along the coast of the Bay of Bengal, the result of a bounteous supply of water and a million tons of nutrient-rich silt carried to the coast by three mighty rivers—the Meghna, the Brahmaputra, and the sacred Ganges. Much of the sediment had come all the way from the majestic Himalayas. Over the centuries these three rivers had combined to build in Bengal the largest delta on Earth, consisting of thousands of islands and hundreds of channels and, at its seaward edge, the largest mangrove forest in the world. It was a land of Bengal tigers, elephants, and crocodiles, but it was also a land of rice paddies and sugarcane fields carved out of swamps by a million Bengalis. It was a land controlled by the British, whose ships filled the port of Calcutta, eighty miles up the Hooghly River in West Bengal.

On this morning Bengalis and British alike were enjoying the quiet, dry weather now that the rainy season was over, unaware that five hundred miles to the south in the center of the Bay of Bengal rain was falling "as in a solid mass."[2] High winds had been churning the seas for the last three days, but no ships had reached the coast to inform the British. Months later, British scientists working for the newly formed Meteorological Department of India would examine the logs of schooners that had been on the Bay of Bengal that October. And by comparing barometric pressure measurements, they would recognize an area with very low barometric pressure. Around this low-pressure area a huge portion of the atmosphere was rotating counterclockwise, the winds reaching well over a hundred miles per hour near the center. Such a rotating mass of air over the Indian Ocean had been recently given the name *cyclone,* although it was sometimes called a *hurricane* like its counterparts in the Atlantic or sometimes called a *typhoon* like those in the Pacific. Scientists did not yet understand how these violent whirlwinds were created, much less know how to predict their path or wind strength. But they were well aware of the damage cyclones caused, with the help of the sea.

On that October 4, unknown to those on land, the turbulent weather being recorded in logs on ships battling to survive on the Bay of Bengal was part of a major tropical cyclone moving in a north-northwest direction—heading directly toward them. Along the Bengal coast the only indication of a possible change in the pleasant weather was a modification in the direction of the light winds. At noon the center of the cyclone was still two hundred miles to the south, but its northern edge finally reached the Hooghly River in West Bengal, bringing with it stormy weather, much to the chagrin of those on shore who had thought they were finally done with rain. Unfortunately, this cyclone would bring more than rain.

In the evening the cyclone's full fury began to make itself felt. By 8 P.M. it was blowing violently, which continued throughout the night.[3] At around 9 A.M. the next morning the eye of the cyclone passed near Saugor Island in the lower Hooghly, producing forty-five minutes of sudden calm that confused the captains of the many ships anchored there. From the barometric pressure readings the British later acquired from ship logs, this cyclone was probably a Category 4 storm on today's Saffir-Simpson Hurricane Scale, reducing to a Category 3 after it moved onto land and up the Hooghly. Its violent winds caused a great deal of damage, especially near the eye of the cyclone. But the real danger came not from the air but from the sea, for this cyclone was bringing with it what British scientists in the mid-1800s called a *storm wave,* what scientists today generally call a *storm surge.*

A storm wave is not to be confused with one of the thousands of short steep wind waves that came crashing down on decks of ships caught in the cyclone. This was a much longer and wider wave that raised the surface of the sea over hundreds of square miles, and in the process it pushed a huge volume of water toward the coast. Wind waves were present also, but they rode on top of the storm wave and added to the havoc the storm wave would wreak when it reached the shore. In the mid-1800s scientists had yet to learn exactly how storm waves were produced, though they believed it had something to do with the fast winds and low atmospheric pressure. During the morning of October 5, the waters along the coast of West Bengal and the nearby rivers slowly rose, but the crest of the storm wave did not arrive at the mouth of the Hooghly River until a little after 10:00 A.M., about an hour after the cyclone's eye had passed.[4]

The timing of the storm wave could not have been worse. Its crest arrived just two hours before the tide reached high water. To make matters worse, it was nearly a full moon, so the height of high water during this spring tide was at its highest. The Hooghly was a shallow river whose width narrowed upstream, so its highest tides formed a steep bore that moved rapidly up the river (as described in Chapter 1). Dikes and embankments along the Hooghly River were high enough to withstand the spring tidal bore but not the bore created by the storm wave. This combination of storm wave and spring tide was overwhelming—not just because its high water levels flooded the land but because the crest arrived suddenly and with great destructive power.

The storm wave picked up the ship *Martaban,* which had been anchored near the mouth of the Hooghly River, and carried it over normally shallow sandbars without the ship ever coming close to touching the bottom. In spite of the havoc on deck caused by the wind, the rain, and the wind waves breaking on the deck, the captain managed to repeatedly check water depth with a lead line. He later wrote, "The ship never shoaled at less than 7 fathoms" (42 feet), and he estimated that "the storm-wave must at least have risen 40 feet to have carried me across these sands."[5] The storm wave also picked up the river steam vessel *Salween,* carrying it up onto the shore and over the tops of trees, and putting it down atop the local telegraph office.[6]

But the storm wave's real measure of destructive power was the devastation it wreaked on the countryside as the deluge washed over or broke through dikes and embankments, at heights of ten, twenty, or even thirty feet. The storm wave swept away entire villages, leaving little evidence that they had ever existed. It flooded the land as far as ten miles from the banks of rivers and channels. When the captain of the *Salween* landed on the telegraph office, he discovered sadly that the post office had also been gutted by the storm wave, killing the postmaster and his wife and children, and that the entire native village of Kedgeree had been washed away with all its men, women, and children drowned.[7]

A similar scene played again and again in village after village over all of West Bengal. To escape the violent winds and the torrential rains, villagers took refuge in their huts. They huddled in the dark with the frightening noise of the roaring winds all around them. But then, "almost in an instant, and without any warning, the water was over the village."[8] The storm wave, appearing like a wall of water, violently swept each hut away, drowning those trapped inside. Of those who somehow were freed from their crumbling homes, a few managed to grab onto floating roofs or pieces of wood, and if they had the strength to hold on until the chaos subsided, they found themselves washed many miles inland. Other lucky ones were able to grab onto trees and hold on for dear life. The great majority of survivors were men, for the women and especially the children did not have the strength to hold onto the trees or floating wood for the hours it took to survive.[9] Those lucky enough to be saved by tree branches observed not only the terrible sight of their families being carried away into the mist by the swirling waters, but sometimes strange sights such as "cattle and tigers...swept into an indiscriminate mass together...the latter...powerless to do any harm."[10] Only the strong-swimming water buffaloes seemed able to survive in the wild currents.

The scene of total devastation revealed the next morning was the true indication of the power of the storm wave. Bodies covered the shores and were spread inland for miles. So many floated in the Hooghly River that relief ships from Calcutta had to go slowly to steer around them. They lay in and around every village. The heartbreaking sight of so many bodies of young children was difficult enough for government officials to bear as they tended to survivors, but soon the awful stench made it even worse. In each village bodies were left to rot by surviving villagers because these were the bodies of strangers from some other village. The bodies of their loved ones had been washed miles away, to places unknown, but probably near another village whose surviving inhabitants had also left them to rot. Those who were still alive had more immediate problems. Their food stores had been washed away, and their water tanks had been fouled by saltwater and putrid vegetation. In desperation they ate bad food and drank impure water, which led to an outbreak of cholera, dysentery, and smallpox. People starved, and people died of disease.

It did not take long for the full calamity to become apparent as British officials tabulated losses. At least 100,000 cattle had been drowned. Rice crops were destroyed or damaged by saltwater, many fields already turning black. Entire regions lost more than three-quarters of their population. At least 50,000 people were killed directly by the storm wave, with another 30,000 dying over the following weeks from disease.[11] The combined 80,000 death toll was probably a conservative estimate, because it was not known how many migrant workers had been in the area for the upcoming rice harvest.

In Calcutta, eighty miles up the Hooghly River, there were fewer deaths. It was far enough from the coast that the storm wave had decreased in size by the time it reached the city, worn down by friction of the river bottom and also reduced in volume by water lost over riverbanks. If the embankments and dikes downriver had held better, the wave would have been larger at Calcutta, and the city would have suffered worse destruction. The damage that did occur was primarily to ships in the Port of Calcutta. Although diminished when it reached the port, the storm wave was still a steep bore that violently lifted ships at their moorings, pulling out anchors, breaking chains, or if anchor and chain held, pulling the vessels underwater. Ships freed from their moorings "were grounded, a mass of confused wrecks, with cargo-boats, lighters, and smaller boats of every description, on the sands of

Goosery, Shibpore, and Cossipore."[12] Of the 195 major vessels in port, only 23 escaped harm, greatly upsetting the shipping companies of London and Liverpool. A witness wrote that "where the eye generally wandered over hundreds of native boats of all sorts and sizes," after the storm wave "not one was visible; but the water and the shore were covered with minute pieces of plank."[13]

The 1864 storm surge was not the first to flood the low-lying Bengal coast and kill so many people, but it was the first to be well documented, in this case by the newly formed Meteorological Department of India. From its reports we get a good picture of the storm surge's destructive power and its terrible impact on the people, since the British scientists recorded not only statistics but also accounts of survivors. What is evident is that the storm surge came as a complete surprise. People had the moon to help them predict tides, and they understood the seasons well enough to predict when the rainy season should occur, but there were no warning signs that might help the people predict when the sea was about to flood their land.

Similar or even greater death tolls due to storm surges had occurred there countless times over the centuries before 1864, though we have only glimpses of each catastrophe from limited written records. As we look further back in history for evidence of great coastal floods, we must rely on archaeological and geological studies. And while no written accounts of specific catastrophic floods exist, myths and religious documents hint of such events. The earliest known catastrophic flood took place four thousand years ago at the Harappan port of Lothal on the Indian coast of the Gulf of Cambay, connected to the Arabian Sea, which is west of the Bay of Bengal. Lothal was destroyed at least four different times, the worst destruction occurring in 1900 BC, when a large region around the port was leveled by a storm surge.[14] Thus it is not surprising that stories of these floods show up in the earliest written religious works of ancient India. The *Satapatha Brahmana,* written between 800 and 500 BC, recounts a story of a worldwide flood.[15] This is India's equivalent of Noah's Ark and the Great Flood (or Deluge) in the Book of Genesis. The Indian version is quite similar. The only twist is that the Indian Noah, named Manu, is warned of the great flood by an Indian god name Vishnu, who takes the form of a fish. But the warning is the same—the flood will cover the world and kill evil people—and Manu is told to build a vessel for him and his family and some animals, but in this story the fish god pulls the vessel to safety as the world floods.

Nearly every region of the world has a great-flood story. While this fact is cited as evidence for a real worldwide flood (an impossibility, of course) by those who believe these stories to be true, it has also led to geological investigations by those looking for a real event (but not a worldwide event) that could have inspired one or more of these great-flood stories. Entire books have been devoted to this subject.[16] Here we will not go into all the arguments for and against many of the theories that proposed a location or a cause of some large ancient flooding event. We will only say that if there was a catastrophic flood in the distant past that served as the inspiration for the story of Noah's flood, there are reasons to believe that such a flood would have been caused by a huge storm surge from the sea.

The oldest known great-flood story was written sometime before 2075 BC in Sumerian, the language of the civilization that lived in the lower Euphrates valley at the north end of the Persian Gulf.[17] In the following millennium Babylonians lived in the Euphrates valley, and they had their own version of the deluge story. The Babylonian version is very similar to the later Hebrew version in Genesis, which apparently had two authors, the combined result written sometime after 586 BC. There is also a Greek version, as well as a Muslim version in the Koran.[18] Many historians and scientists have suggested that if one of these stories was inspired by a real flood, with other storytellers then copying the first story and adding their own details, then the most likely location for that flood was near the mouth of the Euphrates when the Sumerians lived there. Every deluge story involves the prediction of a great flood by some god, and a necessary response by the Noah character in order to save himself and his family, namely, building a vessel of some kind. In this case such "predictions" are prophecies. But the interesting aspect of an ancient gigantic flood occurring in the Euphrates valley at least 4,000 years ago and inspiring the Noah's Ark stories is that a storm surge makes more sense as the flooding mechanism than does the typically proposed river flood.

A river overflowing from heavy rains initially seems to make sense as the basis for the great-flood story. But the Euphrates flooded many times, and not being a rare occurrence, such floods would be unlikely to inspire an extraordinary story. The Nile flooded almost every year, and Egypt does not have a deluge story in its heritage. But in the biblical and other deluge stories, all or part of the flood is described as coming from the sea. The relevant verse from the Bible is in Genesis 7:11: "In the six hundredth year of Noah's life, . . . were all the fountains of the great deep broken up, and the windows of heaven were opened." Here "great deep" means the sea, and the "fountains" are the waters from the sea that contributed to the flood along with the rain. In the Bible, and in the other deluge stories, Noah's Ark was not carried toward the Persian Gulf, as would have been the case with a river flood, but instead was carried to the north toward Armenia, as would be the case with water from the sea inundating the land.[19] Some biblical scholars equated the "fountains of the great deep" with subterranean waters coming from the deepest part of the sea, perhaps released by an earthquake. But neither they nor the authors of Genesis understood that winds can push the waters from the sea onto the land (the storm surge), which is a much better explanation for why such a vast area was flooded. A tropical cyclone in the Persian Gulf would of course bring rain, perhaps many days' rain, but it would be the wind-produced storm surge that would cause the flooding, as we saw in Bengal in 1864 and as we will see in more examples in this chapter and the next.

The reference to the flood coming from the sea is more explicit in nonbiblical versions of the deluge story, and perhaps there was even some understanding of the wind's role. For example, in the Babylonian story, which is part of the *Epic of Gilgamesh* and was originally found on the eleventh of twelve clay tablets pieced together by George Smith in 1872, we have this description: "For six days and nights the wind blew, and the deluge and the tempest overwhelmed the land. When the seventh day drew nigh, then ceased the tempest and the deluge and the storm, which had fought like a host. Then the sea grew quiet, it went down; the hurricane and the deluge ceased."[20] The Babylonian Noah goes on to say, "I looked upon the world, and behold all was sea." If a large area of low-lying land is covered by the sea, the flooding can reach to the horizon, and thus it would appear to cover the world.

The first author to hypothesize that the deluge came from the sea was Eduard Suess in 1885, but to explain the sea rushing in to cover the land, he felt it necessary to propose not only a large storm but also an earthquake-produced tsunami occurring at the same time.[21] However, a storm surge can cause greater flooding than a tsunami. The flooding caused by a tsunami, as destructive as it can be when it first happens, does not last long, but the flooding caused by a large and lengthy storm can cover the land for days. What would be required is a huge tropical cyclone developing over the Arabian Sea and then moving up the Persian Gulf.[22] Large tropical cyclones are extremely rare in this region, but they have occurred, and that rarity makes each of them a noteworthy event. Recently, on June 3, 2007, Cyclone Gonu reached Category 5 status with 160-mile-per-hour winds and became the most intense cyclone on record in the Arabian Sea. It hit Oman, at the entrance to the Persian Gulf, only the third time in twelve centuries that a tropical cyclone made landfall there, the other two occurring in 865 and 1890. Gonu's storm surge caused flooding in Oman and in the United Arab Emirates. As powerful as Gonu was, it could not get far into the Persian Gulf because of the narrow entrance. The northern Arabian Sea narrows down into the Gulf of Oman, which is connected to the Persian Gulf by the very narrow Strait of Hormuz, only thirty miles wide. Gonu was weakened by the land and barely made it into the Persian Gulf. However, four thousand years ago sea level was higher in this region than it is today, and it is likely that the entrance to the Persian Gulf was wider.

Insurance companies often talk about the so-called hundred-year storm—the really big storm that happens on average once every hundred years. But is it possible that around four thousand years ago a thousand-year storm made it into and up the Persian Gulf to the mouth of the Euphrates? Its storm surge would have catastrophically washed over the low-lying lands of the Euphrates valley, producing a great flood that would in the following centuries grow into more than a legend, a flood that some would come to believe covered the world, except for a mountaintop on which an ark finally landed.

Ancient documents, of course, whether religious or historical, mention only floods, not the tropical cyclones that cause the storm surges that produce those floods. It is not until the fifth century AD, according to the earliest known written evidence, that someone recognized tropical cyclones as a special type of violent wind storm that came in from the sea to destructively flood the land. People living along the Pacific coast of southern China called this type of storm *jufeng*,[23] which they described as "a wind (or storm) that comes in all four directions,"[24] an apt description of a cyclone's rotating wind system by someone who did not know that the wind was rotary. In the Northern Hemisphere a cyclone rotates counterclockwise, so that north of its eye the winds blow from the east.[25] West of the eye the wind blows from the north; south of the eye, from the west; and east of the eye, from the south. As a tropical cyclone moves through an area, an observer will therefore see different wind directions at different times.

Because of the violent floods they produced, there was a need to predict when a jufeng would strike the coast. The earliest methods were, of course, not scientific, one common Chinese guidance being that one knows when a jufeng is approaching because "before it comes, roosters and dogs are silent for three days."[26] In later documents *jufeng* was replaced by the more descriptive *da feng yu jia hai chao* ("strong storm driving [a] sea surge"), which are the earliest written references to a powerful storm surge like the one described centuries later by the British in the Bay of Bengal. The oldest known historical document that describes landfall of a jufeng on the Chinese coast is *Jin Tang Shu* (*Old History of Tang Dynasty*), published in AD 816, in which the storm surge produced by a jufeng damaged the city wall of Mizhou.[27] By the end of the ninth century, ideas on how to predict a jufeng and its storm surge sounded slightly more scientific. Some people believed that a preparatory wind (called a *lingfeng*) would occur before the arrival of a jufeng.[28] Others thought that a jufeng would come when "among summer and autumn, clouds sometimes appear to be gloomy but with light like rainbow." These warning signs were called *jumu* ("the source of jufeng"), and it was written that "sailors always watch this so as to take the necessary precautions." They also recognized the important fact that a storm surge was most dangerous when it arrived at high tide.[29] Beginning with the Song Dynasty (960–1279) the central imperial government kept official records of jufeng landfalls and their storm surges, including casualties and property losses.

The Japanese also experienced the destruction of tropical cyclones and their accompanying storm surges, but they had a different word for a cyclone, a name that came into use after one of the most crucial events in their history. In November 1274, Japan was invaded by Kublai Khan, the emperor of Mongolia and the grandson of Genghis Khan. He had a fleet of a thousand Korean ships and forty thousand Mongolian, Korean, and Chinese troops (he had already conquered Korea and northern China). Miraculously, however, his invading fleet was decimated by the high winds, large waves, and high storm surges of a tropical cyclone. The Japanese attributed the great storm and their salvation to three deities, and from that point on they referred to a large tropical cyclone as a *kamikaze*, "divine wind."[30] Five years later Kublai conquered southern China, ending the Song Dynasty and gaining additional resources for a second invasion of Japan. In August 1281, now with a fleet of 4,000 ships and 150,000 men, he again attacked. Incredibly, another tropical cyclone arrived just at the right time to once again destroy Kublai's fleet and save Japan again. That was more than enough to solidify *kamikaze* in the Japanese vernacular.

The only other people in the world at that time who had some understanding of tropical cyclones and the storm surges they produced were the native tribes living in and around the Caribbean Sea. When Europeans arrived at those shores, they found that the locals spoke of a *furacane* or a *huracán* (there were many variations of the word), apparently named after the local god of the winds.[31] Christopher Columbus learned of this after experiencing the high winds and rising sea of a hurricane while on an island in the West Indies in June 1495 during his second voyage to the New World. The Caribbean islanders, as well as the Mayans of the Yucatán, may have been the first to understand that a hurricane was a storm with rotating winds. The Mayans' wind god Huracán was frequently portrayed as a head with two arms spiraling out from its sides.[32] These spiral arms were usually in a direction that could imply a counterclockwise direction (the direction that hurricanes rotate in the Northern Hemisphere), and they looked incredibly like today's internationally recognized symbol for a tropical cyclone.[33]

From the time of Columbus on, there is a long recorded history of hurricanes and their storm surges in the Caribbean, in the Gulf of Mexico, and along the Atlantic coast of North America.[34] European explorers also encountered hurricanes in other regions of the world, though they often referred to them using their local names, which in the western Pacific was *typhoon,* that term having replaced *jufeng* as the word for *tropical cyclone.*[35] Still another word often used for a hurricane or a typhoon was *tempest,* although it was sometimes applied to any big storm.[36] *Cyclone* originated much later than *hurricane* or *typhoon* or *tempest.* In an attempt to find a universally accepted term, the word was invented in 1844 by an Englishman, Henry Piddington,

President of Marine Courts of Enquiry in Calcutta, of whom we will learn more later.[37]

The strong winds that generate a storm surge, however, do not need to come from a tropical cyclone. Away from the tropics, especially along coasts in the northern regions of the Atlantic and Pacific Oceans, storm surges are frequently generated by the strong winds of weather systems that we call *extratropical cyclones* (or *extratropical storms*). Extratropical cyclones usually are not as violent as tropical cyclones, but they cover a much larger geographical area. The northeasters (or nor'easters) that strike New England are a type of extratropical cyclone, and the violent gales from the North Sea that strike the coasts of Great Britain and the Netherlands are the winds of extratropical cyclones. These storms are typically strongest in the winter (unlike tropical cyclones, which typically occur in late summer and autumn). They also have low-pressure centers (weaker than a hurricane's), but these centers are cold rather than warm, as with a hurricane. Their winds also rotate around their center in a counterclockwise direction (in the Northern Hemisphere, but clockwise in the Southern Hemisphere) but are usually somewhat weaker, though still strong enough to produce dangerous storm surges. The Atlantic Coast of the United States has the unfortunate privilege of being assaulted by both tropical cyclones (hurricanes) and extratropical cyclones (northeasters). Hurricanes generate higher storm surges, but storm surges from northeasters can cause more property damage because they cover a much larger geographical area.

Europeans had no real experience with tropical cyclones and the storm surges they cause until the fifteenth century, when they began exploring the world.[38] By this time, however, they had had many centuries of experience with huge floods from storm surges caused by the gales of extratropical cyclones—centuries of experience, but no idea how to predict when the next gale and the next flood would occur. Without that prediction capability, the best they could do was try to build coastal defenses to be ready for the day when a really large storm surge would wash over their lands. The most vulnerable coastal regions of Europe were the Low Countries—the Netherlands, western Germany, Denmark, Belgium, and northern France. Documents from Roman times describe destruction by storm surges produced by violent North Sea gales. In AD 15 Tacitus described how a storm-surge-caused flood swept over two Roman legions.[39] Pliny the Elder had also reported on catastrophic floods, noting that the German people lived on earthen hills they constructed to rise above the highest flood levels they expected to see.[40]

This was one of the earliest human engineering responses against the destructive power of storm surges, but it was just the beginning. By the tenth century some coastal dwellers had begun building embankments and dikes to keep the sea completely away from their homes and farmlands.[41] Then they went further, and after centuries of storm surges eroding their land, they began to take back some of that land. They built dikes around parcels of marshland, cutting them off from the sea, and then drained them to become dry land on which they could live and farm. These tracts of fertile land reclaimed from the sea were called *polders*.[42] Major progress in creating and maintaining polders by the Dutch and their Frisian cousins to the north in Germany and Denmark came when they began using windmills to pump out the water, and soon thereafter windmills covered the landscape.

In spite of the success that windmill power had brought them, there remained a constant battle against the sea, maintaining their dikes against the destruction caused by every storm surge directed at them by North Sea gales. No matter how strong and how high they built their dikes, eventually a storm surge would be high enough and powerful enough to cause breaks in them and, once there were breaks, to scour out huge holes and tear them down. The mightiest of extratropical storms from the North Sea were given names, often from the nearest religious holiday, and became legends. But the lives they took and the villages they swept away were not myths, and to coastal residents the threat was very real and something for which they were always preparing.

One of those legendary floods was the Grosse Männdrenke ("Great Drowning of Men") of 1362. On January 16 the storm surge from a North Sea gale washed away at least half the population of the island of Strand, off the coast of North Friesland in northern Germany, along with many tens of thousands of people in the coastal regions of the Netherlands and Denmark. Since January 16 was the feast day of St. Marcellus, this disaster was also called the Second Saint Marcellus Flood (the "second" designation because back in 1219 on the same day a North Sea storm had caused the First Saint Marcellus Flood, its storm surge drowning an estimated 36,000 people). To the north of Strand in Denmark sixty parishes were swept out of existence; to the south of Strand in Holland the storm surge had further opened the Zuider Zee to the sea; and to the west of Strand, across the North Sea in Yorkshire, England, the port of Ravener-Odd was destroyed. On the island of Strand the wealthy port of Rungholt disappeared into the sea, never to be seen again. Rungholt eventually obtained a "lost city" mythical status, with local mariners claiming they could hear its church bells ringing when they sailed over the waters covering the ruins of the sunken city. Accounts of the catastrophe had interpreted the storm surge as God's wrath incurred by the people's sins and blasphemy and the decay of morality in Rungholt. Only a preacher and two virgins were said to have survived.[43] In England it was also supposed to have been the wrath of God that destroyed the port of Ravener-Odd, because of their piracies and other wicked deeds.

In the centuries following 1362 the dikes had been rebuilt, and the Frisians had again begun building polders and reclaiming land from the sea. In 1630

Duke Friedrich III hired a Dutchman known far and wide for his techniques in building dikes, creating polders, and using windmills to pump out water for more efficient land reclamation. He was born Jan Adriaanszoon, but because of his business, he had given himself the name Leeghwater, meaning "empty water."[44] Leeghwater lived in the village of Dagebüll on the mainland across from the island of Strand while he supervised creation of new polders. In 1633 his workers had closed the last gap in the dike around the Bottschlotter polder, and since then they had been damming the future Kleiseer polder. The duke himself had inspected the work. But on October 11, 1634, those four years of work were about to be undone in one night, for a southwesterly storm from the North Sea was driving a great storm surge toward them. Leeghwater later wrote an account of what happened on that night.[45]

In the evening, Leeghwater wrote, he became concerned about the strong southwesterly winds that had begun to blow, and this led him to return to his house, which sat on a dike eleven feet above ground level. By the middle of the night a storm surge had raised the water level high enough that wind waves riding on top of the storm surge smashed against the dike, sending spray onto the leaky roof. Lying in his bed, Leeghwater's son felt water dripping on his face. With winds howling and water almost over the dike, Leeghwater and his son fled to the safety of the dike master's mansion on higher ground. The winds changed to northwesterly, and the storm surge raised the water level so high that waves smashed in a door on the west side of the mansion, the water extinguishing the fire in the fireplace and filling Leeghwater's kneeboots with seawater. In his account, Leeghwater described his sad feelings when his son asked him, "Oh father, shall we die here?" At the north end of the mansion the water eroded the soil to a depth of a man's height, causing the floor and the hallway of the house to begin to break apart. It looked like the mansion with all its inhabitants would be washed over the dike into the raging sea, but somehow it survived.

The next morning Leeghwater saw that all the houses and tents that had been below them were washed away with all the people inside, including his master carpenter and his foreman and their families. Sea dikes that had been there for a hundred years were all destroyed. On the island of Strand only four or five strongly built parish churches remained from the twenty-four that had been there. He estimated that seven or eight thousand people drowned, including nine preachers. His own house on the dike was gone. In its place he found a large seagoing vessel sitting there. Several ships were also found resting on the highest streets in the nearby village of Husum. The storm surge washed away half the island of Strand.[46] Later calculations showed that of the Strand's nine thousand inhabitants, sixty-four hundred died, and many thousands died elsewhere in North Friesland. At least fifty-nine thousand head of livestock were lost.[47] With the dikes

Figure 3.1 "Die erschreckliche Wasser-Fluth" ("the terrible flood waters"), an etching of the destruction caused by the 1634 storm surge along the coast of North Friesland, Germany. (*From Eberhard Happel's "Greatest Curiosities of the World," published in 1683*)

destroyed, the agricultural fields in the polders were now covered with saltwater.[48]

By 1743 there was still almost nothing understood about the weather or its effects on the sea, and thus almost nothing was understood about how storm surges were produced. There have been many failed attempts to discover some correlation between patterns that people thought they saw in the weather or the ocean and the easily observable patterns in the movements of the moon, the stars, and the planets in the night sky. Scientists and theologians alike saw these predictable astronomical movements as proof that the world they lived in was an orderly, logical, and divine world. But astronomy was of no help in predicting the weather (except for the changing seasons) or in predicting movements of the ocean (except for the tides). Even the tides could not be predicted as accurately as an eclipse of the moon, which scientists could calculate down to the minute. And, as it turns out, it would be an eclipse predicted to occur exactly at 8:30 P.M. on October 21, 1743, in

Philadelphia that would stimulate Benjamin Franklin to take the first big step toward understanding the wind storms that produce storm surges.

More precisely, Franklin's thinking would be stimulated by the bad weather that prevented him from observing that eclipse, to which he had eagerly looked forward. At his home in Philadelphia the moon had been hidden by the thick clouds and heavy rain of a ferocious storm, but Franklin learned, much to his surprise, that the eclipse had been observable in Boston because the storm had begun there at a later hour. His brother Thomas, who was in Boston, said the storm did not start there until two and a half hours later, at nearly 11:00 P.M. Franklin later wrote in a letter, "This puzzled me, because...being a northeast storm, I imagined it must have begun rather sooner in places farther to the northeastward than it did in Philadelphia."[49] Then he learned that the storm had begun even earlier in Virginia than it had in Philadelphia. In another letter he wrote, "The storm did a great deal of damage all along the coast, for we had accounts of it in the newspapers from Boston, Newport, New York, Maryland, and Virginia."[50] In Boston a large storm surge had flooded the streets.

Franklin became the first person to realize that a storm was a moving system of winds and that it moved in a direction that was not necessarily the same as the direction of the winds themselves. He wrote, "Though the course of the wind is from northeast to southwest, yet the course of the storm is from southwest to northeast; the air is in violent motion in Virginia before it moves to Connecticut, and in Connecticut before it moves to Cape Sable [Nova Scotia], etc."[51] Franklin explained this with an analogy of a chimney over a fire. "Immediately the air in the chimney, being rarified by the fire, rises; the air next to the chimney flows in to supply its place, moving towards the chimney; and, in consequence, the rest of the air successively, quite back to the door [the source of cold air from the outside]."[52] He envisioned warm air rising over the Gulf of Mexico and cooler air from the northeast flowing in to replace it, this airflow beginning first near the Gulf, then a little farther away (for example, Philadelphia), and then still farther away (for example, Boston). This first, still very incomplete, explanation of a moving storm system might have been used to predict winds at one location based on winds at another location, but in Franklin's time, communication was not nearly fast enough to take advantage of his insight. Franklin's growing understanding of storms and the difficulty in predicting them did not, however, stop him from putting weather predictions in *Poor Richard's Almanack*, if only to keep up with the numerous competing almanacs of that time. We now see as absurd the idea of predicting weather a full year in advance, but all the Colonial almanacs did that, and at least Franklin, with his usual good humor, made fun of his weather predictions, realizing that, unlike his tide predictions, they had no astronomical basis.

Franklin does not seem to have ever suggested that this October 1743 storm might have had a rotating motion. He never knew that it was an extratropical cyclone, with a calm center over the ocean and air moving counterclockwise around it, which is why the winds along the coast came from the northeast. Even though Franklin had been very interested in and written much about waterspouts and whirlwinds (tornados),[53] those narrow rotating wind systems never inspired him to wonder whether, for example, a hurricane might also rotate, though on a much grander scale. However, others had been inspired to speculate. Mariners had known for centuries that with hurricanes and typhoons winds changed direction, often moving around the entire compass over a fairly short time, and that sometimes the winds stopped completely for a short while. A few recognized this as an indication that the air was rotating around a calm center, the first probably being Bernhard Varenius, a German geographer who in 1650 had obtained much information from Dutch navigators.[54] In 1698 Captain Langford of England's East India Company wrote about five hurricanes in the Caribbean Sea and referred to them as "Whirl-Winds."[55] A hundred years later, in 1801, Colonel James Capper wrote, "All circumstances properly considered,...positively prove them [hurricanes] to be whirlwinds."[56]

It would not be until 1831, however, that someone would use meteorological observations to demonstrate the rotation of a hurricane and thus lay a more scientific basis for the study of storms. It took a major hurricane to provide those data. The data came from the Hurricane of 1821, the only hurricane in recorded history whose eye passed over New York City,[57] arriving there on the evening of September 3. Its storm surge caused a thirteen-foot rise in the water in only an hour, flooding the lower end of Manhattan up to Canal Street. The waters of the Hudson and the East rivers joined to cover the sidewalks of New York. But New Yorkers were lucky; the hurricane hit at low tide. One can only imagine the destruction if it had been high tide. Two days earlier while to the east of Florida the hurricane had been a Category 5, but after passing over land several times it had been reduced to maybe a Category 2 by the time it hit New York City.

But the most important thing about this hurricane was the insight it provided William Redfield after it had passed through Connecticut and Massachusetts. Redfield was a saddle maker turned steamboat captain and a self-taught amateur scientist. He was walking through the woods with his son from their home in Middletown, Connecticut, to northwestern Massachusetts and the home of the family of his wife, who had just died giving birth. In spite of his great sadness, he noticed that all the "fruit trees, corn, etc were uniformly prostrated to the north-west" in Connecticut, but when he reached Massachusetts, he noticed that the trees and corn "were uniformly prostrated towards the south-east." Ten years later Redfield would publish a scientific paper in which he used this information (plus wind data he acquired from all over those two states as well from states along the Atlantic Coast) to demonstrate that "this storm was exhibited in the form of a great whirlwind."[58] That paper thrust him into the scientific limelight and also into what was later called the "American storm controversy."

This "controversy" was an ongoing debate that began when James Espy, a scientist at the Franklin Institute in Philadelphia, put forth a different theory on hurricanes and storms, one that said the air did not rotate around a quiet center (the eye) but instead the air rushes in from all directions directly toward a low-pressure center. This was similar to Franklin's suggestion a century before, except that Espy added to Franklin's analogy of heated air moving up a chimney the important idea that the updraft in the center of a hurricane is stronger if moist air is involved. He used the concept of *latent heat,* which is the heat required to turn a liquid into gas (or the heat released when a gas condenses into liquid). Espy realized that water vapor is lighter than dry air and that, as moist air rises and expands, the latent heat released as the water vapor condenses helps the convection to continue longer.[59] Thus, warm, moist air contributes more to convection in a hurricane than does dry air, which is one reason hurricanes form over warm oceans. This was an important contribution to meteorology, but unfortunately Espy got the wind direction wrong because he did not include the effect of the Earth's rotation.[60] The debate became national, with scientists in New York supporting Redfield and scientists in Philadelphia supporting Espy. Then it became an international debate when scientists in England supported Redfield, but scientists in France supported Espy. The issue was finally cleared up in 1856 in a paper by William Ferrel, who recognized that Redfield and Espy each had part of the answer (as so often happens).[61]

One of Redfield's supporters was Henry Piddington, the British scientist in India who had coined the word *cyclone* in 1844. He wrote forty memoirs and books about cyclones in the Indian Ocean and China Sea, finding support everywhere for Redfield's proposal that cyclones had winds that blow in large circular motions around a quiet center. But more important, Piddington was the first to write about the dangers of the cyclone's storm surge, which he called a "storm wave." He had written, "There really exists an actual *wave* or elevation of the sea,…which rolls in upon the land like a huge *wall of water,* as such a supposed wave should do, causing of course dreadful inundations."[62] Piddington had seen firsthand the havoc that a storm surge could wreak and the death it could bring in India and other low-lying lands around the Bay of Bengal.

At this point, we should make clear the meanings of the different terms used for the variation in the height of the sea's surface, including storm surge, for there has often been confusion, especially in the popular press. The term *water level* is the most general term, meaning the height of the sea's surface, but after averaging out the rapid surface oscillations due to wind waves (which usually have periods from a couple of seconds to about twenty seconds; see Chapter 5). Scientists define a *storm surge* (called a *storm wave* by Piddington and a *sea surge* by others) as a change in water level along the shore caused by winds, and to a lesser extent by changes in atmospheric pressure. Scientists use this term even if the winds are not strong enough to be categorized as a storm. The storm surge has sometimes been called a *storm tide,* but that term is more correctly used to mean the total water level change, namely, the storm surge plus astronomical tide. Here we put the word *astronomical* in front of *tide* because *tide* has often been used to mean simply a change in water level, including the effect of the wind, even though scientifically that word should only be used to refer to changes in water level due to the effects of the moon and the sun (as we talked about in Chapter 1). Others refer to storm surge as a *meteorological tide,* and although this still uses *tide* to mean a water level change, it at least says that the change in water level is of meteorological origin, namely, from the effects of winds and changing atmospheric pressure. The storm surge moves toward and sometimes onto the land as a very long wave, which steepens when it moves into shallow water. This long wave has sometimes been erroneously referred to as a *tidal wave.* To further add to the confusion, *tidal wave* has been often erroneously used for *tsunami,* which is the very long wave caused by an earthquake or volcanic eruption (as we will talk about in Chapter 7). Even today, especially in newspaper accounts, the terms *tide* and *tidal wave* are often misused, as is *sea level.* The term *sea level* implies an averaging out of all water level oscillations, including the tide and storm surge. It is used when we are talking about very slow long-term changes. These longer-term sea levels (usually yearly sea levels) are used in El Niño and climate change studies (see Chapter 10). Researchers studying sea level rise due to global warming try to average out all other oscillations that might bias the calculation of the trend, that trend hopefully only being determined by very slow changes in the heat in the upper ocean and the amount of melted land ice entering the sea.

For Piddington, the destructive power of a storm surge was first dramatically illustrated at Coringa on the Bay of Bengal coast of India in December 1789. Coringa and its 20,000 inhabitants disappeared in a single day after being hit by a succession of three great storm surges.[63] Fifty years later another 20,000 people died when the rebuilt town was assaulted by another storm surge. After witnessing so much destruction and loss of life caused by the flooding from these and other storm surges, Piddington became very sensitive to potential vulnerabilities at coastal locations. Although as yet in no position to predict when a storm surge would strike a coast, he believed he knew enough to predict which stretches of coast and which harbors would be most seriously affected when a storm surge finally came. In 1853 Piddington was worried about the vulnerability of a new port that was proposed for the

Mutlah River as an eventual replacement for the Port of Calcutta (because someone believed the Hooghly River was silting in).[64] He sent an open letter to Lord Dalhousie, the Governor-General of India, about the proposed Port Canning, explaining how storm surges could flood it. He wrote, "Everyone must be prepared to see the day when in the midst of the horrors of a hurricane they will find a terrific mass of salt water rolling in or rising upon them with such rapidity that in a few minutes the whole settlement will be inundated to a depth of from five to eighteen feet of water."[65] Lord Dalhousie and others ignored his warning and built Port Canning on the Mutlah. On November 2, 1867 (nine years after Piddington's death), a cyclone produced a storm surge that did just what Piddington had predicted, passing "over the town with fearful violence."[66] At the same time a storm surge in the Hooghly River did no damage to the Port of Calcutta. Five years later Port Canning was abandoned.

Although Piddington was best known by mariners who used his *Sailor's Horn-Book for the Law of Storms* as a guide for avoiding the dangers of the eye of a tropical cyclone when at sea,[67] he also had ideas about predicting storm surges from cyclones making landfall. Though the science was at that point incapable of such prediction, he suggested as early as 1842 that telegraphs might provide storm surge warnings.[68] A prediction of the time of arrival of a cyclone and its accompanying storm surge at certain coastal locations might be possible, he thought, if real-time data via telegraph were received from locations where the cyclone had already been. Piddington hoped that "our children may see this done."

But Piddington himself lived to see real-time telegraph capability enabled, though not in India, nor along the southern coast of China, where he thought tropical cyclone patterns were conducive to such a warning system. The first system was built in the United States and was not used for tropical cyclones. By 1850 Joseph Henry, first head of the Smithsonian Institution in Washington, DC, had established a network of telegraph weather stations across the United States, which were used to provide weather warnings.[69] The huge map in the lobby of the Smithsonian showed in almost real time the weather conditions across the country.[70] By watching this map over time, Henry, and many interested onlookers, could clearly see the eastward or northeastward movement of storms, as Franklin had first suggested, and from that movement he could estimate where each storm was headed. Eventually a storm would reach the Atlantic Ocean, where its winds might generate a storm surge along the coast, and it would be possible to make a rough prediction of its size and timing. Unfortunately, the Civil War and a fire at the Smithsonian put an end to the telegraph network. It would not be until the 1870s that another telegraph weather network would be created in the United States, this time run by the U.S. Army Signal Service.

In 1860, two years after Piddington died, a telegraph network of coastal observers was begun in Britain under the direction of Robert FitzRoy, the first director of Britain's first Meteorological Department.[71] Being a Navy captain, later an admiral, FitzRoy believed that coastal areas were most in need of storm warnings because of storm-wave-caused floods and the threat to mariners heading to sea in their boats. Two years earlier he had begun to distribute specially built barometers to dozens of fishing villages, initially in Scotland. He used telegraphs to receive meteorological data from twenty-two villages. In February 1861 he began to send them back predictions, first storm-warning signals and then *daily weather forecasts* (a phrase coined by FitzRoy). These forecasts were calculated using all the weather information he received by telegraph, and was based on an understanding of the rotary nature of extratropical cyclones (that understanding having been advanced by his careful analysis of the Royal Charter Storm in October 1859, which sank 343 ships, including the *Royal Charter*).

But FitzRoy's weather forecasts were not appreciated by all, especially astro-meteorologists (astrologers, in actuality) who made a great deal of money selling almanacs with weather predictions erroneously based on movements of the moon, the sun, and the planets. Unfortunately FitzRoy was equally criticized by scientists in the Royal Society, who said that his practical science, which was meant to be used by everyone, was merely popular science that played into the hands of the astrologers. FitzRoy ended up killing himself on April 30, 1865, driven to that sad end, his wife believed, by anxieties over forecasts that failed and the criticisms aimed at him from seemingly all directions. His daily weather forecasts and his storm warnings were stopped the next year. The public reaction to the stoppage was so great that the storm warnings were brought back the following year, but it would be a decade before the daily weather forecasts were resumed.

As worthless as the predictions of the astro-meteorologists were, all it took was one perceived success to raise their stature in the minds of the public (and in their own minds as well) and thus to cause problems for real scientists who were trying to save lives with prediction methods based on ocean physics. Such a perceived success in predicting a storm and its destructive storm surge using an astro-meteorological technique occurred in October 1869 at the Bay of Fundy, with the Saxby Gale and its accompanying Saxby Tide (which more correctly could have been called the Saxby Storm Surge).[72] The northern end of the Bay of Fundy was an unlikely place for a destructive storm surge. With an astronomical tidal range that can reach forty-eight feet, a storm surge arriving anytime except near high water will not flood the countryside, because the countryside is above the highest tidal high-water line. But the timing turned out to be just perfect on October 5, 1869. To begin with there was a perigean spring tide, which (as we saw in Chapter 1) means the moon was the closest it could be to the Earth and the sun and the moon were working together, it being new moon. And moving up the Atlantic Coast was a hurricane that proceeded up the west coast of the Gulf of Maine and then into the Bay of Fundy, a rare occurrence. That combination of perigean spring tide and six-foot storm surge produced the largest one-day tidal range in history, fifty-four feet at Burntcoast Head Lighthouse, Nova Scotia (although that record is somewhat misleading, since the people lived with forty-eight-foot astronomical tidal ranges, the world's largest). Still, a six-foot storm surge can do a lot of damage. The storm tide flooded much of the land that the Canadians had reclaimed from salt marshes using dikes, much as the Dutch and Frisians had done in Europe, those dikes being overtopped. Salt water remained trapped behind the dikes for days after the storm.

Although the casualties due to the Saxby storm tide were less than in other disasters (about one hundred people died), the stories were no less tragic. A local paper, the *Moncton Times*, reported the story of the O'Brien family, caught by the storm surge. They "were awakened in the night to find their dwelling partly filled with water and all means of reaching dry land apparently cut off. In this dreadful emergency their only chance of escape seemed to be by means of a raft, and hastily constructing such a one as the drift timber within reach enabled him to make, he and his family got upon it and committed themselves to the mercy of the waves. The wind blew them across the river but unfortunately during the journey the raft parted and the four little boys drowned...in the midst of the fearful storm and thick darkness of Monday night."[73]

What had given the Saxby Tide its name was that some believed it had been correctly predicted ten months earlier in England by Lieutenant Stephen Martin Saxby. Saxby was an acting Instructor of Engineering in the Royal Navy, who had his own system of weather prediction based on the moon.[74] In 1868 Saxby had written a letter to the editor of *The Standard* in London, which was published on Christmas day. In that letter he wrote, "With regard to 1869, that at seven A.M. on October 5,...the new moon will be on the Earth's equator when in perigee and nothing more threatening can, I say, occur without miracle."[75] On September 16, 1869, a second letter by Saxby was published in *The Standard,* in which he wrote that although his "warnings apply to all parts of the world; effects may be felt more in some places than in others."[76] Given the lack of specificity with regard to where the threatening event would take place, and given that at any moment somewhere in the world there would be storms, and also given that there would be a perigean spring tide on October 5, then it was a good bet that somewhere in the world there would be flooding caused by a storm surge. Saxby got lucky in the Bay of Fundy.[77] On October 8, 1869, the editor of the *Daily Morning News* in Saint John wrote that he was inclined to believe that "Saxby is in this instance a true prophet," but he also admitted that Saxby would instead be a "charlatan" if in the following days they did not hear of more lives lost, property destroyed, and vessels sunk all over the world.[78] Of course, they never did.

It was then true and always will be true that to predict a storm surge, one must first be able to predict the storm. In the late 1800s meteorological science was still not advanced enough to develop a statistical or dynamical prediction technique for storms. The only method available was the use of real-time weather data from a telegraph observation system letting people know that a storm was heading in their direction—the methods of Henry and FitzRoy and the ideas of Franklin and Redfield still being the best approaches available. Such systems were begun again in the United States and in England, as well as in the Netherlands and France. These systems, being in the north, had to respond only to relatively slow-moving extratropical cyclones. The first successful telegraph network used for predicting hurricanes, and their very dangerous storm surges, was established in the Caribbean by Spanish Jesuits at Real Colegio de Belén in Havana, Cuba.[79] Father Benito Vines, the director of Belén from 1870 to 1893, carried out an ongoing study of the characteristics of hurricanes in the Caribbean, publishing two books and becoming an internationally recognized expert. By 1888 Vines had established a telegraph meteorological observation network with seven regularly reporting stations (Trinidad, Barbados, Martinique, Antigua, Puerto Rico, Jamaica, and Santiago de Cuba) and with as many as thirteen other stations also reporting at times. Using real-time data from these stations and knowledge that he had gained from studying so many hurricanes over the years, Vines made the first hurricane forecasts, most very successfully. After Vines's death, Father Lorenzo Gangoiti became the director and continued Vines's program and its hurricane forecasting.

A national weather service with a telegraph weather network was started in the United States in January 1871, this time under the auspices of the Signal Corps of the U.S. Army. In 1891, after twenty years of disagreements between meteorologists and some military officers about the need for scientific research to improve weather forecasting, this weather service was transferred to the Agriculture Department and became known as the Weather Bureau. A scientist named Mark Harrington was selected to be its first civilian chief. However, a year later, the new Secretary of Agriculture, Julius Sterling Morton, brought in by the newly elected President Cleveland, was opposed to scientific research, even attacking scientists within the Weather Bureau, and preferred to ask a major in the Signal Corps for weather advice. Morton eventually convinced the President to fire Harrington on July 1, 1895, and he found a replacement to his liking within the Bureau, Willis Luther Moore, a man with more political ambition than scientific prowess.[80] As we will see in the next chapter, this selection started a chain of events that five years later would lead to the greatest natural disaster, in terms of deaths, in U.S. history—a disaster caused by the large storm surge produced by a hurricane of which the residents of Galveston should have been warned, but were not.

CHAPTER 4

Defending Our Coasts

Flooded Cities

In September 1900 Willis Luther Moore, head of the newly formed U.S. Weather Bureau, was permitted to establish a U.S. weather office in Havana, because the United States was then administering Cuba, since winning the Spanish-American War. In two years Cuba would receive its independence, but in the meantime Moore could use his new weather office in Havana to acquire weather data for hurricane forecasts to be made in Weather Bureau offices in Washington, DC. Moore was jealous and resentful of the forecasts that were already being distributed by the much more experienced observatory at Real Colegio de Belén in Havana. To squash these competitors, he managed to convince the War Department to ban the distribution by telegraph of weather warning cables from Belén—even though five years earlier the U.S. Navy's Hydrographic Office had published a translation of a book about West Indian hurricanes by Father Vines, Belén's director, referring to him as "the accomplished meteorologist."[1] The War Department apparently accepted Moore's questioning of Belén's weather forecasting skill and motives—at least until complaints led to a rescinding of the ban. But by then it was too late, as we shall see.

A hurricane crossed over Cuba on September 4 and 5, 1900. At Belén, Father Gangoiti forecast that it would increase in strength and move into the Gulf of Mexico with Texas as its most likely target, but because of Moore's action, no one in the United States knew of Father Gangoiti's forecast.[2] All that was known was that Moore's forecast from Washington said that the hurricane would not enter the Gulf of Mexico and instead would move up the Atlantic coast of Florida. On the evening of September 5 at Key West, at Florida's southern tip, the gale-force winds blowing from the northeast abruptly quieted and then resumed from the south at gale force, indicating that the eye of the hurricane had gone by. But this information was ignored by Moore and the Washington office. They were busy warning New Jersey fishermen to stay in port. The next day there was no evidence of a storm along the Atlantic coast of Florida, but there were wind reports from the Gulf Coast that pointed to a cyclonic circulation. Moore finally telegraphed his Gulf Coast offices from Pensacola to New Orleans to raise storm warnings. At the same time, however, more urgent hurricane warnings were being raised from Cedar Keys, on Florida's west coast, south to Miami and then north along the Atlantic Coast as far as Savannah, Georgia. On the following day the storm warning along the Gulf Coast was extended westward to Galveston, Texas, where it caused scarcely any concern among the population. That less urgent warning remained in effect through Saturday, September 8, the day the hurricane and its catastrophic storm surge hit Galveston.[3]

In Galveston even Isaac Cline, the Weather Bureau's Local Forecast Officer and Section Director for the city, did not know about Father Gangoiti's forecast. It might have surprised him—assuming he believed it. For Cline believed that no serious hurricane would ever hit Galveston, and he had stated that unequivocally when asked for his opinion by the *Galveston Daily News*.[4] There had been a debate over whether Galveston needed a seawall as protection from a storm surge produced by a hurricane. Proponents for the seawall pointed to the once thriving port of Indianola, a hundred miles southwest of Galveston, which had been hit twice by hurricanes, in 1875 and 1886, the first storm surge destroying most of the city and the second wiping the rebuilt city off the map. Both hurricanes had crossed Cuba and gone by Key West, precisely as the 1900 hurricane had just done.[5] In the end, the seawall was not built, probably to a large degree because of the opinion of Cline, an opinion based on his total lack of understanding of hurricanes and storm surges.[6]

And yet on the morning of Saturday, September 8, even Cline was concerned enough to go down to the beach to measure the period of the swells, the long, slow waves coming in from the sea. He sent a telegram to Washington: "Unusually heavy swells from the southeast, interval one to five minutes, overflowing low places south portion of city, three to four blocks from beach. Such high water with opposing winds never observed previously."[7] Under normal conditions those "opposing winds" would have lowered the water level, not raised it. The information Cline put in the telegram should have indicated to him that a ferocious storm was somewhere beyond the horizon, but he never mentioned the word "hurricane."[8] Another person down at the beach that morning, Cline's neighbor Dr. Samuel Young, came away with a more alarming interpretation: the higher-than-normal tides and the large waves meant that a cyclone was coming. Young was Secretary of the Cotton Exchange, but also an amateur meteorologist who enjoyed studying the weather map drawn each day at the Cotton Exchange by one of Cline's employees. Young rushed from the beach to the Western Union office and sent a telegram to his wife, who was returning on a train with their children from a holiday. He told her to stay in San Antonio until he sent her word.[9]

Unknown to Cline, the hurricane was moving toward a landfall forty miles west of Galveston, the worst possible location, for that meant winds on the storm's right (to the east) would blow toward Galveston and drive water onshore. For those lucky enough to be farther west on the other side of the eye, the winds would be blowing offshore and the water level would be lowered, not raised. Even though the winds at Galveston did not reach storm speed until 1:00 P.M., the water level rose throughout the day because of water being pushed ahead of the hurricane. By 3:30 P.M. half of Galveston's streets were covered in water. By 5:00 P.M. wind speeds reached hurricane level. By 5:30 P.M. Isaac Cline had left his weather office and was at home, three blocks from the sea, with his pregnant wife, three young daughters, his brother Joseph (who also worked for the Weather Bureau), and about forty-five neighbors who had sought refuge in his house, which was supposed to be the strongest building in that part of the city. At 6:30 P.M., when Cline looked out his door, he saw that "where once there had been streets neatly lined with houses there was open sea."[10] The water level rose steadily, but at 7:30 P.M. it increased another four feet almost instantaneously, and everyone in Cline's house rushed to the second floor. By 8:30 P.M. the water reached twenty feet above the usual high-water line. The scene on that second floor had already played out in hundreds of houses now gone and would be repeated in hundreds more all over the city. People huddled in the dark or perhaps in the glow of a single candle, on the second floor or in an attic. They were surrounded by frightening sounds—the howling of the wind, the drumming of the rain, the water splashing on the floor below them, the creaking of beams as the waters pushed and twisted the house, the explosion of plaster off the walls. People screamed or cried or prayed or sang or held each other tight.

Cline's house had so far withstood the violent rushing waters, but other floating houses or pieces of houses slammed into it like battering rams. Finally, something very large crashed into the house, pushing it off its foundation, and it collapsed into the sea. Just before this, Joseph Cline had broken through a window and jumped out with two of Isaac's daughters. Everyone else went under with the house as it capsized and sank. Somehow Isaac Cline escaped the submerged house and made it to the water's surface, as did his youngest daughter, but his wife was lost, along with most of the people in his house. Joseph and Isaac with the three girls climbed onto a floating roof. There were similar escapes in other parts of Galveston, the escapees then having to face a world only slightly less scary than the one they had just left—a harrowing ride on floating debris in almost total blackness, with 150-mile-per-hour winds throwing all manner of projectiles at them, its shrieks hiding the screams coming from the houses they floated past.[11]

Figure 4.1 A photograph of some of the destruction from the 1900 Galveston storm surge. (*National Oceanic and Atmospheric Administration*)

The sun rose the next morning on a landscape of wreckage, as far as the eye could see, from over 3,600 collapsed buildings (see figure 4.1), among which 6,000 bodies were found.[12] That number of deaths was the most ever from a natural calamity in the United States. Many, if not most, of those deaths probably would have been prevented if Willis Moore had allowed the Cuban forecast to be telegraphed or if Isaac Cline had not discouraged building a seawall. Cline paid a tragic price, losing his wife and unborn child. The whole experience changed his life, and from then on he devoted himself to studying hurricanes and storm surges, eventually becoming a nationally known expert and writing two books and many papers on the subject.[13] In a report published right after the storm surge, Cline admitted, "I believe that a sea wall...would have saved much loss of both life and property."[14] A seawall was completed in 1904 and proved its worth when a hurricane hit Galveston on August 15, 1915. Although of similar strength as the 1900 hurricane and making landfall at almost the same location, and thus producing a comparable storm surge, this time, with proper warnings and the seawall, the deaths in the same area of the city of Galveston amounted to only eleven.[15]

As horrific as the 1900 storm surge was to the people of Galveston, it paled in comparison to the horror that storm surges have produced in countries around the Bay of Bengal, where deaths over the centuries are estimated to be in the millions. The 1864 West Bengal storm surge that we looked at in Chapter 3 killed at least 80,000 people. Another storm surge hit East Bengal (now Bangladesh) on the night of October 31, 1876, and killed a staggering 215,000 people.[16] Storm surges killed comparable numbers in 1839 and 1897, although exact numbers are not agreed upon. In the twentieth century, storm surges would cause other large death tolls, which we will look at shortly. And in between the really large catastrophes many more smaller storm surges resulted in fewer casualties, but still much larger than at Galveston.

It is difficult to fully comprehend the human misery that such large death tolls represent. We are moved by the personal accounts given by survivors in Galveston and Bengal, telling of loved ones lost and of the power of the sea as it overwhelmed them. The scenes of fear and desperation in Bengal had similarities to those in Galveston, but in Bengal families held on to each other in single-room huts made of bamboo, mud, and thatch, frightened by the screaming winds outside, but ultimately being overwhelmed by a wall of water. Their huts broke apart more quickly than the two-story wooden houses of Galveston, and they provided less floating debris to be grabbed onto by those few who survived the destruction of their homes. Tree branches were their only real salvation, but we can never know how many desperate fathers released their lifesaving branches to follow after the crushed huts swept away into the night with their wives and children inside. We may try to extrapolate those stories to the other faceless thousands or tens of thousands whose lives must have ended in equally harrowing ways. To appreciate the fear and sorrow throughout Galveston on the night of September 8, 1900, we imagine 3,600 houses with the same fear and sorrow that was inside the Cline house. In Bengal on the night of October 31, 1876, we would have to imagine the fear and sorrow in 40,000 bamboo huts. Can we fully grasp the idea of 40,000 families washed away in the night and the sudden horror inside those dark huts when the storm surge struck?

What we do begin to grasp is what the British began to understand in the mid-1800s, that the Bay of Bengal was a tragically "special" place. Even in more recent years, for example, the hundred-year period from 1897 to 1996, there were 117 significant storm surges in the Bay of Bengal.[17] Of all major hurricane disasters around the world, with human death tolls of ten thousand or more, 85 percent occurred along the coast of the Bay of Bengal—in India, in Burma, and especially in the area now called Bangladesh (formerly East Pakistan and before that East Bengal). But why the Bay of Bengal? Fewer tropical cyclones actually form there compared with other tropical oceanic regions of the world. What the Bay of Bengal has is a special situation, geographical and human, that increases the deadliness of storm surges generated there by tropical cyclones. First, the Bay of Bengal is a confined area with land on all sides except the southern boundary, virtually assuring a landfall somewhere for every cyclone. Second, the funnel shape of the Bay of Bengal tends to amplify a surge's height, and as a cyclone moves northward, this effect becomes stronger. Hydrodynamic storm surge models show that a storm surge's height is doubled by a coastline with a right-angled corner, as in Bangladesh.[18] The models also show that the wide continental shelf and slowly decreasing depths at the northern end of the bay amplify the height of storm surges as they move toward the coast. Third, and very important, the coast along the Bay of Bengal is very low lying. There are no high lands to which the people can escape. Much of the coastal land in Bengal is river delta that is not only easily flooded, but whose rivers and channels provide pathways up which the storm surges can travel, becoming larger and steeper as channel widths narrow, often becoming a moving wall of water (a bore). And fourth, the most important factor causing the high death rates is the large population that lives near the shore—people who until very recently were not provided with proper warnings and did not have anywhere safe to go. As we will see, it would take another hundred years before a warning capability would be provided and before concrete storm shelters on stilts would be built.

In the more developed United States and Europe, in the early twentieth century, telegraph-based and later radio-based warning systems were implemented, which along with stronger buildings and more developed transportation systems greatly limited deaths due to storm surges. With a hurricane the emphasis was on predicting its track across the ocean and especially the location of its landfall. Since hurricane winds rotate counterclockwise in the Northern Hemisphere, the storm surge would be most destructive to the right of the eye (looking toward shore), and people were told to evacuate if they lived east of a predicted landfall on the Gulf Coast or north of a landfall on the Atlantic Coast. The predicted time of landfall was very important for scheduling evacuations. It also indicated whether the storm surge would arrive at high tide, when the flooding would be much worse.[19] A different approach was necessary for storm surges produced by extratropical cyclones, such as northeasters along the U.S. Atlantic Coast and North Sea gales that assaulted Great Britain and the Netherlands. They covered a much larger geographical area, and they usually had no strongly developed eye with its extremely high winds, as did hurricanes. A telegraph warning system was still important, although additional analysis had to be done to determine how much water the winds would move toward the coast.

Although storm surges are generated by both extratropical cyclones and tropical cyclones (hurricanes), the two mechanisms are different.[20] For a hurricane approaching the coast, the water is pushed up onto the land by wind blowing directly toward shore. For an extratropical cyclone, which covers a much larger area and can have winds far out over the ocean, it is the wind blowing parallel to the coast that generates the storm surge. In this case, these alongshore winds produce an ocean current that is acted on by the Coriolis force (due to the rotation of the Earth, as discussed in Chapter 1), transporting water to the right of current direction (in the Northern Hemisphere).[21] For example, a northeaster off New England generates a current that moves from the northeast toward the southwest along the coast, and the Coriolis effect pushes water toward the coast, raising the water level. The storm surges from both hurricanes and extratropical cyclones are also affected by changing atmospheric pressure, but this has a much smaller effect than the wind.[22]

To make a storm surge prediction, and thus be able to tell if a flood is coming, scientists tried to correlate changes in water level with changes in wind speed and direction. They formulated some general rules for changes in water level at specific locations on the coast based on the wind's speed, direction, and length of time it blew.[23] Some formulas were based on differences in atmospheric pressure over a geographical area, which indicated likely wind speed and direction. But such predictions were generally poor. There were, however, a few coastal regions where because of a fortuitous dynamic situation, a reasonable storm surge forecast at one location could be made based on recent water level measurements at another location. One example was the east coast of Great Britain facing the North Sea. Arthur Doodson, the British tidal expert at the Liverpool Tidal Institute mentioned in Chapter 2, discovered that storm surges were typically generated by storms at the northern end of Britain, after which they traveled southward along the east coast and then moved up the river Thames, sometimes reaching London. This is what happened with the storm surge that surprised London on the night of January 6, 1928. It first measured a foot and a half above the normal tide, at Dunbar, Scotland, at 3:00 P.M. It traveled south along the coast, and eight hours later it reached the mouth of the Thames, by then having increased to over five feet above normal tide. The surge's arrival coincided with a high tide and so caused a great deal of destructive flooding. Part of the storm surge continued up the Thames to London, still growing. Another part crossed the narrow Dover Strait, at the eastern end of the English Channel, and traveled northward along the coast to Germany and Denmark, causing more damage.[24]

Having seen this progression along the coast with other storm surges, Doodson felt that storm surge predictions along the southern part of Britain's North Sea coast and up the Thames might be made based on water level data telegraphed or radioed from Scotland. In the following decades, however, no government funding could be obtained to implement such a real-time-data-based storm surge warning system. In the meantime, oceanographers in Great Britain, the Netherlands, and Germany continued to study the problem. They looked for more sophisticated statistical techniques for predicting storm surges based on relationships between water level data and wind and pressure data. But to get the government to increase support for such research and to implement the needed warning system it was going to take a disaster. That disaster would arrive in 1953.

On the cold Saturday night of January 31, 1953, an extratropical cyclone with gale winds averaging ninety miles per hour developed north of Great Britain and moved into the North Sea.[25] The winds generated a large storm surge that traveled down the east coast of Britain, up the Thames, and across the English Channel to the Netherlands. The storm surge coincided with a spring high tide and thus overcame embankments and sea defenses, taking both nations by surprise.

In Britain the storm tide overwhelmed everything. Seawalls and dikes along the east coast of Britain and up the Thames River were breached in more than twelve hundred places.[26] More than 160,000 acres of farmland were inundated with salt water, making them unproductive for a long time, and at least forty-six thousand head of cattle were killed. More than twenty-four thousand homes were damaged or destroyed in the hundreds of towns that were flooded. Phones were knocked out over large areas. Not only had coastal defenses failed, but there had also been no warning provided to the people. It did not help that it was a weekend and government offices were closed, or that it was at night and people were in bed when the storm surge flooded their homes.

Trapped inside their homes with freezing water rising higher and higher, they climbed onto chairs and then onto tables or hung on doors. After fighting for hours to stay out of the water, some collapsed and slipped from their place of safety into the deadly water. Some pulled themselves onto floating furniture. Mothers tried to keep their children above the water even as they themselves stood knee deep and then waist deep in the numbing water that was slowly sapping their strength. "Children died quietly of exposure in their parents' arms as they tried to hold them, hour after hour, above the water: [one mother saying,] 'After a while he did not speak any more and appeared to go to sleep.'"[27] The lucky ones escaped to higher floors or attics, which though cold kept them above the freezing waters and shielded them from the frigid gale winds outside. They could not, however, keep out the sorrowful sounds of neighbors screaming from inside houses next door. Those without an attic or second floor desperately tried to smash holes in their ceilings so they could climb out onto their roofs before the waters reached them. Once on the roof they died from exposure unless they were rescued quickly. And their fellow citizens were trying to rescue them. One was a twenty-two-year-old American serviceman who managed to save twenty-seven people clinging to rooftops. Reis Leming could not swim, but he still managed to push a small rubber raft through the freezing water, fighting roaring winds and high waves.[28] He made three trips before being overcome by exposure. A British nurse massaged him back to life, and nine days later the Queen gave him the George Medal for bravery, the first American in peacetime to receive that honor. This was just one of hundreds of heroic rescues in Britain, most by neighbors rescuing neighbors.

Because of this heroic effort there were only 307 deaths. There were, however, huge economic losses, and this massive flooding is considered the worst natural disaster in Great Britain in modern times. It occurred only seven years after the end of World War II, which is one reason why coastal defenses were vulnerable, there not having been resources or time to maintain them during the war. Now nature had let them know it was time to begin repairing them. Over thirty thousand workers from all over the country, including thousands of British and American troops and their valuable equipment, quickly began repairing seawalls and dikes, trying to complete the work before the next spring tide in mid-February. More than fifteen million sandbags were used along with countless truckloads of gravel, clay, and other materials.[29]

As it turns out, London had been lucky. The deaths and damage caused by the storm surge could have been much worse. The heart of London had avoided being flooded by literally inches. Very early in the morning on February 1 the water level at London Bridge was more than six feet higher than the predicted tide and higher than ever measured there before. The embankments, which had been built higher than the highest possible tide, were over-topped by the storm surge, but luckily only by a couple of inches, so only a little water rushed into central London. Breaks in dikes and embankments all along the Thames had saved London by allowing millions of gallons of water to flood the lowlands instead of reaching the city. The next time the British might not be so fortunate. If a larger storm surge ever reached London, not only would more lives be lost, especially in the subways, but the economic consequences of knocking out the nerve center and economic center of the country would be staggering. As one politician put it, such a storm surge would achieve what the German Blitz had been unable to accomplish.[30]

A short time after the storm surge hit England, it was the people of the Netherlands who were being assaulted. The maximum storm tide exceeded all previous measured high waters, and there were at least 150 breaches in the sea dikes that protected the polders.[31] Unlike the British, the Dutch had a storm surge warning service, and on January 31 at 11 A.M. a warning was telegraphed.[32] A second message was telegraphed at 6 P.M. and also broadcast on the radio, warning of dangerous high waters at Rotterdam and two other locations. But it had been many years since a warning like this had been issued, and telegrams were sent to only thirty people instead of the hundreds who should have received them. And it being a weekend, many were not read. Also, there were no radio broadcasts between midnight and early morning in the Netherlands, so when the storm tide hit its peak between 3 A.M. and 4 A.M., the radio had been off the air for at least three hours. As a result, the storm surge breaches of the seawalls and the flooding of villages and towns came without warning, most people wakened in the middle of the night by freezing water. As in Britain, the national government made no immediate response to the flood. All rescue attempts were made by local groups and individuals who woke neighbors and mobilized people to save the dikes.

At least 750,000 people lived in the areas where the 150 breaches in the dikes took place, and 100,000 of these people had to evacuate. More than 1,800 people drowned. More than fifty-five thousand acres were inundated, and forty-seven thousand livestock were killed. At least three thousand houses were swept away as well as three thousand farms. However, as in Great Britain, it could have been much worse, for Rotterdam came very close to being flooded. If Rotterdam had flooded, not only would there have been a huge number of casualties, but the economic consequences of shutting down the world's largest port would have been enormous. The protection of the three million people living below sea level in North and South Holland provinces came down to a single dike.

Breaches in a dike always begin with a small opening, but water rushing through that hole quickly erodes it into a larger and larger opening until millions of tons of water are rushing through. Thus the story of the Dutch boy plugging a hole with his finger is not far-fetched, although that story is actually American in origin and not that well known in the Netherlands.[33] At 5:30 on the morning of February 1, water broke through a section of dike called the Groenendijk, along the river Hollandse IJssel to the east of Rotterdam, that did not have a protective stone layer over its base of clay. The mayor of Nieuwerkerkaan de IJssel had anticipated problems with the dikes and had been up all night rallying volunteers. At the Groenendijk breach, volunteers worked in the dark, with cold, pouring rain, winds wailing, and waves crashing over the dike. But no matter how hard the volunteers worked, they could not close the breach, which grew to a width of forty-five feet. In desperation the mayor commandeered a sixty-foot grain boat and ordered its captain to ram it into the dike from the seaward side to plug the hole. The volunteers roped the ship in tightly and filled in all around it with sandbags. The Netherlands thus avoided what probably would have been the worst natural calamity in its history.[34]

As a result of the 1953 storm surge Great Britain and the Netherlands both made major changes in storm surge prediction and storm surge protection. In Britain a Storm Tide Warning System was set up in 1954 at the Meteorological Office with the help of the Liverpool Tidal Institute, which provided real-time water level data from six tide gauges.[35] Following the close call for central London, there were calls for a movable flood control structure or barrier to be installed downstream of London on the Thames. A movable barrier would leave the river open for the huge amount of commercial sea traffic into the Port of London but could be closed when flooding was expected from storm surges coinciding with high tides.[36] Work on the Thames Barrier began in 1974 and was completed nine years later. The barrier was built across a one-third-mile-wide portion of the river downstream of central London, leaving four navigable channels that could be closed off when a storm surge propagated up the Thames toward London. Each thirty-five-hundred-ton circular floodgate rotates into position when the river must be closed.

Only seventeen days after the storm surge hit, the Dutch responded by establishing the Delta Committee, and then implemented its recommendations, the Delta Plan, over the next few decades.[37] They made organizational changes at all government levels, assigning clear responsibilities for providing warnings and for maintaining sea defenses. A huge program was started to build up the sea defenses. Storm surge barriers were completed in 1986, in the process reducing the exposed coastline of Holland by 430 miles. The last two-and-a-half-mile-long section, the Oosterscheldekering, had huge sluice-gate doors that could be closed when a storm surge threatened.[38] An even larger movable storm surge barrier, the largest in the world, was completed by the Dutch government in 1997. That barrier can close off the Nieuwe Waterweg, the waterway leading to the Port of Rotterdam, in thirty minutes. Hung from sixty-foot-high retaining walls, two arc-shaped movable gates swing together from the opposite shores. The first time both the British and the Dutch barriers were closed at the same time was on November 8 and 9, 2007, when a gale coinciding with a spring tide appeared to threaten both countries. Luckily, the surge maximum did not coincide with high tide, and the barriers easily handled the combined storm tide.

In other nations as well, storm surge prediction was finally beginning to make some progress, using two primary approaches.[39] The first approach was to use statistical techniques, obtaining long records of data (water levels, winds, atmospheric pressure) at many locations and using these data to formulate correlation or regression equations that could then be used to predict water level rise when wind speed and direction were fed into them.[40] The second approach was to use numerical hydrodynamic computer models based on the physics of the water movement.[41] These models used equations similar to those first developed by Laplace (as we saw in Chapter 1), but they included meteorological forcing at the sea surface, specifically, wind pushing the water along horizontally (via friction) and atmospheric pressure pushing down or lifting up the sea. These new models were now possible because of the computer power that was just beginning to develop in the 1960s.[42] For either storm surge prediction approach a reasonable weather prediction capability was also needed. The winds and atmospheric pressure had to be known for the near future in order to predict what the storm surge would look like. That had often been a big stumbling block, until the 1960s when weather satellites were first put into orbit.[43]

How satellites were used in storm surge prediction was slightly different for tropical cyclones (hurricanes and typhoons) than it was for extratropical cyclones (northeasters and North Sea gales). For storm surges produced by hurricanes, it was most important to accurately predict the hurricane track and especially the exact location of its landfall. Since the path of a hurricane could be followed using satellites, landfall predictions could be made from the last position of the hurricane. For storm surges produced by extratropical cyclones, the satellites allowed determination of the position and geographical extent of, for example, a northeaster. High- and low-pressure systems and warm and cold fronts could also be followed as they moved across the country. This greatly aided in determining the wind directions and speeds along the coast and over the ocean that would generate the storm surges.

The improvements in storm surge prediction made in the United States and Europe by 1970 were beginning to be used in other parts of the world. And yet, on November 12, 1970, a storm surge produced at the northern end of the Bay of Bengal by only a Category 3 tropical cyclone was to become the most deadly storm surge in modern history. As often as this region has been hit with destructive storm surges, it still takes the right combination of factors to produce a real catastrophe. On the night of November 12 a cyclone headed for East Pakistan (formally East Bengal), bringing with it a twenty-foot storm surge that would hit the coast just as the tide was reaching spring high water due to the full moon. The rice harvest was only two weeks away, so at least another hundred thousand migrant workers were camped near the rice paddies in the low-lying delta regions. This was in addition to the two million very poor inhabitants who lived near the coast in mostly bamboo huts.

A critical factor in past storm surge calamities in the Bay of Bengal had been the lack of a warning, but now weather satellites orbited the Earth and sent back pictures of cyclones, hurricanes, and typhoons from around the world. The tropical cyclone heading for East Pakistan was seen in pictures taken by two satellites operated by the U.S. Environmental Science Services Administration (ESSA), the predecessor to NOAA. ESSA passed on these pictures to the Indian Meteorological Office and the Pakistan Meteorological Department in Dacca. But two big questions still remained. How would the warnings get to the millions of people living on the low-lying islands and coasts of the Meghna River delta region of East Pakistan? And once they had the warning, what could they do about it?

Recommendations with respect to both these questions had been made nine years earlier in a sixty-three-page report to the Pakistan government from the then head of the U.S. Hurricane Forecast Center, Gordon Dunn. Dunn had been invited to East Pakistan after two tropical cyclones had killed 16,000 Bengalis in October 1960.[44] He had recommended implementing a storm warning system, including coastal radar systems and instruments to receive satellite pictures, and building earthen mounds high enough for people to escape to. One of the islands in East Pakistan that Dunn had visited had 250,000 people but no electricity and only one battery-operated radio. By November 1970 two coastal radars were in operation, and the Pakistan Meteorological Department was receiving satellite photographs. A network of tide gauges had also been set up to measure storm surges. But communications were no better, and no earthen mounds had been built.

On the morning of November 8, 1970, the two satellites sent pictures of a depression over the southern central portion of the Bay of Bengal. By the following morning the satellites showed that this depression had moved northward and intensified enough to be called a tropical cyclone. It continued moving northward, intensifying over the warm waters of the bay. On November 11 it reached its peak strength with sustained wind speeds of 115 miles per hour. By that time it had crossed into the shallow waters of the continental shelf and turned northeastward, heading into the narrowing northern Bay of Bengal. On the night of November 12 it would be East Pakistan's turn to bear the brunt of a cyclone and its storm surge.

As the cyclone neared the coast, the Pakistan Meteorological Department issued a "great danger signs" warning over Pakistan Radio, but this designation apparently confused those who heard it, because they were used to the previous numerical warning system, in which "No. 1" indicated the greatest danger. Some Bengalis never heard the warning because they didn't have radios and they lived too far from someone who did. Others heard the warning but ignored it because several previous warnings, including one in October, had been for cyclones that ended up not causing much damage or many deaths. Because the Pakistan Meteorological Department had no method to determine how powerful a cyclone was, and thus no way of predicting how large the storm surge would be, they had put out warnings for most cyclones.[45]

No one had a radio in Chatlakhali, a village of forty-five people on a twenty-five-square-mile island with a total population of about eleven thousand. There was, in fact, no radio on the entire island. The one-room tin-sheet hut of Nur Hassain Faram Faraji's family housed him and his wife, his parents, his three sons, and his two daughters. November 12 had been a fairly typical day, but by 9 P.M. Nur Hassain and his family were cowering in their violently shaking hut, as the wind howled and his children cried. It was now too dangerous to try to take his family to the two-story concrete community center a quarter mile away, where five hundred people had squeezed into the forty-by-sixty-four-foot upper floor when the lower floor filled with water. Even an hour's warning would have made it possible for the Hassain family to reach the community center. Now all they could do was hold onto the wooden beams inside the house, in the dark, listening to the sound of the wind. Around midnight the wind noise was suddenly overwhelmed by a sound which was "like nothing I had ever heard," Nur Hassain said later. "Like the sea was coming to us."[46] Others in nearby huts looked out into the dark night after hearing "a great roar growing louder and louder." In the distance they could see a glow coming nearer and getting bigger, which they realized too late was the crest of a huge wave.[47] In an instant the twenty-foot storm surge smashed Nur Hassain's house, the roof caving in and the sides falling away. Suddenly there was quiet, but only because he had been pushed underwater. He felt a hand, which he grabbed. It was his ten-year-old daughter. The two of them were swept inland, but then the raging waters pulled her away from him, never to be seen again. He was knocked senseless, and awakened to find himself next to his fourteen-year-old son, but then he too was swept away. Somehow Nur Hassain managed to grab a tree and later found his son again, holding onto another tree about a half mile away. No one else in his family survived. There was also no house, no rice, nothing for him and his son to start over with—when they finally got over the tragedy of losing everyone they loved.

Nur Hassain's lost family members were seven of at least 300,000 Bengalis who died that night. Some think the death count is closer to a half million after including the unknown number of migrant workers who were there for the upcoming rice harvest and survivors who died afterward from starvation and disease. Many had been sleeping when the storm surge hit, and those not immediately drowned awoke with the shock of being swept away by cold, turbulent water. A great majority of those who died were children, the elderly, and women, for only the trees were there to save people, and only the strongest people could last in the trees while being lashed at by rushing water and fierce winds. The arms and chests of survivors were covered with scratches from holding onto the tree trunks and branches for hours.[48] In some villages not a single child survived. Bhola Island had the highest number of deaths—the eye passed just to the west of it, and the storm surge did maximum damage there—which is why this cyclone is sometimes referred to as the Bhola cyclone.

The people who survived wandered around aimlessly for days, numb and expressionless. When relief workers finally showed up, Americans, British, and many others, the survivors temporarily came out of their shock as though they felt that maybe these foreigners could find their missing loved ones, but as hope faded a second time, they slipped back into a stupor. Decomposing bodies floated in the waterways and were strewn over the land, creating a stench so foul that relief workers had to put cloths over their noses. Burying of bodies by the Pakistani army was hampered by survivors who would dig them back up looking for their loved ones.[49] Many bodies had been swept out to sea, but no one knew how many.

As survivors came out of their grief, they found that their means of producing food and making a living were also destroyed. At least 400,000 homes were demolished or badly damaged. More than 500,000 poultry and 280,000 cattle were lost, and the value of the rice crop damage was estimated to be at least sixty-three million dollars. At least 9,000 marine fishing boats were wrecked, affecting 90 percent of the marine fishermen. Tens of thousands of inland fishing boats were also lost, and 46,000 inland fishermen died, more than half the region's fishermen. The storm surge destroyed 65 percent of the annual fishing capacity of East Pakistan, which provided 80 percent of the animal protein consumed by the population of East Pakistan. In all, at least 3.6 million people were affected in some way by the storm surge.[50]

But one more major effect of the 1970 East Pakistan storm surge was still to come. East Pakistan was once part of British-controlled India (when it was called East Bengal). British India had been dominated by two religious groups, Hindus and Muslims. Most of the Muslims were in either the western part of India or the eastern part. Those in the west had a political connection with the Middle East and spoke Urdu, while those in the east were Bengalis who had converted to Islam and spoke the Bengali language. In 1947 when the British colony of India received its independence, it became two nations. The central portion, whose majority was Hindu, became India. The western and eastern ends became Pakistan, two culturally distinct areas separated by a thousand miles whose only common bond was Islam. In the ensuing years, West Pakistan controlled the Pakistan government, generally treating the Bengalis of East Pakistan as second-class citizens and taking economic advantage of them. Resentment in East Pakistan had grown considerably by 1970.

When the storm surge devastated East Pakistan, the Pakistan government was slow to respond to the tragedy, while foreign nations, including the United States, Britain, and even India, an enemy of Pakistan, sent aid immediately. The "president" of Pakistan (actually an unelected military dictator ruling under martial law), General Agha Muhammad Yahya Khan, had flown over the stricken coastal regions at three thousand feet, too high to see the many thousands of bodies, while drinking imported beer. He was overheard to say, "It didn't look so bad."[51] This was a sign of things to come.

Food and supplies began to arrive in Dacca from all around the world, but what was needed most were helicopters to get the food and supplies to survivors, since there were no roads and the waterways were clogged with bodies and debris. Nine days after the storm surge, only one Pakistan government helicopter was in use for relief. In the meantime six American helicopters and some British helicopters were in full operation.[52] The first public statement of a growing anger at their government by the Bengali population was printed in *Holiday,* an influential English-language newspaper printed in East Pakistan. The front-page headline read "Regime Fails the East Wing. Our People Can Bear No More."[53] The lack of a decent relief program by the Pakistan government upset the political leaders in East Pakistan and put new life into demands for local autonomy. The Red Crescent, Pakistan's Red Cross, began operating independently of the government-led relief commission.[54] Sheik Mujibur Rahman, leader of the Awami League, East Pakistan's dominant political party, warned of a possible struggle for the secession of East Pakistan.

To the low-flying relief helicopters the carnage left by the storm surge was appalling. Even the smell reached them as they flew over. One eyewitness wrote later, "The shoreline... was littered with blackened and bloated bodies washed up by the tide. It was a terrible sight. Amid the scene of death and devastation, what stood out prominently was the Awami League's election symbol—a boat riding the crest of a wave—all over the place, even in the remotest village and habitation."[55] National elections were held on December 7 and resulted in a landslide victory for the Awami League. As a result of that election Sheik Mujibur Rahman should have become the next prime minister of Pakistan. Instead of being sworn into office, he was arrested, and General Yahya set the Pakistan army on an operation of terror and genocide, which killed many thousands of Bengalis. The Bengalis fought back as best they could. The Bangladesh Liberation War had begun. India was brought into the war when the Pakistan army began killing Hindu Bengalis. By the end of 1971 India had soundly defeated Pakistan, and East Pakistan then became the independent nation of Bangladesh—a nation whose birth was stimulated by the destruction of a storm surge.[56]

The deadly storm surge that hit East Pakistan in November 1970 would not be the last. On April 29, 1991, Bangladesh was hit at high tide by another destructive storm surge, produced by Cyclone Gorky, which had winds of 150 miles per hour. At least 138,000 people died, which some saw as an improvement over the 1970 storm surge, especially since this cyclone was more powerful than the one in 1970 and the storm surge caused greater property damage. Some of the reduction in deaths was due to three hundred large cyclone shelters (concrete buildings on stilts) having been built, so some people had a place to which to escape. But this storm surge hit a part of Bangladesh where there was higher ground to escape to behind the city of Chittagong, although those living on the offshore islands did not have that option. Another reason was the improved system for warning people, using radio and 20,000 volunteers from the Red Crescent Society of Bangladesh, who went through neighborhoods with megaphones. The storm surge predictions had been accurate, making use of data from two National Oceanic and Atmospheric Administration (NOAA) satellites and a Japanese satellite. There were also reduced deaths after the storm surge among the survivors due to starvation and disease, because a U.S. amphibious task force of fifteen ships and 2,500 personnel returning from the Gulf War was diverted to Bangladesh. This was followed by 7,000 more U.S. soldiers in what came to be called Operation Sea Angel.

But 138,000 dead is still a tragically large number. Some of these deaths were due to the fact that many warned people did not evacuate, or only women and children evacuated because the men returned to protect their homes from looters. In some cases people did not leave because the warnings did not make the seriousness of the situation clear enough to them. Others did not believe the warnings, because they remembered past warnings that had been exaggerated. While prediction of a cyclone's landfall had improved, even small errors in the predicted location of the eye can produce significant errors in the predicted storm surge heights along the coast near the eye.[57]

During the following decades storm surge prediction steadily improved, especially through the use of numerical hydrodynamic models, which could be run on computers with ever-increasing power.[58] One of the earliest and most important operational storm surge models was the SLOSH model formulated by Chester Jelesnianski and used in the U.S. National Weather Service.[59] *SLOSH* stands for "sea, lake, and overland surges from hurricanes," but it was also used for extratropical cyclones. It was first developed in the late 1960s and was greatly improved following Hurricane Camille, a Category 5 hurricane with 190-mile-per-hour winds, whose 24-foot storm surge killed 259 people along the Mississippi coast on August 17, 1969. Camille was the largest recorded storm surge in U.S. history until surpassed in 2005 by the storm surge from Hurricane Katrina.

For hurricanes, as we have seen, the destructiveness of a storm surge depends on where the eye of the hurricane makes landfall. For the northern Gulf Coast of the United States, for example, locations to the east of an eye's landfall will experience the largest storm surge heights. For locations to the west of the eye, the water will go down and recede from the shore. So the approach taken in using the SLOSH model was to run numerous simulations for many different landfalls. Each coastal region had simulations run for a variety of hurricane tracks, wind strengths, and the forward speeds of the hurricane and for particular tide levels. Then during an actual hurricane, the most appropriate composite of these simulations was used for the forecast hurricane values at landfall and the predicted tide.[60] The model simulations were done for dozens of coastal regions for all sections of the Gulf and Atlantic coasts of the United States. Thousands of simulations were run, and composite products useful for emergency management were generated.

To predict storm surge heights at a location where a hurricane was predicted to make landfall, forecasters would pull out from the archive the appropriate composite of SLOSH simulations for that location with the same wind speed and pressure and track as the actual hurricane, observed from satellites. If the hurricane changed its path or strength, a different composite of SLOSH simulations would be used. This reliable routine is still carried out at the U.S. National Hurricane Center in Miami, Florida, even as it has used other more sophisticated model systems that run with real-time data.[61] The predictive skill of SLOSH and other newer storm surge models depends on having accurate information on the bathymetry (water depths) leading to the shore and the topography (land elevations) where the flooding will take place. Both can significantly affect how high the storm surge will be and how far inland the flooding will reach.

SLOSH and other numerical hydrodynamic storm surge models have also been used for predicting the storm surge produced by extratropical cyclones, with various simulations being run for different categories of wind speeds and directions and atmospheric pressures along a section of coast. These models must cover a much larger section of ocean, for example, the length of the entire East Coast of the United States stretching out beyond the edge of the continental shelf, or the entire Gulf of Mexico, or the entire North Atlantic, including the Caribbean and Gulf of Mexico. Such models are also run in real time, driven by forecast winds and pressure from weather models. Today with the internet and modern communication capabilities and with satellite and state-of-the-art instrumentation for making oceanographic and meteorological measurements, it is a much simpler task to bring real-time data to the storm surge model and to the weather models that support it.[62]

These greatly improved storm surge prediction capabilities have come in handy during the first decade of the twenty-first century, which has seen several Category 4 and 5 hurricanes produce large storm surges. Of these destructive storm surges, the one that made the most news and had the most impact on the United States was, of course, the one generated by Hurricane Katrina, which on Monday, August 29, 2005, hit the coast just east of New Orleans at the Louisiana-Mississippi border. Katrina was a hurricane whose track, area of landfall, intensity, and resulting storm surge were very accurately predicted by the National Hurricane Center (NHC), in the National Weather Service of NOAA. The unfortunate consequences of Katrina that received so much worldwide attention resulted not because of an inaccurate prediction but because of the government's inadequate preparation for and response to this predicted natural calamity.

On August 25, four days before Katrina's landfall, NHC began giving the first of what would end up being hundreds of television and radio interviews on the danger of Katrina. Two days before Katrina came ashore near New Orleans, NHC released to the public a storm-track prediction and a predicted wind speed at landfall that would turn out to be amazingly accurate. On Saturday, thirty-two hours before Katrina's landfall, NHC's *Hurricane Katrina Advisory Number 19* included a prediction for "coastal storm surge flooding of 15 to 20 feet above normal tides... locally as high as 25 feet along with large and dangerous battering waves... to the east of where the center makes landfall."[63] The next morning in *Advisory Number 23* the storm surge prediction was increased to 18 to 22 feet above normal tide levels, locally as high as 28 feet. NHC made these storm surge predictions using the SLOSH model.[64] That afternoon, fourteen hours before Katrina's landfall, *Advisory Number 24* included the ominous statement "Some levees in the Greater New Orleans area could be overtopped."

To make sure that no one was uninformed about the seriousness of Katrina's threat, NHC distributed its predictions widely. To coordinate communications between NOAA and the emergency management community at the federal and state levels, NHC had activated the Federal Emergency Management Agency (FEMA)/National Weather Service (NWS) Hurricane Liaison Team several days before Katrina's landfall.[65] The NOAA Homeland Security Operations Center had also begun providing situation reports directly to the White House. Max Mayfield, Director of NHC, personally called the governors of Louisiana and Mississippi, the mayor of New Orleans, and the Chief of Operations at the Alabama Emergency Management Agency on Saturday evening to make sure they understood the gravity of the impending Katrina landfall. On Sunday during a Hurricane Liaison Team teleconference, Mayfield briefed President Bush, Homeland Security Secretary Chertoff, FEMA Director Brown, and various White House staff.[66] By this time Katrina had become a Category 5 hurricane. That morning the NHC advisory released at 7:00 A.M. described Katrina as a "potentially catastrophic" hurricane.[67] Mayfield mentioned the potential overtopping of the levees at the Saturday and Sunday briefings. To make sure the media was aware of the urgency of Katrina, NHC gave 471 television and radio interviews from August 25 through Katrina's landfall in Mississippi on the morning of August 29. There were nine hundred million hits on the NHC website during Katrina. The day before Katrina hit the coast, the National Weather Service office in Slidell, Louisiana, included in its warning the statement "Most of the area will be uninhabitable for weeks... perhaps longer... human suffering incredible by modern standards."[68]

NHC's hurricane and storm surge predictions turned out to be very accurate. The storm-track prediction, released to the public more than two days before Katrina came ashore, was only 15 miles off the actual track. The predicted wind speed for landfall, released two days before landfall, was off by only 10 miles per hour from the measured wind speed.[69] The measured storm surge, based on high-water-mark observations, since most water level gauges were damaged by Katrina, was 24 to almost 28 feet along the western half of the Mississippi coast and 17 to 22 feet along the eastern half, very close to the NHC predictions. A U.S. record storm surge of 27.8 feet was measured at Pass Christian, Mississippi, 50 miles northeast of the center of New Orleans.[70] In Mississippi the storm surge penetrated at least 6 miles inland, and moved up bays and rivers as far as 12 miles. Even Mobile Bay in Alabama, 90 miles east of New Orleans, had a storm surge of 10 to 15 feet. The large swells that Katrina had generated while still offshore as a Category 5 and then a Category 4 hurricane caused a significant wave setup that increased the water level along the entire Gulf Coast.

The levees were overtopped. This led to breaches when the overtopped water eroded the backside of the levees. Some levees were breached without being overtopped, due to later-discovered design flaws. New Orleans began to flood very soon after Katrina made landfall. Ultimately, more than 1,800 people died, one of the greatest losses of life from a natural calamity in U.S. history. This death toll may be much smaller than death tolls that have occurred in the Bay of Bengal, but it is still higher than anyone would or should expect in the United States in the twenty-first century. Over a million people became homeless and had to be moved to other states. Katrina also became the costliest storm in U.S. history, inflicting many billions of dollars of damage and having a major economic impact on the whole nation, including crippling the nation's largest oil-producing region. The Gulf region accounts for a third of U.S. domestic oil production; a fifth of U.S. natural gas production; 60 percent of imported oil, which enters through its ports; and almost half the nation's refining capacity. All the area's production facilities were shut down by Katrina's storm surge, and the entire nation suffered. Gasoline futures had their highest jump in history.[71]

The meteorologists and oceanographers had done their job. Without those accurate hurricane and storm surge predictions and their wide dissemination, tens of thousands could have died. It was the delayed and inadequate government response to the natural catastrophe that led to the sad and embarrassing calamity that followed Katrina's landfall. In their 2006 report "A Failure of Initiative," the congressional Select Bipartisan Committee wrote, "We repeatedly tried to determine how government could respond so ineffectively to a disaster that was so accurately forecast."[72] There had also been a lack of preparation for the inevitable "big one" that everyone knew would someday hit New Orleans, warnings from scientists and engineers having been repeatedly ignored.[73] New Orleans would have been helped immeasurably by adequate evacuation plans and storm centers, better levees, and preservation of natural barriers such as wetlands, which reduce the power of a storm surge. Some of the levees built to protect New Orleans from the Mississippi River kept sediment from reaching the wetlands, and so the wetlands began to disappear. In addition, the extraction of water and oil from below the land surface had increased the sinking of New Orleans. The city was below sea level, surrounded by water on three sides (by the Gulf of Mexico, the Mississippi River, and Lake Pontchartrain), and protected by inadequate levees. Thousands of acres of wetlands were also submerged, allowing the saltwater to intrude farther inland, leading to more wetland loss.

Katrina was not the only large tropical cyclone to produce a destructive storm surge in the last decade, but as with Katrina the predictions of those storm surges were much more accurate than in decades past, even in the Bay of Bengal. On October 29, 1999, a Category 5 cyclone with greater than 155-mile-per-hour winds hit Orissa on the Indian coast, producing a nineteen -foot storm surge that reached ten miles inland. About 9,800 people were killed, but this was a small number by Bay of Bengal standards. This reduction in casualties was primarily due to an accurate early warning three days before landfall by the Indian government and the forced evacuation of tens of thousands of families from the area of predicted landfall. The natural protection of many of the villages provided by mangroves along that coast also appears to have helped save lives.[74] But the true test as to whether improved prediction and communication capabilities were making a difference in the Bay of Bengal was when the next large cyclone hit Bangladesh. That test occurred two years after Katrina, when Bangladesh was hit on November 15, 2007, by a storm surge generated by Cyclone Sidr, a Category 5 cyclone that reduced to a Category 4 when it hit land. The storm surge, ranging from twelve to twenty feet depending on the location, caused severe damage. Although this cyclone was just as powerful as the cyclone that hit this area in 1991, and more powerful than the 1970 cyclone, the confirmed deaths from it were only 3,400, a huge reduction from the 138,000 deaths in 1991 and the at least 300,000 in 1970.[75] The difference is that Cyclone Sidr had not only accurate predictions of the cyclone and the storm surge but also a much more active warning system. Over 40,000 volunteers from the Cyclone Preparedness Program, including members of the Bangladesh Red Crescent Society, rode bicycles and used megaphones to convince residents to go to almost 4,000 emergency shelters that now existed in fifteen provinces. The Bangladesh government reported that 3.2 million people were evacuated (40 percent of the coastal population). In addition, many miles of embankments had been built since 1991, some along the coasts and some around a network of 124 polders (similar to those the Dutch built four centuries earlier in Holland). The polders not only saved lives but also protected houses, animals, and crops.[76]

With the great success in protecting the people of Bangladesh in 2007, there was reason for optimism six months later when Cyclone Nargis developed in the central part of the Bay of Bengal on April 27, 2008. However, on May 1 Nargis took an unusual turn, and instead of continuing in a typical northerly direction, it swerved and headed east, intensifying as it moved, toward a country that has been visited by cyclones much less often than India or Bangladesh. Nargis was a Category 4 cyclone by the time it and its twelve- to eighteen-foot storm surge reached the coast of Burma (now officially called

Myanmar) on the evening of May 2 and passed over the low-lying and densely populated Irrawaddy Delta. A day later at least 146,000 people had died and at least 2 million were homeless. Three-fourths of the livestock had been killed, a million acres of rice paddies had been flooded with salt water, half the fishing fleet was sunk, and at least 700,000 homes were gone.[77] The horrors of 1970 and 1991 in Bangladesh were seen everywhere—bodies floating in rivers and channels and lying all over the land, as were animal carcasses and debris from thousands of smashed houses. The people had received no warnings from their government, a military junta that was not known for treating its people well. There had been no lack of accurate forecasts from other nations. The Indian Meteorological Department regularly sent forecasts and warnings to Myanmar's Department of Meteorology and Hydrology.[78] After the calamity, nations around the world immediately offered aid, but the military junta initially rejected the offers while doing nothing to help the storm surge survivors, and then for a long time made it very difficult for efficient relief operations to take place.

Hurricane Katrina and the more deadly Cyclone Nargis sadly show us that, now that storm surge prediction has finally reached a high level of accuracy, the most important influence on preventing deaths is how governments respond to those predictions. With satellite and other real-time data now easily provided via the internet to numerical hydrodynamic storm surge models running on supercomputers, driven by the forecast winds from numerical weather prediction models, there should never be an unpredicted storm surge again. These instruments and models are all part of the Global Ocean Observing System. In most countries today the internet and other communication media, especially television and radio, enable widespread dissemination of warnings, but as we have seen, this is not the case in some developing countries, such as Myanmar, where a responsible government is required for that communication to be effective. And as we saw with Katrina, even with such communication, efficient evacuations of large segments of the population will not happen unless there is a coordinated government response.

Today, storm surge prediction, though not as accurate or as easy as tide prediction (Chapter 2), is an area where we have been very successful.[79] In addition to damage done by the flooding caused by a storm surge, there is also damage done by large wind waves, also created by the storm, which ride the storm surge higher and farther inland. But large wind waves do not need the help of a storm surge to be destructive. As we shall see in the next chapter, predicting where waves will occur and how large they will be has been even more difficult than predicting storm surges. Although the prediction of average sea state is now reasonably successful, we still cannot predict the occurrence of hundred-foot rogue waves, which can break oil tankers in half and capsize ocean liners.

CHAPTER 5

Stormy Seas

Predicting Sea, Swell, and Surf

On December 7, 1942, exactly one year after the attack on Pearl Harbor brought the United States into World War II, the RMS *Queen Mary* floated on the waters of the Hudson River at Pier 90 on the west side of Manhattan. That entire day a constant stream of young American soldiers, along with nurses and other personnel, slowly made their way up her long gangplanks. By the time she left New York harbor the following day, a total of 16,500 troops were squeezed into every corner of the gigantic ship, ready for the cross-Atlantic voyage to Gourock, Scotland.

Prime Minister Winston Churchill had offered President Franklin Roosevelt the two superluxury liners, the *Queen Mary* and the *Queen Elizabeth,* as a way of speeding up the transport of desperately needed U.S. troops to Britain. *Mary* was the fastest ocean liner in the world and could reach any British port in just five days.[1] It took convoys of Liberty ships and other troop-transport vessels two weeks to deliver Americans across the Atlantic. The *Queen Mary* was also a safer way to transport troops because she could outrun any German U-boat. She was so fast that a German U-boat could not keep her in its sights long enough to accurately aim a torpedo at her. *Mary* could therefore cross the Atlantic unescorted. She still zigzagged every few minutes, a tactic the other troop-carrying vessels were forced to use to minimize a U-boat's chances of torpedoing them, but it hardly seemed necessary for *Mary.* Over the entire war *Mary* and *Lizzie* would transport more than 765,000 soldiers and not lose a single one.[2] Hitler would become so frustrated that he would offer an Iron Cross and $250,000 to the U-boat that could sink a *Queen.* But neither ship would ever come under fire. Only once did each *Queen* run into trouble, but it had nothing to do with German U-boats.

Docked at Pier 90, the *Queen Mary*'s eighty-one thousand gross tons and thousand-foot length dwarfed the battleships and destroyers at neighboring piers. She was more than a hundred feet longer than the *Titanic* had been, with almost twice the gross tonnage.[3] Although painted gray and stripped of all her luxurious amenities, unmistakable signs remained of the superluxury liner she had been before the war—her gargantuan size, her three huge smokestacks, and especially her sleek lines. But to the thousands of American troops now onboard, the living conditions would be far from the luxury that previous *Queen Mary* passengers had enjoyed. Most of *Queen Mary*'s more than twenty-one hundred cabins now slept eight soldiers at a time. About half the soldiers onboard were double-bunked, meaning that the beds operated on twelve-hour shifts. Food was to be served twice a day, and was certainly not going to be gourmet, the British staple of sticky oatmeal for breakfast and leathery mutton for dinner being typical.[4] The American soldiers had not yet learned this as they lined all seven decks of the *Queen Mary* when she left New York Harbor on a sunny December 8. Bands played, crowds cheered, and fireboats shot streams of water high into the air as she cruised past the Statue of Liberty and headed for the British Isles, three thousand miles across the Atlantic Ocean.[5]

By the fourth day at sea the weather had turned foul. A squall developed, and its high winds generated huge waves that made *Mary* pitch and roll.[6] She was not as adept as her passengers might have hoped at handling these large North Atlantic wind waves. With seven times the normal number of passengers and all the extra equipment, weapons, and ammunition, she was top heavy. Her stability problems increased as the voyage went on, because the oil she consumed came from fuel tanks located below the waterline. The zigzagging also did not help, the sudden change in heading sometimes initiating a deep roll. So *Mary* pitched and rolled, and thousands of landlubber soldiers paid the price, overcome by seasickness, their bunks immersed in the odor of vomit. Many became so dehydrated that they had to go to the ship's hospital.[7] By the next day the howling winds had reached hurricane strength. The *Queen Mary* shook viciously each time she climbed up an enormous wave crest and fell forward into a deep wave trough, occasionally exposing her propeller blades, and then was pummeled by the next wave. She was surrounded as far as the eye could see by an angry ocean with white, frothing peaks and deep, dark valleys. Water was everywhere on the ship, from the spray as *Mary*'s bow crashed down on the next wave and from the wind shearing off the tops of waves and throwing the water at the ship. And through it all, the *Queen Mary*'s passengers remembered what they had been told. Don't go near the railings, because if you fall over, you are lost. For even if you survived the fall (sixty-five feet from the main deck), and then survived the icy Atlantic waters (numbness and incapacitation would set in within a few minutes), the ship would not stop to look for you. With

German U-boats seemingly everywhere, sixteen thousand lives would not be risked to save one.[8]

But the worst was still to come. Seven hundred miles from Scotland, the *Queen Mary* suddenly fell into an almost bottomless trough and was then broadsided on her port side by a monstrous wave crest that was at least twice the height of any she had encountered. This mountain of water shattered windows on the bridge, ninety-five feet above the waterline. It tore away all the lifeboats on the port side of the top deck. It broke through portholes, sending water rushing into hundreds of cabins. But most seriously, the weight of this stupendous wave, thousands of tons of water, slowly pushed the *Queen Mary* over farther than she had ever rolled before.[9] The lifeboats on the starboard side swung down with the ship and almost touched the sea. Soldiers on the lower decks of the starboard side looked out of their portholes and saw dark seawater. Many were thrown out of their bunks and found themselves lying not on the floor but on a wall of their cabin. Passengers broke arms and legs or suffered concussions. In the dining rooms, chairs slid from one side to the other, and dishes and glasses shattered. For the soldiers at antiaircraft positions, a hundred feet above the waterline, the scene they witnessed was especially harrowing. Those on the port side stared in disbelief as a mountain of water appeared to come at them, and those on the starboard side were one minute a hundred feet in the air and the next almost touching the ocean itself.[10]

One soldier, who had just finished his watch at a 20-mm gun on A deck when the *Queen Mary* began to roll, was slammed against the bulkhead. He could hear pans and dishes crashing in the galley below him. When he finally reached his cabin, he saw members of his gun crew come sliding out the doorway riding torrents of water from broken portholes.[11] Because of all the water, many soldiers threw on life preservers, convinced the ship had been torpedoed.[12] When *Mary* had listed over farther than anything the gunman had ever experienced before, later determined to be fifty-two degrees from vertical, and when she seemed to stay there for an eternity, the gunman, like so many others on the ship at that frightening moment, figured that the ship would never right itself again. In fact, according to later calculations, if the *Queen Mary* had listed over only three more degrees, she would have capsized. If she had capsized, she would have almost certainly sunk, taking with her up to 16,500 lives and instantly becoming the largest maritime disaster in history. And perhaps also the greatest maritime mystery in history, for she might have sunk without a trace, although some Nazi U-boat would have surely claimed her as its greatest prize. A tragedy and a mystery were apparently avoided, due to the "exceptional seamanship on the part of her bridge officers . . . a quick turn of her helm [so that] her bows were brought dead on to this exceptional wave."[13]

The near calamity on the *Queen Mary* was not reported in the newspapers until almost a year later.[14] By that time her sister ship the *Queen Elizabeth* had also encountered a wave more than twice the size of all the waves around it. In February 1943 that wave "struck down the forward end of the [*Queen Elizabeth*'s] superstructure and damaged guns on the exposed forward decks."[15] Windows on the bridge were shattered ninety-two feet above the waterline, but she too was not fatally damaged or capsized. Decades later such uniquely monstrous waves would come to be called *rogue waves* or *freak waves.* Other disasters caused by rogue waves would be reported, but some of those other ships would not be as fortunate as the two *Queens.*

When World War II began in 1939, mariners had no way to predict the average size of waves produced by winds blowing over the sea, much less a way to predict when the most monstrous of such waves would appear. But the war created a special need for a wave prediction capability, a need more important than predicting when a ship might run into a single very large wave. When the Allies carried out amphibious landings on enemy beaches, their chance of success depended on the absence of large wind waves. Waves do more than make a rough ride for ships approaching the enemy shore. They break on the beaches, creating high surf that makes it extremely difficult to land boats of men without drowning them. Predicting high waves is not just a matter of predicting the winds; it is also a matter of knowing what wind directions produce the largest waves.

It had only been in the decades preceding World War II that wind waves on the surface of the ocean had begun to be understood, but they were so complex that no one was close to predicting them. Like storm surges, scientists could not see any correlations with the very predictable movements of the moon and the sun, as there were with the tide. But the demands of war can speed up science when the lives of hundreds of thousands of men depend on it. And it would be up to two oceanographers at the Scripps Institution of Oceanography in California to come up with a method of wave forecasting that the Allied navies could use for their amphibious landings—in North Africa in 1942, in Sicily and mainland Italy in 1943, and in Normandy in 1944, as well as throughout the Pacific as the United States captured one island after another from the Japanese.

To fully appreciate the method those two men devised, and the impact it had on the war as well as on future wave prediction methods, we must first take a look at the somewhat sparse history of our knowledge of wind waves. It is not that the earliest awareness of waves on the ocean did not go as far back as the earliest awareness of the tides; it did. But philosophers and then scientists could find few patterns to help them turn their knowledge into a prediction capability. Whereas the tide was periodic and regular, wind waves were random and irregular. Whereas the tide was caused by very predictable astronomical forces (the periodically changing gravitational attraction of the moon and the sun), wind waves were caused by unpredictable winds blowing over many areas of the ocean. Whereas tide prediction involved predicting a single very long wave, wind wave prediction involved predicting countless short waves that added to each other in ever-changing ways, the result so complex that it could only be treated by statistics (of average heights and wave periods). It was no wonder that progress came so slowly.

The first descriptions of huge wind waves on the ocean damaging or destroying ships can be found in the earliest books from ancient civilizations. A vivid narrative in the first book of Virgil's legendary epic poem the *Aeneid* (19 BC) describes enormous waves that devastated the Trojan fleet under the command of Aeneas. After the disastrous loss to the Greeks at Troy, described in Homer's *Iliad,* Aeneas's fleet was hit by a storm as it headed for Italy (where his descendants will found the Roman Empire). Though the *Aeneid* is fiction, Virgil's description of gigantic waves must be based on real experiences. His description could have been applied two thousand years later to the waves that assaulted the *Queen Mary* and other modern ships:

> a roaring gale from the North
> struck full on the sail, *dashing up waves to the stars.*
> A snapping of oars; then the prow swung round, and the ship
> was broadside to the waves; *in a sheer cliff*
> *the water piled up.* Some aboard were poised on a wave-top,
> while to others the deep yawned wide, disclosing the land
> between wave and wave.[16] (emphasis added)

Although the ships of 19 BC were much smaller than the *Queen Mary,* so that the waves did not have to be as high to make a terrifying impression on the sailors, phrases like "waves to the stars" suggest great heights, and "in a sheer cliff the water piled up" suggests a steep rogue wave.[17]

A century later Plutarch wrote about a rare event in history that involved wave prediction. This event was rare in that it literally changed the course of world history, and it was even rare from an oceanographic standpoint in that it involved an accurate prediction of wind waves, something that would not happen again for almost two millennia. As unlikely as it might seem, an accurate wind wave prediction determined the outcome of the naval battle of Salamis in 480 BC.[18] A Greek naval force had been formed by Themistocles to defend their homeland against invasion by the Persian forces of Xerxes. The Persian fleet had 1,200 ships, including ships from Egypt and Phoenicia, while the Greek fleet had only 368 ships. But Themistocles had a plan that took advantage of his local knowledge of the wind waves in the Strait of Salamis, the strait between the Greek mainland and the island of Salamis. Themistocles knew from local fishermen that each morning, usually sometime between 8 and 10, a strong wind from the open sea would blow down the strait.[19] The channeling effect of the narrow strait increased the wind's speed, which generated high waves.[20] The Persian ships, whose high sterns and decks made them clumsy and unwieldy, could not deal with these waves, while the low, sleek, and solidly built Greek ships, called *triremes* because they had three levels of rowers, could cut through the waves with no difficulty.[21]

Themistocles positioned his ships so as to lure the Persian ships into the Strait of Salamis. He managed to delay the battle until the breeze coming from the sea was strong enough to begin generating big waves. At one point he even had his Greek ships row backward to put off the confrontation with the Persian ships until the waves were high enough.[22] The rough sea caused the Persian ships to pitch and roll, affecting their maneuverability so that they could not protect their sides from attacks by the Greeks, who thrust bronze-covered rams attached to their bows into the Persian ships. The pitching and rolling also greatly reduced the accuracy of Persian archers. In the end the mighty Persian fleet was shattered and Greece was saved. It was one of those moments in history whose repercussions were world changing. If the Persians had conquered Greece, Western civilization might never have developed—or if it had, it would have turned out quite differently.[23]

The Battle of Salamis was a rare example of mariners being able to predict wind waves, but only because this was a special situation in which they could predict the winds in a narrow strait. No wind is more predictable than the daily change in wind direction that we now call the *sea breeze–land breeze,* the wind blowing from the sea onto land during the day and reversing during the night.[24] In this case the Greeks knew that a strong wind would produce high waves in the Strait of Salamis. As we will see, there would be few other special situations that would allow such easy wind wave prediction.

Sailors recognized the dangers that large wind waves posed to their ships, but were unable to predict when such waves would appear. The philosopher-scientists of those early civilizations were not much help, though some were interested enough to ask relevant questions. Aristotle wondered, "Why is it that sometimes craft traveling on the sea in fine weather are sunk and completely disappear, so that no wreckage even comes to the surface?" He correctly speculated that large ocean waves could travel along the sea's surface beyond the region where winds had generated them and arrive at places where there was no wind.[25] Thus a ship did not have to be in a storm to be hit by a large wave. Aristotle had seen great waves striking the coast, and he noticed that ocean waves were larger than the waves he saw on lakes. He also realized that waves breaking against the shore broke up rocks to produce sand, and he noticed that an ocean shore had much more sand than a lake shore, another indication that ocean waves were larger than lake waves.

In trying to understand how waves were generated by the wind, some philosophers noted a strange occurrence that they sensed provided a clue—if one put olive oil or fish oil on a sea surface with waves, the waves would disappear and the sea surface would become flat. In the first century Plutarch asked, "Why does pouring oil on the sea make it clear and calm?"[26] He wrote that divers working underwater sometimes swam with olive or fish oil in their mouths. If there was not enough light (because too many waves at the surface kept the light from penetrating to a great depth), they would spit it out. The oil floated up and calmed the water surface, thus allowing more light to penetrate. Plutarch wondered "whether the sea, which is terrene and uneven, is not compacted and made smooth by the dense oil."[27] And he quotes Aristotle as saying that "the wind, slipping over the smoothness so caused, makes no impression and raises no swell." He was getting closer to the answer, but that was as far as anyone would get for a long time. Pliny the Elder in his *Natural History* repeats the account of divers underwater with oil in their mouths and states that the "whole sea is made still with oil," but he added no more understanding to the cause of this almost magical effect.[28] During the Dark Ages, oil-on-water incidents were sometimes believed to be miracles. In the seventh century the Venerable Bede, a monk in England, wrote about a bishop giving holy oil to a priest who was to bring back by sea a future wife for the local king. Sure that there would be a storm, the bishop told the priest that when the storm over the sea begins, he must throw the holy oil on the water so he could have "pleasant calm weather, and return home safe," all of which did occur as he had predicted.[29]

The stories of enormous waves destroying ships would continue throughout the Middle Ages, but there would be no advances in scientific understanding of such waves, much less in methods for predicting when they would appear. It was not until the fifteenth century that a scientist again paid serious attention to the problem of waves. Leonardo da Vinci watched and made sketches of wind waves on the ocean. He tried to make the problem scientifically more manageable by studying simple waves produced in a pond when a rock is tossed in. The waves in his experiment did not, however, stay simple for long. Leonardo noticed that the waves radiated in all directions from where the rock entered the water, eventually hitting the edge of the pond and being reflected back in the opposite direction.[30] These reflected waves passed right through the incident waves, the crossing waves sometimes adding together to become larger waves and other times subtracting from each other to become smaller waves or even to momentarily disappear. The complicated pattern produced was an indication of the complexity that scientists would unsuccessfully try to understand over the next five centuries.

The waves we observe on the sea are an ever-changing combination of waves that propagated there from different storms around the ocean. But the wave crests that one sees are not the crests of the waves that traveled there. Each wave crest that one observes is a short-lived combination of those long-traveled wave components. To understand this we must first look at the simplest case, in which only two waves of equal height are traveling in the same direction but at different speeds, for example, in an experimental wave tank, the only place where such a simple situation can be created. At certain moments two crests from the two wave trains are at exactly the same location at the same time, that is, they are *in phase* with each other, and they add together to produce a larger wave. But then as the faster wave train passes the slower one, their combination decreases. When the crest of the faster wave train reaches the trough of the slower wave train, they exactly cancel each other out, and there is momentarily no wave observed at all. But this is only for two interfering wave components. What we end up seeing on the sea's surface is a moving group of waves whose heights are greatest in the middle of the group and smallest at the front and rear of the group. With dozens of wave components adding and subtracting to each other, it becomes a much more complicated pattern of waves. This is why, when we try to follow a particular wave crest as it moves along the water surface, we see its height slowly change, eventually decreasing and disappearing, with other wave crests growing to take its place. Only in shallow water near a beach can we follow the larger waves for any distance, for reasons we will talk about later.

With no hope of scientists being able to predict when waves will come and how large they will be, mariners did the best they could based on their own observational experiences. They learned that some times of the year, typically winter, were worse for storms and waves. They also learned that certain regions of the ocean had rougher seas. One sea captain wrote in 1699 of the very high waves that he experienced in the Florida Current (the beginning of the Gulf Stream): "the Wind blowing against the Current makes an extraordinary Sea, and so thick come the Waves one after another that a Ship can't possibly live in it."[31] Very high waves were also experienced in other great ocean currents, such as the Kuroshio off Japan and especially the Agulhas off the southern end of Africa.

Scientific progress continued, but very slowly. In 1670 the Irish scientist Robert Boyle realized from talking with divers that water movement due to waves decreases as one goes deeper in the water. Waves could be ten feet high at the sea's surface during a storm, but ninety feet below there was no movement at all, no indication of the violence occurring at the surface.[32] Divers also noticed that wave action below the surface was circular, water particles moving in orbits that get smaller with depth. This was mentioned by Isaac Newton in 1687 in his landmark book *Philosophiae Naturalis Principia Mathematica*. Newton also showed mathematically that a simple wind wave moves along the surface of deep water with a speed that is proportional to its wavelength (the horizontal distance from one crest to the next)—in other words, longer waves travel faster than shorter waves.[33] A century later Pierre-Simon, Marquis de Laplace, used this to show that as wind waves travel long distances, they *disperse,* namely they begin to separate from each other, the longer waves leaving the shorter waves behind.[34]

A longer wind wave, traveling beyond the winds that produced it, eventually becomes lower, more rounded, and more symmetrical and is referred to as a *swell.* Swells can travel thousands of miles across an ocean without losing much energy.[35] At any particular coastal location we see the combined effect of numerous swells that originated in different parts of the ocean, some of them quite distant. The wavelengths of swells typically vary from 180 feet to 1,300 feet, with corresponding wave periods varying from six to sixteen seconds. They travel across the ocean at speeds ranging from twenty-one to fifty-six miles per hour.[36] Of course, swells can be even longer, higher, and faster if produced by a very large storm. For example, a very long swell measured off the British coast of the English Channel in 1899 had a wavelength of half a mile, a period of 22.5 seconds, and traveled at a speed of seventy-nine miles per hour.[37]

Swells move in groups, but, as described above, what appear to be individual waves within a group are actually the crests and troughs that result from adding together wave components that have almost the same wavelengths and periods. The highest wave crests and the deepest troughs come in the middle of the group, where these wave components are in phase with each other and thus add together to maximum effect. At the front of the wave group and at the rear are the smallest wave crests. It is here that the wave components are out of phase with each other, canceling each other out. A group of swells travels at half the speed of the individual wave components that combined to produce the group. This is the speed at which *wave energy* travels across oceans. The so-called *ninth wave,* sometimes called the seventh wave, of legend and surfer lore, is the center wave crest of a group of swells.[38] Whether it is the ninth, seventh, or some other number of waves depends on how far the swells have come and how that journey has affected the period of the swell components. Once swells reach shallow water near a beach, however, the speeds of its wave components no longer depend on their wavelengths, but instead depend only on water depth, and the waves of each group then travel at the same speed. In shallow water, decreasing depth steepens the wave, eventually leading the heightened crest to break.

Around the same time that Laplace was studying waves, across the Atlantic Benjamin Franklin was also investigating them, but with a more observational and experimental and less mathematical approach. He gained insights on how wind generates waves by studying the effect that a little oil has on waves, the same effect that Aristotle, Plutarch, and Pliny had wondered about two millennia earlier. In 1757 Franklin was at sea on a vessel that was part of a British fleet of ninety-six ships sailing to Nova Scotia to capture the French fortress at Louisbourg. Franklin "observed the wakes of two ships to be remarkably smooth, while all the others were ruffled by the wind, which blew fresh." When he asked the captain about it, the captain replied, "The cooks have, I suppose, been just emptying their greasy water through the scuppers."[39] This answer Franklin perceived to be given with some contempt, as though this was something all mariners knew. This incident reminded Franklin that he had once read Pliny's account of oil poured on water to still the waves in a storm. From that point on, Franklin gathered more stories of oil used to calm troubled waters, but it would be several years before he would carry out an experiment himself.

One day in London when Franklin was standing by Clapham Pond, he noticed that the waters were very rough, due to the wind. He pulled out a cruet of oil and dropped a little on the water. The oil spread out over the water surface with surprising swiftness, but it didn't seem to smooth the waves, which were very large since Franklin was standing on the leeward side of the pond. He went over to the other side of the pond, the windward side, where the waves were first produced, then growing larger as they moved over to the leeward side. Again he dropped a teaspoon of oil on the water but with a very different and remarkable result. The oil "produced an instant calm over a space several yards square, which spread amazingly, and extended itself gradually till it reached the lee side, making all that quarter of the pond, perhaps half an acre, as smooth as a looking-glass."[40] Franklin wrote that the oil spread "so thin as to produce prismatic colours, for a considerable space, and beyond them so much thinner as to be invisible, except in its effect of smoothing the waves."

From looking at waves and thinking about why oil would make them go away, Franklin gained an understanding of how waves are produced by the wind. He wrote, "air in motion, which is wind, in passing over the smooth surface of water, may rub, as it were, upon that surface, and raise it into wrinkles, which if the wind continues, are the elements of future waves." He then goes on to say that "the small first-raised waves, being continually acted upon by the wind, are, though the wind does not increase in strength, continually increased in magnitude, rising higher and extending their bases, so as to include a vast mass of water in each wave." He explains this growing process, what we now refer to as *resonance,* with the analogy of using a finger to produce a large swinging motion in a weighty suspended bell. The bell is too heavy to push into a large swinging motion with your finger. But you can begin with a small push, and when the bell makes one small oscillation and begins moving again in the direction it was initially pushed, you can give it another small push, imparting some additional energy, so that this time the oscillation goes a little farther. After doing this repeatedly, the heavy bell now swings over large distances. And in a similar manner, small water waves become larger water waves, a little bit of added energy at a time.

But how did this very thin film of oil suppress the waves? To explain why all the waves in Clapham Pond disappeared, we quote Franklin one more time: "For the wind being thus prevented from raising the first wrinkles that I call the elements of waves, cannot produce waves, which are to be made by continually acting upon and enlarging those elements, and thus the whole pond is calmed."[41] But how does the oil keep the wind from producing those initial wrinkles, which mariners called *cat's paws* and scientists now call *capillary waves*?[42] That part of the problem Franklin didn't figure out, and it would be at least a century before someone did. With a typical wind wave (but not a capillary wave), the restoring force is gravity. If the water surface is pushed up, gravity will pull it back down. But as this water moves down, inertia carries it beyond the equilibrium position (the position when the water surface was initially flat). At some distance below the equilibrium position, the pressure from the weight of the higher water all around it pushes the water back up, and it again moves past the equilibrium position. This oscillation continues for some time, but it is not just a vertical oscillation. The change in shape of the water surface (which *is* the wave) moves horizontally along the water surface.[43] For very small capillary waves, however, the restoring force is not gravity, but rather *surface tension,* the attractive force between the water molecules at the sea's surface. Surface tension resists expansion of the surface, so if wind pushes the water surface down, surface tension pulls it back up, again overshooting the equilibrium position and starting an oscillation. The oil film on top of the water surface is only a single molecule thick, but it resists expansion much more than does the surface tension of the water. Thus, when there is an oil film, the wind cannot create capillary waves, and without capillary waves no larger waves can grow.[44]

Franklin enjoyed doing tricks and could be quite a showman. He liked to keep a little oil in the hollow joint of his bamboo cane. On one occasion, walking with some fellow travelers, he came upon a stream with wind waves on it. He stepped up to the water's edge and declared that he would make the waves disappear. With all the grandeur of a sorcerer he shook his cane three times over the water. Drops of oil fell unseen onto the water, and the waves quickly diminished, the water surface almost instantly becoming as smooth as a mirror, while his audience looked on in stunned amazement.[45]

Numerous other scientists, some of them very well known (Lagrange, Airy, Stokes, Lord Rayleigh, Russell, Lord Kelvin, to name a few), discovered other characteristics of water waves, usually through mathematical treatments. None of their insights helped immediately in the prediction of wind waves, but their work would be important decades later when hydrodynamic wave models began to be developed. These scientists had much more success with the easier problem of the motion of very long waves, such as tide waves, storm surges, or tsunamis (the definition of a *long wave* is a wave whose wavelength is much greater than the water depth). One characteristic that made these long waves much simpler to deal with mathematically than wind waves was the fact that their speed across the ocean was determined simply by the water depth. The shallower the water, the slower a long wave traveled. Long waves with different periods traveled at the same speed and did not disperse like wind waves.[46]

But when wind waves enter shallow water, they begin to take on some long-wave characteristics, and scientists were better able to understand and predict their behavior close to shore. Changes occur in wind waves when they begin to feel the sea bottom, at a depth comparable to their wavelength. As the water depth become shallower, wind waves become larger, because the mass they carry with them is forced into a smaller area. The shape of the wave profile also begins to change because the wave speed is now being affected by the water depth. The depth under a crest is greater than under a trough, so the crests move faster than the troughs. The wave becomes steeper, and eventually the crest begins falling over and the wave breaks onto a beach or over a sandbar offshore.[47]

The sea bottom also affects the direction that wind waves travel in shallow water. When waves approach the shore, their crests are often fairly wide, sometimes running some distance along the length of a beach. When the depth of the bottom varies, different parts of a wave crest travel at different speeds. The part of a wave crest over deeper water will travel faster than the part of a wave crest over shallower water. The wave crest will therefore bend. This is called *wave refraction.* Wave energy can therefore be focused by the shape of the sea bottom. Some bottom shapes, such as an underwater ridge or the decreasing depths around a point of land, focus the wave energy toward one location and thus increase the size of the waves (such a ridge would then be referred to as a *waveguide*). Other bottom shapes, such as an underwater canyon or the depths leading into a deep and widening bay, spread out the wave energy and decrease the wave size.[48]

For waves in deep water, very little water is actually moved by the waves. As a wave moves by a floating object, the object is moved up and down and back and forth, but after the wave has passed by, it ends up in almost the same location as where it started. The water particles themselves move in circular orbits and are transported only slightly forward, in the direction of the wave propagation. In shallow water, however, the circular orbits become elliptical and much more water is moved toward the beach, as is especially noticeable close to shore when a wave breaks and foamy water rushes up the beach.[49] When wind waves approach the beach at an angle, some of the water they carry is forced sideways by the beach, creating a *longshore current,* which can carry sand along the coast and erode the beach. Sometimes two longshore currents from opposite directions meet head-on, in which case strong *rip currents,* or *riptides* (a poor name, since it has nothing to do with the tide), are formed, which flow straight out to sea, sometimes carrying panicked swimmers with them.[50]

For centuries surfers have delighted in the huge waves that break over sandbars when large swells from a distant storm cross the ocean and arrive at their beach. Tahitians had been surfing long before Captain Cook witnessed them totally enjoying that activity in 1777, and so had the Hawaiians and the Polynesians, and possibly even the Incas. Such a sport probably could only have developed where the long and large swells of the Pacific arrived at tropical beaches inhabited by people who were children of the sea. These sea peoples were so knowledgeable of the swells of the ocean that they could steer their boats on the open ocean based on the direction that a specific swell was traveling and their knowledge of what direction that swell had always come from in the past.[51]

Europeans also knew large waves, but primarily from local storms fresh from the North Sea or from the North Atlantic. Their main concern was the loss of their merchant ships to those waves. In both regions there were rare occasions when an incredibly monstrous wave would suddenly appear out of nowhere. European coasts may not have been hit more often, but it seemed that way because they had better record keeping—in the form of logbooks from the many lighthouses that were built on headlands or small islands or offshore reefs or any place considered treacherous to shipping. The problem was that those were the types of places where wave refraction would focus the wave energy onto the lighthouses. Now and then the logbook of a lighthouse keeper would report the assault of a single truly enormous wave. But worse, sometimes the evidence for such a monstrous wave was the fact that the lighthouse and its keeper were suddenly no longer there.

The Eddystone Light, on a treacherous group of rocks in the English Channel fourteen miles from the Plymouth Breakwater, is perhaps the best known of the many lighthouses that fell victim to an extremely high wave. The first Eddystone Light, built in 1698 by Henry Winstanley, was originally an 80-foot-high octagonal wooden tower. Large waves frequently washed over the Eddystone's lantern, so Winstanley rebuilt the lighthouse in 1699 to a height of 120 feet with a ring of stone 4 feet thick up the first 20 feet of the base. He was so confident of its strength that he said nothing would please him more than to be in his lighthouse during a great storm. Four years later he unfortunately got his wish, when he and his lighthouse experienced the great storm of November 26, 1703, one of the largest storms ever recorded in the British Isles. The entire lighthouse, with Winstanley and five others inside, was washed away, never to be seen again.[52]

Winstanley was but one of many lighthouse builders celebrated in Britain. Most famous were four generations of the lighthouse-building Stevensons of Scotland (the family of Robert Louis Stevenson, author of *Treasure Island*),

Figure 5.1 Gigantic wave assaulting the second Eddystone Lighthouse, built in 1759 by John Smeaton. (*The Family Magazine or Monthly Abstract of General Knowledge, 1843*)

who battled enormous waves through two centuries, continually upgrading their lighthouse designs to try to withstand them.[53] The patriarch of the family, Robert Stevenson, built at least fifteen lighthouses, and his three sons, Alan, David, and Thomas (father of Robert Louis), built dozens more. It was

Thomas, though, who made significant contributions to the understanding of wind waves. He invented the *dynamometer*, an instrument that measured the force of a wave striking a solid surface.[54] He made many measurements demonstrating the overwhelming power of waves. The most powerful wave he measured pounded the 157-foot-high Skerryvore Lighthouse during a westerly gale on March 29, 1845. It produced a force of over 6,000 pounds per square foot.[55] But even without a dynamometer, the incredible power of large waves was evident from the objects they moved. Waves lifted a twenty-six-hundred-ton stretch of reinforced masonry off the foundations of the Wick Harbor breakwater. Waves from the English Channel tore up the stone breakwater at Cherbourg on the coast of France, hurling blocks that weighed three and a half tons sixty feet into the air.[56] And, of course, the tremendous height that these waves could reach was also a clue to their power. In 1831 Robert Stevenson built the 66-foot-high Dunnet Head Lighthouse atop a 345-foot cliff at the most northerly point of the Scottish mainland. During its building the workers were shocked to discover that occasionally a gigantic wave would scale the cliff face and reach the lighthouse. Even allowing for enhanced run-up due to a sloping cliff face, this was extremely high for any wave to reach. In 1872 at the 125-foot Dubh Artach Light built by Thomas Stevenson, a wave reached 92 feet up the tower, yanking a copper lightning conductor off its lee side.

Thomas Stevenson was the first person to undertake a serious measurement study of waves. Motivated by his desire to build lighthouses that could survive their assault and to build harbors that could protect ships from them, he measured wave heights every day in 1852. From these data he developed what was probably the first wave prediction formula, although in this case it predicted only the maximum wave height expected to occur at a coastal location at some unspecified time. As a lighthouse designer, he did not need to be able to predict exactly when an extreme wave would come, just that it would come. He came up with a simple formula that said the maximum height of waves at a coastal location was proportional to the square root of the distance over which the wind blew (called the *fetch*). A larger body of water such as an ocean could thus produce larger waves than a bay or a lake.[57]

Thomas Stevenson also realized that strong tidal currents and river currents affect the size of waves. When swells from the ocean run into an opposing strong current, their heights are increased. The Pentland Firth, between the northern tip of Scotland and the Orkney Islands, was notorious for its *tidal races* (or *roosts*), narrow waterways with strong tidal currents (with speeds up to eighteen miles per hour) that turned ocean swells into large, dangerous waves.[58] Thomas realized that the dangerous surf often seen at the mouths of some rivers was not due solely to a shallow sandbar but could also be due to the strong outward river current meeting the waves. The current arrested the motion of the waves and decreased their wavelengths, causing them to become higher and steeper and eventually to break. Thomas suggested that the current slowed the wave's motion so much that the following wave could overtake it, with both waves coalescing into one larger wave, which acquired more speed and overtook waves in front of it, becoming even larger—an incorrect theory, but one that stimulated further thinking for years.[59]

While the extremely large waves that have destroyed or damaged lighthouses are clearly examples of the sea's wave power, every day millions of smaller waves are demonstrating their combined power by eroding shorelines and beaches. Sometimes it doesn't take a ninety-foot wave to bring down a lighthouse. Given enough time, the millions of smaller waves can do the job by slowly eroding the land around a lighthouse. Such is the case for the Montauk Point Lighthouse on Long Island, which George Washington authorized after Congress appropriated the land for it in 1792. The lighthouse was originally built 297 feet from the edge of a bluff. Over the years the waves of the Atlantic have eaten away at the base of the bluff until that distance has shrunk to only 50 feet. Unless the erosion is stopped, the lighthouse will someday fall into the sea.[60]

By the time World War II began, there was still no capability to predict the height of waves produced by local winds (referred to by mariners as *sea*, as in "a heavy sea"), or the size of the swells that crossed hundreds of miles of ocean from distant storms, or the size of the surf produced when the sea and swell broke onto the shore. But less than a year after the United States had been pulled into the war by the attack on Pearl Harbor, an urgent need arose for the capability to predict the sea, the swell, and the surf. On July 25, 1942, President Roosevelt and Prime Minister Churchill agreed to carry out Project Torch, the first Allied attack on Nazi-held territory, the assault on North African territories held by the Nazis and Vichy French. By the end of the month General Dwight Eisenhower had begun the plan to capture North Africa, involving three amphibious landings tentatively planned for November 1942. The first landing was to be at Casablanca on the Atlantic Coast of Morocco, and the second and third at Oran and Algiers, respectively, on the Mediterranean Sea. With thirty-five thousand American soldiers to land at Casablanca, thirty-three thousand British soldiers to land at Algiers, and thirty-nine thousand at Oran, these were to be the largest amphibious landings ever attempted. There was, however, a serious problem.

In the late fall and winter Casablanca was always hammered by large Atlantic swells coming from the northwest. The swells would break on the beaches in what General Eisenhower called a terrifying fashion that was "forbidding from the standpoint of small-boat landings."[61] Large waves would make it dangerous to transfer men from the battle cruisers to the Higgins landing boats that would take them to the beaches. The waves would cause the landing boats to bob about wildly and crash repeatedly into the battle cruisers. Men trying to get aboard would fall into the sea or be crushed between the two vessels. Even worse, once the landing boats reached shore, they would face those same waves grown even larger because of the shallow water, breaking on the beach so violently that the boats would be swamped and the men with their weapons and heavy backpacks would be pulled helplessly underwater and drown. Huge swells could appear suddenly even when the weather was clear and the local winds were weak. Dangerous wave conditions were expected at Casablanca four days out of five in the fall and winter.[62]

As the date approached for this crucial opening act in the European theater of World War II, one that would have a critical effect on the rest of the war, Churchill, a former First Lord of the Admiralty, worried about those waves. He asked Roosevelt, "What happens if, as I am assured is 4–1 probable, surf prevents disembarkation on Atlantic beaches?" Roosevelt answered, "Bad surf conditions on the Atlantic beaches is a calculated risk."[63] In the meantime, both the Americans and the British were working on methods to calculate that risk for each day in November. They were trying to predict the size of the waves that would be approaching the North African coast and the size of the surf when those waves broke on the beaches.

That same July, Walter Munk, a graduate student at Scripps Institution of Oceanography in La Jolla, California, was brought to Washington to work in the newly formed Oceanography Section at the Army Air Force Directorate of Weather. He was brought there by a first lieutenant in the Army Air Force who had attended an oceanography training course at Scripps taught by Harald Sverdrup, Director of Scripps and Munk's mentor.[64] At that time Sverdrup was the leading physical oceanographer in the world (a recognition that Munk would also achieve years later).[65] At Washington, Munk was briefed on the plans for Operation Torch. A method was needed to predict surf conditions at the Casablanca beaches so that two or three days could be selected when the surf would not be too high to threaten the success of the landing. To do this Munk would have to predict the generation and decay of wind waves from the information available on weather charts, and then estimate the subsequent transformation of those waves into breaking surf.[66]

The U.S. Army had already begun practice landings on the Atlantic beaches of North Carolina. Those practice landings had to be halted whenever breaking waves reached six feet, because the waves swamped the Higgins boats. This did not bode well for Operation Torch, since six-foot seas were exceeded most of the time during fall and winter along the Atlantic beaches of Morocco.[67] Before coming to Washington, Munk had been working on a similar problem along the California coast, where he had compiled wave and wind statistics. Using these data, he had been trying to develop a formula that could predict expected wave heights based on wind speed. In his formula he was incorporating hydrodynamic wave theory and not simply relying on statistical correlations.[68] Munk spent September in Washington assembling wave and wind data from wherever he could find it, which he hoped would improve his formula for predicting sea, swell, and surf.[69]

At the end of September, Munk felt that he was close to success, but to finish he told the Army that he would need help from Dr. Sverdrup. Sverdrup came to Washington and spent most of October working with Munk.[70] They compared wave predictions (made using their forecasting formula) against wave observations, including wave records from the Azores made in support of Pan American Airways seaplane landings. They "hindcast" waves for the Azores using wind derived from weather maps from the same time periods as the wave data.[71] By the end of October, Sverdrup believed their method was ready to predict wave conditions for the amphibious landings at Casablanca. Their method using weather maps consisted of three steps. First, the wave height and the wave period of the storm-generated sea were calculated using three parameters—the wind speed, the time duration of the storm, and the fetch (the distance over which the waves traveled while the wind was blowing). Second, after the waves moved beyond the storm or the storm stopped, these swells would slowly attenuate, and this decrease in size was estimated.[72] Third, when the waves approached the beach, they would be distorted by shallow water until they broke, and this was computed from principles of conservation of energy.[73]

As the proposed November landing date approached, the problem of the sea, swell, and surf worried Major General George Patton, who was in charge of the attack on Casablanca. It also worried Rear Admiral Henry Hewitt, who commanded the 107 vessels that were bringing Patton's green American troops from the United States directly to the African shores for their first encounter with war and death. Although they hoped that the French would not fight, since they had been allies of the United States before being conquered by Nazi Germany, something told them this was wishful thinking, and they focused on every potential problem as though they were fighting the Germans themselves. The problem of the surf on the landing beaches was made even more of a problem because the tide (with a ten-foot range) would be ebbing, so if landing craft were delayed in unloading their men because of the surf, they would be stranded there until the next high tide. Because of the tides, Hewitt had even recommended delaying the landings a week, but Patton rejected that idea, since a week later the chances of bad weather would be greater and there would be more moonlight, making surprise more difficult.[74]

The meteorologist aboard Hewitt's flagship the *Augusta*, Lieutenant Commander Richard Steere, was asked by Hewitt and Patton to find three days when ocean swells would be below eight feet and accompanied by clear skies for effective air support.[75] Steere received forecasts from Munk's special wave forecasting team, which was operating at Joint Weather Central in the U.S. Weather Bureau building and included Navy meteorologists and personnel from the Weather Directorate and the Weather Bureau.[76] Munk's group took the weather forecast maps produced by the meteorologists at Joint Weather Central and produced wave forecasts for the same time period. The accuracy of those wave predictions depended not only on Munk and Sverdrup's formula but also on the winds calculated from the weather maps. Steere received those same weather forecasts from Washington, so he could see what Munk based his wave forecast on, as well as weather forecasts from London and weather and sea state reports from the Army Air Force Weather Service forecaster at Eisenhower's headquarters at Gibraltar.

Early on November 7, Munk's team predicted surf that would be too high for landing on November 8. They also predicted that those conditions would last through at least November 12.[77] Weather reports from a submarine off the Moroccan coast confirmed that sea conditions were not encouraging. Eisenhower was considering diverting the American ships to Gibraltar and giving up on Casablanca, but he left the last-minute decision up to Rear Admiral Hewitt based on Steere's recommendation.[78] Steere, being on-site and with the most weather information coming in to him, recognized a late change in the synoptic weather situation near Morocco and, after applying the Sverdrup-Munk formula to this new data, recommended that the Casablanca landing go forward.[79] And indeed, late on November 7 the seas began to subside. At 2 A.M. on November 8 General Patton stood on the deck of the *Augusta* in the dark looking out at the lights of Casablanca and Fedala in the distance. He wrote in his diary, "Sea dead calm—no swell. God is with us."[80]

As it turned out, the morning of November 8 was the only time the amphibious landing could have successfully taken place at Casablanca. That afternoon the wave heights began to increase again, and they continued that way through November 12, as had been predicted by the Munk group in Washington.[81] The landings began before dawn at three locations along the Moroccan coast—at Fedala (12 miles north of Casablanca), at Port Lyautey (60 miles farther north), and at Safi (140 miles south of Casablanca). Even with wave conditions better than hoped for (breakers two to four feet), the landings were not easy. The surf caused many landing boats to lose control, some being smashed against the reefs and others capsizing.[82] There were three assault waves on Fedala. The experience described by a newsman in the third assault wave was typical. "We plunged up to our armpits into the surf and struggled to the reef. Waves washed over our heads, doubling the weight of our sixty-pound packs with water, but sweeping us nearer safety. . . . Twice the surf pulled me loose and twice it returned me. My strength was ebbing fast when another soldier pulled up the man before me and lent me a wet hand to safety.... When I could stand again, I saw about me scores of dripping soldiers, their legs weary and wide-braced."[83]

If the waves and surf had been any higher (as they had been on the previous days and would be on the following days), most of those exhausted-but-alive soldiers would most likely have drowned. As it was, even with those relatively good wave conditions, 64 percent of the 370 landing craft used at Fedala were damaged by the surf.[84] Patton went ashore at 1:20 P.M., and later wrote of "getting very wet in the surf."[85] The waves were definitely increasing again, but by this time most American objectives had been met. Before night fell, the American flag had been planted in Northwest Africa, and the largest amphibious operation in history up to that time had been a success.[86] But an even larger amphibious landing lay on the horizon. Operation Torch had been only the first step in the war against Nazi Germany and its ally Italy. If all went well and the Allies captured North Africa and then Sicily and Italy, the plan was then to carry out Operation Overlord, which would send an even larger force across the English Channel to attack the Nazis in northern France.

In the next chapter we will see the important role played by Allied wave predictions at the Normandy beaches on D-Day and during the weeks after. Then, after decades of successful sea, swell, and surf forecasts, we will see why we still cannot predict the occurrence of gigantic rogue waves like the one that assaulted the *Queen Mary,* and we will see what is being done to develop such a capability.

CHAPTER 6

"Holes" in the Surface of the Sea

Rogue Waves

By mid-August of 1943 the Allied command had decided that they should land on the beaches of Normandy in the spring of 1944.[1] The success of the operation to drive the Nazis out of northern Europe, Operation Overlord, depended first on winning the beachhead at Normandy on D-Day and then on efficiently moving large numbers of troops and supplies across those beaches. The Overlord planners estimated that it would take at least fifteen weeks to move enough Allied divisions across the English Channel and the Normandy beaches to equal those that Germany already had in northern France and Belgium. That fifteen-week estimate assumed that there would be good weather and good wave conditions the entire time, which was highly unlikely.[2] Accurate predictions of sea, swell, and surf were therefore going to play a vital role in determining the success of both the initial amphibious landing and the subsequent transport of troops and supplies for months thereafter.

By this time Walter Munk and Harald Sverdrup were back at Scripps in California. Sverdrup had set up a training course on wave prediction, and Munk was assisting. Over two hundred officers from the Army Air Force, Navy, and Marine Corps were trained, those officers later participating in the planning and execution of amphibious landings at Sicily and Normandy, and then all landings in the Pacific theater of war, including Iwo Jima, Okinawa, and the Philippines.[3] The course was constantly being modified as improvements were made to the wave prediction technique. The trained officers helped test the Sverdrup-Munk method, and each had to adapt it to the particular coastal region to which he was assigned. To further improve their method, Munk and Sverdrup tried to calibrate the predicted wave heights and periods by comparing them to observed heights and periods during each amphibious landing when it occurred. These were only visual estimates made by the coxswains on the landing boats, but they were still useful. This led to the definition of *significant wave height* as the average of the highest one-third of the waves, which seemed to be what the coxswains were seeing. *Significant wave height* would become a critical term in all future wave studies.[4]

In the fall of 1943 a Swell Forecast Section was established in downtown London at the Admiralty weather center, two floors underground because of German bombing. This was one of three centers in Britain involved in making weather forecasts for the D-Day landings at Normandy; the other two were the British Meteorological Office at Dunstable and the U.S. Strategic Air Force center at Widewing. In the Swell Forecast Section were three wave forecasters: two American officers who had been trained at Scripps in the Sverdrup-Munk wave prediction method and one British officer who was more familiar with another technique developed in Britain.[5] The British technique was based on correlations between a number of wave observations and wind observations, which had led to rules of thumb and crude forecasting graphs. Upon comparing the two techniques, the section decided to use the more sophisticated Sverdrup-Munk technique. The section had to first determine the height and period of the swells entering the English Channel from the North Atlantic and determine how high they would be when they hit the five designated landing beaches at Normandy (Utah, Omaha, Gold, Juno, and Sword). Then they had to predict the height and period of waves generated by local winds over the Channel itself. Finally, they had to determine the size of the surf on the Normandy beaches, which involved understanding the effect on waves of the shallow water, the strong tidal currents, and the coastal configuration, the latter including effects of waves hitting the beaches at various angles.[6]

To carry out this work the Swell Forecast Section needed to obtain wave data, so they organized a synoptic network of fifty-one wave reporting stations along the British coast of the English Channel. The data came from visual observations made by His Majesty's Coast-guard lookouts, who usually made three observations per day, each consisting of a wave count over three minutes and an estimate of the height of each wave breaking during that time. There were also underwater pressure recorders at four of the stations. Wave conditions on the Normandy beaches were estimated by studying aerial photographs of the beaches. One fortunate result of the section's work was the discovery that the Cotentin Peninsula usually blocked Atlantic swells from the west so that they did not reach the Normandy beaches. Thus for the beaches they could concentrate on waves and surf generated by local winds. The swells were, however, still important for the long trip across the English Channel to reach Normandy.[7]

The ultimate objective was to predict waves and surf at the Normandy beaches based on weather forecasts. Each weather forecast was a synthesis of the weather forecasts put out by the three weather forecast centers. The enormous responsibility of synthesis fell to Eisenhower's Chief Meteorologist, Group Captain James Stagg. It was made all the more difficult because the Dunstable and the Widewing meteorologists regularly disagreed with each other.[8] The three weather forecast centers held regular conferences via secure telephone lines that included Stagg and his deputy at Supreme Headquarters Allied Expeditionary Force (SHAEF) and the meteorological officers for the Allied fleet and the Air Force. The Swell Forecast Section listened in on the same phone line as the Admiralty meteorologists. When agreement was reached on the forecast winds for that day, the section would generate sea, swell, and surf forecasts. Stagg took the weather and wave results of each conference to Eisenhower's twice-a-day staff meetings with his commanders. By early April 1944, SHAEF was requiring five-day wave forecasts for the English Channel and adjoining sea areas.[9]

Based on tide and moonlight conditions, D-Day was planned for June 5, with June 6 and June 7 still acceptable if bad weather prevented going on June 5 (see Chapter 2).[10] Starting and stopping such a massive operation could not be done at a moment's notice. Some of the ships involved in Project Neptune (the naval portion of Operation Overlord) were coming from as far away as Scotland and the Irish Sea and so had to begin their voyages days before the expected landing. Thus, weather and wave forecasts had to be made for several days into the future.[11] June was normally a month of good weather, but in 1944 the June weather did not cooperate. Although as usual there had been disagreements between pessimistic meteorologists at Dunstable and optimistic meteorologists at Widewing, the forecast that Stagg brought to Eisenhower's 9:30 P.M. staff meeting on Saturday, June 4, was gloomy.[12] High winds, high waves, and heavy cloud cover were expected on June 5. There could be no air cover for the operation, a critical requirement. The waves and surf would make landing extremely difficult, and perhaps impossible. The waves would also reduce the accuracy of the big guns on the Allied ships. At the 4:00 A.M. Sunday staff meeting the bad situation had not changed, and Eisenhower decided to postpone D-Day for twenty-four hours. Hundreds of ships that had already left port had to be turned around. Even ships that had not left port had already boarded their men, and they could not be allowed to disembark. So everyone just waited on board crowded vessels, rocked by the rough seas. But at the 9:00 P.M. Sunday staff meeting on June 5 there was a ray of hope. Stagg had a more promising forecast, the prediction of a short break in the bad weather that might allow the landings to take place on June 6. After discussion with his commanders, Eisenhower made the decision to go. And so the largest amphibious landing ever undertaken was under way once again—more than 160,000 soldiers on five thousand ships and landing craft were headed for Normandy. Yet, even with all that firepower, it was not clear that it would be enough to overcome the German forces entrenched along the French coast.[13]

The bad wave conditions of the previous day had not subsided much when the ships began crossing the English Channel.[14] It was a long, rough ride. Seasickness was rampant. Soldiers on the ships that had left port early and had been temporarily turned around had been seasick for days. The heavily laden low-powered landing craft were especially affected by the rough seas, even as the six-to-eight-foot waves slowly reduced to three-to-five-foot waves. Many troops were in their crowded landing boats for at least eighteen hours, in air heavy with sweat and vomit, feeling the constant rolling of the short steep seas and wet from the salt spray that hit them each time their vessel crashed into the next wave. If there was a benefit to the widespread seasickness, it was that it kept the soldiers from thinking about the dangers that would meet them at the beaches and about their loved ones at home whom they might never see again. Many were not scared, because hours of seasickness made them feel like they were dying anyway. Their greatest desire was to get off those boats, even if it meant facing German gunfire.[15]

As they landed on the beaches, the waves and surf had reduced to heights considered just barely manageable by Eisenhower's commanders. Unloading the men onto the beaches was made difficult by waves swamping some of the landing boats and pushing the heavily loaded men around. Landing was already a difficult proposition, even without the waves, because the boats had to avoid the underwater obstacles that Rommel had planted between the low-water and high-water lines, and the men had to avoid being hit by German gunfire. The waves caused many of the specially designed floatable tanks to sink as soon as they were launched. It also made it more difficult for the demolition teams trying to blow up Rommel's obstacles.[16] But in all respects, on the Channel and then landing on the beaches, it could have been so much worse, even disastrous, if the landings had taken place during the higher wave conditions of the day before.

The final contribution of wave prediction to the overall success of Project Overlord came after the Normandy beaches had been secured by Allied forces. Transporting Allied troops and supplies across the English Channel and across the Normandy beaches as quickly as possible was critical to the success of the battles that followed D-Day as the Allies moved toward Germany. But the weather was often not cooperative. Two weeks after D-Day a huge storm hit Normandy that totally disrupted movement of troops and supplies.[17] But even smaller storms had serious effects. The Swell Forecast Section kept records of the relationship between surf heights at Normandy and the tonnage of supplies unloaded from ships each day and found a dramatic correlation. A noticeable decrease in the supplies unloaded occurred whenever the surf increased above two feet. The amount of supplies unloaded decreased by as much as 80 percent as wave heights neared five feet. Above seven feet all unloading operations ceased.

The Swell Forecast Section had two representatives on Omaha Beach who prepared twice-daily twenty-four-hour forecasts of the sea state and the surf in support of the cargo operations.[18] These forecasts saved precious hours of unloading, since they allowed the cargo loading to continue up to the very last moment before the surf became unsafe. Without such accurate forecasts, those unloading the cargo could be caught by surprise without enough time to safeguard the ships, ferry craft, barges, and amphibious trucks (called DUKWs). They also allowed loading to begin promptly when conditions had again become safe. When the waves began getting smaller, there was a reluctance to refloat beached craft and begin unloading operations again in case those smaller waves turned out to be a temporary lull, so an accurate wave and surf forecast ensured that the work would begin again as soon as it was safe. Wave forecasts enabled cargo transport to continue into November instead of quitting in September as originally planned. The Germans tied up thousands of troops at European harbors, thinking the Allies would need to capture them for their transport needs, something that was not necessary because of the high efficiency of troop and cargo transport through Normandy, much of that efficiency due to accurate wave predictions. Those men and supplies made the future Allied successes in the war possible.[19]

After World War II, other demands for wave prediction came to the forefront—safe ship routing for commercial vessels at sea, the protection of oil platforms and other offshore structures, preventing the erosion of beaches, and the safety of recreational boaters and the ever-increasing numbers of seaside tourists. The improvements in wave prediction that had been developed by Sverdrup and Munk sparked a renewed scientific interest in the subject and attracted other researchers. Unlike earlier classical wave work, the new wave science tended to be more observationally based, utilizing new kinds of instruments to measure waves and new mathematics to statistically describe the complex wave data. A crucial part of this work was the calculation of a *wave spectrum,* which showed, usually on a graph, the amount of wave energy at dozens of different wave periods. However, instead of *wave period,* oceanographers typically used *wave frequency. Frequency* is the inverse of *period.* It is the number of wave crests that pass by in a given amount of time. For example, a wave with a period of ten seconds has a frequency of one cycle per ten seconds, or 0.1 cycle per second. Lower-frequency (longer-period) waves (swells) have longer wavelengths than higher-frequency waves.

The mathematics for wave analysis and prediction became especially complicated, because not only did one have to use hydrodynamic equations that describe the complex physics of wave motion but one also had to use sophisticated statistics that could meaningfully describe the combination of hundreds or thousands of random individual waves. And to make it even more complicated, all these waves are interacting with each other.[20] The classical theories dealt with individual waves or simple combinations of a few waves, but one cannot measure such simple waves in the ocean. The only thing one can measure in the ocean is the ever-changing combination of hundreds of waves.

As we saw in the last chapter, exactly how waves are generated and made to grow by the wind has been debated for centuries. In 1957 two theories were proposed that are still used in some fashion today. The theory and mathematical treatment by Owen Phillips of the initial generation of capillary waves through a resonant mechanism involving the turbulent pressure fluctuations of the wind (similar to Franklin's suggestion that we looked at in the last chapter) is still mostly embraced for that initial generation but not for their further growth. The theory and mathematical treatment by John Miles says that once capillary waves are present, the wind has a form to push against, providing energy to the waves both by pushing the wave (wind pressure) and by frictionally rubbing on the water surface (wind shear).[21] To these two theories were added other factors, such as mechanisms by which energy and momentum are transferred among different waves, and the sheltering effect of the wave shape, whereby the wind has a different effect on the front side of a wave compared with the backside. Today there still does not seem to be a completely accepted total explanation for wave generation by the wind, but that did not stop the development of wave prediction models.[22] Models for wave prediction slowly evolved over the decades from solely statistical techniques, using analyzed wave data, to techniques using hydrodynamic equations that involve the physics of the wave motion. Since many of the processes were not completely understood, formulas were used in these models to represent those processes, and particular coefficients in these formulas were selected to make the model match data measurements. This is referred to as *parameterization*.

One problem with wave prediction had been that to predict waves at a particular location one needed to predict the winds that generated them, not just locally but all over the globe, because longer waves (swells) can travel thousands of miles from storms over many areas of the sea. Those waves come from different directions, all of them crossing each other and momentarily adding to or canceling each other to varying degrees that make them so irregular and random as to seemingly defy prediction. This problem was solved with the development of global weather prediction models, the predicted winds from which could drive global wave models (both models now being possible because of increased computer power). Such wave models had to be calibrated and verified with measured wave observations. These data were supplied by networks of wave buoys (such as the more than seventy operated by NOAA's National Data Buoy Center) and other sensors for measuring wave heights and directions (such as wave staffs, bottom pressure sensors, land-based radar and laser systems, and rapid-sampling water level gauges). These are another part of the Global Ocean Observing System. Wave measurements over the entire globe are also made using radar on satellites. All these data are very useful for testing and calibrating the global hydrodynamic wave models and can also be provided as real-time information to mariners and coastal residents.

Currently, the most advanced global wave models are the third-generation models, such as NOAA's WAVEWATCH III®, developed and used in the National Centers for Environmental Prediction, and WAM, used at the European Centre for Medium-Range Weather Forecasts.[23] Third-generation wave models explicitly represent all the physics relevant for the development of the sea state. Some physical processes are still parameterized, but this is done much closer to first principals than in previous wave models. Areas where some of the physics is still not fully understood include processes such as wave growth and decay due to the wind, nonlinear interactions between waves, the effect of bottom friction, the effect of surf breaking at the coast, and especially dissipation due to steep waves breaking at sea (called *white capping*). The coefficients selected for the various parameterizations seem to have been well chosen because these wave models do reasonably well in predicting average wave conditions. They have also improved considerably as they received more accurately predicted wind fields from improved weather forecast models. These wave models can also handle the refraction of waves (described in the last chapter) due to variations in bottom depth and variations in ocean currents.

There remains, however, one area of wave prediction that these models cannot handle. None of these models can predict the really big one—the wave that suddenly appears and is two or three times larger than any of the waves around it, a wave like the one that almost capsized the *Queen Mary*. Such a wave has been called a *rogue wave* or a *freak wave*. It is huge, but just as important, it is very steep. Rogue waves can reach heights of one hundred feet from trough to crest. The troughs of rogue waves drop so precipitously that mariners call them "holes in the sea." When a ship suddenly drops into one of these seemingly bottomless pits and her crew looks up at a hundred-foot wall of white water towering over them (which the captain of the *Queen Elizabeth II* said looked like the white cliffs of Dover), there is little they can do but hold on for dear life.[24] Rogue waves have capsized passenger liners and broken oil tankers in half. Over the centuries, reports of these monster waves had accumulated, but many scientists still viewed with skepticism these "sea stories." That skepticism was primarily due to the scientists' own inability to mathematically determine how such huge waves could exist. And since they could not understand the physics of how such monstrous waves could be created, many decided there must be something wrong with each rogue wave account, no matter how reliable the person reporting.[25]

Scientists did believe stories about extreme waves that hit lighthouses. Undeniable evidence backed that up, since lighthouses were washed away (as we saw in Chapter 5). And their mathematical models could explain the physical mechanisms that increased the size of waves in the shallow water near lighthouses. Even so, scientists were not quite sure about the large wave that came out of nowhere at 11:00 on the night of July 3, 1992, and washed over a part of Daytona Beach, injuring seventy-five and damaging beach property. It appears that the wave was produced by a squall line that moved at the same speed and in the same direction as the wave, so that the wave kept growing as it approached the beach, with the sloping sea bottom further amplifying it.[26] Scientists also believed the accounts of extreme waves experienced in areas with strong ocean currents. Here again there was plenty of evidence, but most important, their mathematical models showed how energy could be passed from an ocean current to waves moving in a direction opposed to the current flow. Especially notorious was the Agulhas Current off South Africa (see figure 6.1).[27] Between South Africa and Antarctica there is an unobstructed stretch of stormy ocean waters that goes completely

Figure 6.1 Painting of the East India Company's iron war steamer *Nemesis* among huge waves in the Agulhas Current off the Cape of Good Hope. (*1841 Reeve engraving of Leatham painting, Mariners Weather Log*)

around Antarctica (called the *roaring forties* since it is south of latitude 40° south). Wind blowing over this endless fetch produces long, powerful waves called *Cape rollers* (they occur near the Cape of Good Hope) that travel toward the Agulhas Current, the strong opposing current further increasing their size.[28] Gigantic waves also hit ships near other major ocean currents such as the Gulf Stream off the United States, where numerous ship losses probably due to rogue waves helped build the infamous reputation of the so-called Bermuda Triangle.

But away from major ocean currents and shallow waters, most scientists didn't think waves could grow much larger than fifty feet, and then extremely rarely. Other scientists felt larger waves were possible, but they did not know of any mechanism for generating such a wave in the open ocean. With no prediction capability for rogue waves, mariners were on their own. They could select ship routes that tried to avoid storms, knowing that in or near a storm the probability increased significantly for extreme waves. But that does not always work—in 1995 the captain of the *Queen Elizabeth II* had changed her route to avoid the high winds of Hurricane Luis, but she still was hit by a rogue wave. Mariners could also avoid strong ocean currents, but riding those currents is beneficial because it decreases the duration of their voyage and saves fuel. Considering the large number of vessels that have sunk or just disappeared year after year, mariners obviously needed much better information.

In the eyes of most oceanographers, reports about ships being struck by rogue waves lacked verifiable height measurements of those waves. Of course, ships that mysteriously disappeared left no survivors to provide wave height estimates. In 1963 the SS *Marine Sulphur Queen* became one of those ships that simply disappeared, along with her thirty-nine crew members. Her last known position on February 4 was in the Gulf of Mexico west of Key West. She had intended to ride the Gulf Stream through the Straits of Florida, where debris from the ship was later found, and then sail north to Norfolk, Virginia. The *Marine Sulphur Queen* was the first missing ship mentioned in a 1964 article in *Argosy* magazine called "The Deadly Bermuda Triangle," the first time the phrase "Bermuda Triangle" was used. The author dramatically and misleadingly wrote that the *Marine Sulphur Queen* "sailed into the unknown."[29] There were, in fact, rough weather conditions along the likely route of the ship, with sixteen-foot waves coming from the north.[30] Upon later investigation it was discovered that the *Marine Sulphur Queen* was not seaworthy enough to survive large waves. She was a nineteen-year-old, 500-foot-long, 7,200-ton "T-2" oil tanker that had been converted in 1960 to carry 15,000 tons of liquid sulfur (heated to 255 degrees) in a 300-foot welded steel tank. The tank gave her a high center of gravity and made her more likely to roll over in heavy seas. T-2 tankers also had a weak back, making them more likely to split in two under the strain of heavy seas. Three other T-2 tankers had broken in half and sunk in the previous eleven years. This was the *Marine Sulphur Queen*'s sixty-fourth trip delivering molten sulfur from Beaumont, Texas, to the East Coast.[31]

Shortly after this last voyage of the *Marine Sulphur Queen,* another ship, a sixteen-thousand-ton bulk carrier, followed the same route from Texas through the Florida Strait, riding the Gulf Stream northward. One beautiful night under a full moon, just after they had passed Miami and were traveling on a gently rippled sea with a light breeze, the crew felt the ship suddenly lift forward. Running to the window, they saw an immense foaming wave that crashed over the bow, the water, alive with phosphorescence, flooding the entire deck and bending the ladder that led to the mast house. That wave had apparently been generated by an intense low-pressure system northeast of Cape Hatteras and had grown in size, because it was traveling southward against the northward-flowing current of the Gulf Stream. If a similar rogue wave had hit the *Marine Sulphur Queen,* she would have been a sitting duck, gone in minutes with no time to send an SOS.[32]

Another ship that mysteriously disappeared was the container ship M/V *München,* the 650-foot "unsinkable" pride of the German merchant navy. None of her twenty-seven crew lived to tell what had caused her to sink. After a garbled Mayday at 3 A.M. on December 12, 1978, the *München* vanished during a storm in the North Atlantic. More than a hundred ships searched for her, but all they found was a single badly battered lifeboat with a severely mangled davit pin. That lifeboat had been stored sixty-six feet above the waterline of the *München,* and it would have taken a very great force to bend that davit pin (the davit is a winch system that holds the lifeboat). The cargo ship *Cordigliera* and its thirty crew sank off the southeast coast of Africa on November 13, 1996. Rescue ships found only empty lifeboats, life rafts, and some floating wreckage. And the bulk carrier *Leros Strength,* with its crew of twenty, bound for Poland after loading iron ore in Murmansk, sank twenty-three miles off the Norwegian coast on February 8, 1997. These are only a few of hundreds of missing ships.

A few ships survived assaults by rogue waves, but their eyewitness accounts were generally viewed by scientists as exaggerations, although a few of those accounts must have been difficult to dismiss. In February 1933 the USS *Ramapo* was in the middle of the northern Pacific Ocean en route from Manila to San Diego. The *Ramapo* was a naval tanker that delivered petroleum products from the United States to the Philippines, but she also acted as a survey vessel during her voyages and collected oceanographic data for the U.S. Hydrographic Office. So her crew knew how to make accurate measurements. She was caught in the middle of a violent storm that covered the entire North Pacific, providing an unobstructed fetch of thousands of miles with high winds blowing constantly in the same direction for about seven days. The *Ramapo* was thus sailing through mountainous seas that dwarfed it, but the waves had very long wavelengths (1,000 to 1,500 feet), long enough so that

Figure 6.2 The splitting in half and sinking of the supertanker *World Glory* in 1968 off South Africa after being hit by a rogue wave. (*South African Sailing Directions*)

the 478-foot-long ship could glide down the lee slope of each wave and then move up the windward slope to the next crest, at which point its propeller would momentarily leave the water. Sometime after midnight on February 7 the *Ramapo* slid down the leeward slope of the largest wave her crew had ever seen and was in the following trough, just starting to move up the next windward slope. The moon was out and visibility was good. Using triangulation relative to known lengths on the ship, they carefully measured the height of the wave crest behind them, which was clearly seen from the bridge. The height was 112 feet, making it the highest ocean wave ever measured.[33]

Although such precise height measurements were not taken when a rogue wave hit the supertanker *World Glory* on June 13, 1968, there was little doubt from the accounts of the ten exhausted and oil-smeared survivors that a wave of at least 70 feet was involved. The 737-foot-long, 46,000-ton supertanker was carrying 334,000 barrels of Kuwait crude oil. Oil tankers from the Persian Gulf often avoided the stretch of ocean near the Cape of Good Hope where the Agulhas Current moved around South Africa. But with the Suez Canal closed by the Israeli-Arab War, they had little choice. The *World Glory* had been handling the large Cape rollers reasonably well. She had her bow aimed toward the oncoming waves when at around 3 P.M. they saw an enormous wave heading toward them. The ship buried her nose in the wave, which then moved under the ship's midsection, lifting her high in the air and leaving the bow and the stern hanging unsupported. The weight of the oil in the bow and the stern bent the ship, producing a crack across the main deck. The crack opened and closed with the movement of the ship. Then another wave raised the bow sharply skyward and released it. The ripping metal as the *World Glory* split in two made a terrifying sound. Both sections remained afloat awhile, oil pouring from them. The oil caught fire and burned until another huge wave doused the flames. The thirty-four crew were split between the floating sections, which were repeatedly rocked by more waves. The stern sank after about two and a half hours. The bow section lasted another couple of hours and drifted southwest with the Agulhas Current before sinking. Only ten of the crew were rescued.[34]

West of the Cape, in the South Atlantic, two other rogue wave incidents occurred in 2001 involving two tourist-filled ocean liners. This area of open sea is without strong ocean currents. Within one week and 620 miles of each other, at the end of February and beginning of March, the *Bremen* and the *Caledonian Star* were hit by ninety-eight-foot waves that in both cases broke windows on the ship's bridge. The description by the first officer of the *Caledonian Star* had been heard before—the single enormous wave looked like a mountain. The *Bremen* lost power and navigation tools for two hours. Both were able to limp back to port. The ships had been in major storms, and although they were far from the Agulhas Current, it was speculated that waves from the storms might have been enlarged by traveling through a large eddy that spun off the Agulhas Current and drifted into the South Atlantic. Such eddies (or *rings*) are large circular flows formed when an ocean current has a strong meander (an S-shaped bend) that pinches off, creating a rotating ring whose currents can be almost as strong as the original ocean current.

These are but a few of the countless stories of ships being hit by rogue waves, but they were not enough to convince most oceanographers that rogue waves were real.[35] For them it was going to take hard data obtained with scientific instruments to prove that their wave models were inadequate. Only very recently were data finally obtained that forced these oceanographers to admit that the mariners had been correct. They might argue with the story told by a sea captain, but they could not argue with wave height data obtained from a radar system on an oil platform or on a satellite, data that not only showed that rogue waves were real but that there were large numbers of these monsters of the deep out there on the oceans of the world. The first scientific evidence that began to change their minds was from offshore oil rigs. On January 1, 1995, during harsh weather, a downward looking laser device on the Draupner oil rig in the Norwegian sector of the North Sea was measuring many 40-foot waves, when out of the blue, literally, appeared an 85-foot wave, more than twice as tall.[36] At the Goma oilfield in the Danish sector of the North Sea, a radar had been in operation from 1981 through 1993, attached under a bridge between two oil platforms. When this twelve-year data record was carefully looked at in 1997, they found that at least 446 rogue waves could be identified.[37] Other wave data sets were reexamined, revealing rogue waves that had been edited out of the data as instrument errors. In 1998 during Hurricane Bonnie, wave heights of greater than 60 feet were measured 300 miles east of Abaco Island in the Bahamas using a radar altimeter on a NOAA aircraft.[38]

But it was satellite data that really caused a change in everyone's thinking.[39] Several countries began using satellites with radar altimeters or synthetic aperture radar (SAR) systems to measure waves over all the oceans of the world.[40] In 2003 a great deal of media attention was focused on a group of scientists at the German Aerospace Center when they announced that they had analyzed a three-week period of satellite SAR data and had found ten rogue waves that were more than 85 feet high.[41] However, many scientists had serious doubt about their processing algorithm and their lack of groundtruth. But other scientists began finding rogue waves in a variety of places. On September 15, 2004, Hurricane Ivan passed over wave gauges on moorings that had been deployed by the U.S. Naval Research Laboratory in the northeastern Gulf of Mexico. At one of the moorings a 91-foot wave was measured. This mooring was 47 miles from the eye of Ivan, where the waves may have been even larger (one estimate was 132 feet).[42] In 2007 during Typhoon Krosa, a wave height of 106 feet was measured at a buoy off the northeast coast of Taiwan.[43]

And rogue waves continue to assault ships. On April 16, 2005, a 70-foot rogue wave struck a luxury cruise ship, the 965-foot *Norwegian Dawn,* as it sailed in the Gulf Stream off the coast of Virginia. The ship survived, although sixty-two cabins were flooded, forcing the ship into port.[44] On June 23, 2008, a Japanese fishing boat, *Suwa-Maru No. 58,* was a couple of hundred miles from the Japan coast fishing in the powerful ocean current called the Kuroshio Extension. Wave conditions had been reported as moderate, but the ship was hit by two sets of rogue waves, the second set capsizing and sinking the ship. Only three of twenty crew members survived.[45]

To predict rogue waves we must understand how they are created. When rogue waves were proved to exist, scientists went back to the drawing board to determine what physical mechanism causes them to grow to incredible heights. Such growth must involve a mechanism in which energy is transferred into the rogue wave (either from other waves or from an ocean current).[46] But when oceanographers looked at their hydrodynamic equations, they were not sure what that mechanism might be, except when water was shallow or there was a strong ocean current in the area.[47] In the meantime, prediction methods have been developed for special situations, such as for rogue waves in a strong ocean current. Near the Cape of Good Hope, the South African weather service has been able to provide ships with warnings when rogue waves might be expected, based on when weather conditions are likely to bring in swells from the south or when the speed of the Agulhas Current increases.[48] Considering the proven effect of strong currents on waves, it is likely that the horizontal water motion accompanying long swells could have a similar effect. Thus, a scientific approach that shows that specific swells have converged on an area where a rogue wave sighting was made could be very valuable. This would probably entail using global wave model outputs and analyzing them to determine what swells were produced and where they traveled. Something similar has already been tried on a smaller scale. Before the Japanese fishing boat *Suwa-Maru No. 58* was sunk by a rogue wave in the Kuroshio Extension, wave conditions were reportedly moderate. But a group of Japanese oceanographers carried out a study to try to determine how the sea state developed on those days in June 2008 leading up to the ship's capsizing and sinking. They used an improved third-generation wave model (forced with winds and currents for that time period) and were able to show that long-period swells interacting with wind waves could produce rogue waves.[49]

The ultimate goal, of course, is to develop a global wave model able to predict when and where the conditions appear right for rogue waves to be generated. Another approach, without using a wave model, is to use the real-time analysis of global satellite data to quickly spot those regions where rogue waves have occurred, since other rogue waves would likely occur in the same areas, as long as wind and sea conditions remained conducive to their generation. Real-time satellite data can also be used to track ocean swells around the world, those swells potentially leading to either rogue waves or high breaking waves at a coastline. In specific research projects, this has already been accomplished with results that point out its practical value. In May 2007 the French Research Institute used real-time SAR data from ESA's Envisat satellite to track large swells for three days across the Indian Ocean over a

Figure 6.3 The famous "Great Wave" woodblock print by the Japanese artist Katsushika Hokusai created in 1832. Though many have called this a tsunami, it is actually a large wind wave, which could be considered a rogue wave if its height is twice that of all the waves around it. (*Library of Congress*)

distance of 2,500 miles. The swells eventually hit Réunion Island, where thirty-three-foot waves demolished piers in the port of Saint Pierre and killed six people. Although the oceanographers had predicted that the swells would hit Réunion, they did not realize how much the swells would be amplified when they hit the island. But this was a valuable first step toward a prediction capability using satellite wave data.[50] The same researchers have since tracked swells across the Pacific Ocean and brought the methodology close to routine operation.[51] Satellite swell data, combined with a global wave model to predict where swells are headed, will be valuable because the swells might hit a coast and cause damage (as happened at Réunion Island) or they might cross wind waves produced by a local storm and generate rogue waves. Providing wave forecasts from models and providing real-time wave data from satellites are both operational activities that can be accomplished as part of the Global Ocean Observing System.

If anything symbolizes the power of the sea in the mind of the average person, it is the picture on the TV news shown during every hurricane of a huge wave crashing down on a ship or a pier and spray rocketing into the air. To recreational boaters, waves are more personal, with memories of the jarring rhythm of their boat's bow rising up and smashing down on the next set of wave crests, the spray hitting their faces, as they head for hoped-for comfort and dryness at a distant marina. To commercial mariners waves are even more personal, and more dangerous, with memories of tons of water crashing down on the deck of their cargo ship and their fear as the captain changes course in fifty-foot seas to keep the vessel positioned so that a rogue wave will not capsize their ship or break it in two. We can now reasonably predict all waves except rogue waves, and we are close to being able to predict them.

Rogue waves are terrifying, but as we will see in the next chapter, there is another type of wave in the sea that can cause much more death and destruction. These waves are all the more dangerous because they are almost completely unpredictable, because the earthquakes and volcanic eruptions that produce them are as yet unpredictable. Only when a tsunami crosses an ocean can its arrival be predicted, if the proper real-time observation sensors have been deployed throughout the ocean. Tragically, such a system was not in place in the Indian Ocean on December 26, 2004. But even if it had been, it would not have helped those who lived along the coast of Sumatra, where almost 230,000 people died less than thirty minutes after a submarine earthquake generated a tsunami.

CHAPTER 7

The Sea's Response to an Unpredictable Earth

Trying to Predict Tsunamis

The morning of November 1, 1755, in the Portuguese capital of Lisbon was as beautiful and tranquil as one could hope for at the beginning of winter—sunny, blue sky, no wind, and unseasonably warm. There were barely even ripples on the waters of the Tagus River (Rio Tejo), which widened out into a bay in front of Lisbon and served as a protected harbor for the numerous merchant ships and frigates usually anchored there, six miles upriver from the Atlantic coast. But this serenity was merely the calm before the storm, although not a storm with which the people of Lisbon were familiar. If an actual storm were coming, its rain and winds would eventually provide a warning, however late in coming. Even if unpredicted, a storm or hurricane would at some point no longer remain hidden over the horizon, and there would be time, however short, to try to prepare for the arrival of the flooding storm surge and the destructive wind waves—a last-minute chance to take shelter or move inland to higher ground. But what if the impending disaster from the sea was not brought by a storm? What if that unseen danger were a rapidly moving force of destruction directed at the coast from a distant unknown cataclysm, a danger that sends no warning ahead, a danger that in fact entices people down to the shore to witness strange changes in the sea, the more easily to overcome them with its power?[1]

November 1 was not only a perfect day; it was All Saints' Day, the most important religious holiday in Portugal. So when the ground first began to tremble around 9:30 A.M., the dozens of majestic stone cathedrals and hundreds of churches in Lisbon were filled with worshipers. Lisbon in the mid-eighteenth century was the center of the Portuguese Inquisition, and no Catholic would dare stay home on All Saints' Day. It was standing room only in all places of worship, large or small, the crowds overflowing into the streets. The hymns and sermons and robust prayers drowned out the strange noises that began coming from deep within the earth. But those outside on the streets, along with the many foreigners still in their homes, heard the frightening sounds. Some later said it sounded like hollow distant thunder, while others compared it to the rumbling of heavy carriages driven hastily over stone pavement. The ground's trembling ended after only a minute, but moments later the earth shook with such violence that the upper stories of houses instantly crumpled. Towers and steeples on cathedrals and churches swayed ominously, causing hundreds of bells to ring before tumbling down, as roofs caved in and walls fell inward, crushing many of those who had been hearing holy Mass inside. Narrow streets and alleys disappeared, filled in with large fallen stones, under which people cried for help but were unreachable. The earthquake brought down Lisbon's finest buildings—St. Paul's church, the majestic Opera House, the Palácio dos Estaus (headquarters for the Inquisition), the Customs Exchange, and the Royal Palace itself. And to add to the atmosphere of terror, the sunny, bright day suddenly became as dark as night as falling buildings raised clouds of dust and lime up into the sky, blocking the sun.

The people who survived the first jolts ran down to the Tagus River, since its open shoreline was a good distance away from buildings that might still collapse and away from the smoke of fires caused by falling church candles and shattered stoves and hearths. The many thousands who made it safely to the shore stood in shock, convinced that the Apocalypse had finally come. Most were on their knees, striking their chests and crying out incessantly, "Misericordia, meu Dios" ("Mercy, my God"). Priests wandered through the crowds exhorting everyone to repent, for their sins must have been very great for God to have brought down his own houses of worship on the most important religious day of the year. Other more practical-minded but still-panicking people crowded onto the new marble quay (the Cais de Pedra, or Stone Pier), to get even farther away from the falling city and to look for a boat that could take them to the other side of the river.

But the earthquake was only the first terrifying surprise for Lisbon that morning. Ninety minutes after the first earthquake shock, the water in the river suddenly began "heaving and swelling in a most unaccountable way." The crowds on shore stared in disbelief as ships rocked wildly and were tossed about as though in a storm, even though there was no wind. Transfixed by these sights, the people were jarred out of their confused trance when, as one British eyewitness described it, "there appeared at some small distance a large body of water, rising as it were like a mountain; it came foaming and roaring, and rushed towards the shore with such impetuosity, that we all immediately ran for our lives as fast as possible." Masses of screaming people were now running back into the city. Eyewitnesses later wrote that this first deadly wave was forty feet high. It swept thousands of people away, including as many as three thousand on the Cais de Pedra. The massive quay itself was easily flipped over by the powerful force of the water. The wave carried people up the shore and dashed them violently against buildings. Those who were far enough from the riverbanks to survive the initial onslaught still ended up standing in water up to their waists. In many places the wave went a half mile beyond the shoreline, overturning walls, breaking bridges, shattering houses, and then transporting immense piles of rubble (see figure 7.1). Then, just as suddenly as it had come, the water returned to the river, but with such force that it pulled many into the river, most never to be seen again. The waters receded so far that ships were left sitting on a dry river bottom that only moments before had been covered by forty feet of water. The sight of the bare river bottom, something never seen before, transfixed the confused onlookers, who came closer and were all the more vulnerable when a second wave came with even greater force than the first.

The second wave came ten minutes after the first, charging even farther up the shore, then receding, and followed by a third wave. An eyewitness across the river on high ground looked westward, toward the Atlantic Ocean, and could see each gigantic wave come rushing up the river "like a torrent, tho' against the wind and tide." The captain of a ship on the Atlantic Ocean about forty leagues (140 miles) from the coast later wrote that he felt a strong

Figure 7.1 Etching of the 1755 tsunami at Lisbon, Portugal. (*From G. Hartwig's "Volcanoes and Earthquakes," published in 1887*)

shock and thought his ship had struck a rock, until he threw out a lead line and found no bottom.[2] Other eyewitnesses on the Atlantic coast watched the sea pull back and the sandbar at the mouth of the Tagus go dry before the sea rolled in like a mountain. It propagated up the Tagus, with the water rising fifty feet in an instant near Belém Castle, the royal retreat of King José I and Queen Maria Ana Victoria. The royal couple had heard Mass in the Royal Palace in Lisbon early that morning and were at Belém with their four daughters when the earth first shook at 9:30. Ninety minutes later they looked on in disbelief as the fifty-foot wave battered the shore near their castle.[3] The wave moved so fast that a group on their way to Belém along the banks of the Tagus had to gallop their horses at full speed, barely making it up a hill and escaping the wave.

But it was not only Lisbon that suffered from the wrath of the huge long waves that came after the earthquake. They wiped out entire cities to the south, along the coasts of southern Portugal, Spain, and Morocco and on Madeira (an island 600 miles southwest of Lisbon).[4] They also traveled to the north, wiping out other coastal cities and eventually reaching the British Isles and the Netherlands, Germany, and even Norway, though the waves were much smaller by then. Not knowing of the earthquake near Lisbon, the English were puzzled by the strange water movements they witnessed—ships agitated without there being any wind, or twirled around, or left sitting on the dry sea bottom after the sea suddenly receded.[5] But the waves traveled even farther than that, to the Caribbean, 3,700 miles across the Atlantic Ocean from Lisbon. In Antigua, nine hours after the earthquake off Portugal, the sea surface rose twelve feet several times and then returned almost immediately. At Martinique the sea overflowed lowlands and then returned so quickly that it went past its former boundaries, leaving some seabeds dry for a mile. At the island of Sabia (south of Antigua), the water rose twenty-one feet, and "at St. Martin's, a sloop that rode at anchor in fifteen feet of water was laid dry on her broadside."[6] Through the long waves sent out across the Atlantic in all directions, the earthquake near Lisbon became history's first contemporaneously recognized global natural disaster.[7]

Huge long waves like the ones that overwhelmed the citizens of Lisbon and that traveled to North Africa and the Caribbean are now called *tsunamis*. This name is relatively new to the Western world, although it has been used in Japan since the early 1600s.[8] The switch to the term *tsunami* was motivated mainly by the fact that scientists believe the term popularly used for this type of wave, *tidal wave*, was misleading. This wave has nothing to do with the tide caused by the moon and the sun. Their only similarity is that both waves have a very long wavelength, namely, a distance from one crest to the next of hundreds of miles. A tsunami wave is caused by the sudden vertical displacement of a huge quantity of seawater. That displacement can be caused by the vertical movement of the seafloor during a submarine earthquake, or by a volcanic eruption, or by a landslide under or into the ocean, or even by an asteroid hitting the ocean surface.[9] What is important in producing a tsunami is that the disturbance that moves this mass of water vertically must have a horizontal extent that is greater than the water depth; that is what makes the tsunami a long wave. The area of seafloor moved vertically during a submarine earthquake can cover hundreds of miles, and the average depth of the ocean is only a little more than two miles, which is why such earthquakes cause the largest tsunamis with the longest wavelengths. Disturbances of small extent horizontally, such as a small landslide, can produce very large waves, but they will not be very long waves, and they will have only a local effect, dying out before traveling far from the point of impact. Thus, they are referred to as *local tsunamis.* A volcanic eruption can also displace a large quantity of water, via several mechanisms, either through the expulsion of lava (usually a small effect), or by a sudden caldera collapse into the sea (a larger effect), or by a huge explosion when seawater comes in contact with hot magma (perhaps the largest effect).[10]

And yet the term *tidal wave* probably made sense to the mariners and the coastal residents who used it, since they called all changes in water level "the tide," no matter what caused the change, and in this case the tide seemed to be rising quickly into a large wave. They thought that this "tidal wave" acted like the tide, in that the water receded to a low water and then rose to a high water. It just changed a lot faster than the usual tide, with a complete cycle taking only twenty to forty minutes versus taking over twelve and a half hours for the astronomical tide.

The scientific community had been using a different term, *seismic sea wave,* but this term had its own problems. "Seismic sea wave" is easily confused with "seismic wave," one of several types of elastic waves generated by an earthquake that travel through solid earth. One type of seismic wave can also travel in seawater, even traveling from the solid earth into the ocean and up to the sea surface, where it can jolt a ship, making the ship's crew think they have run aground or collided with a rock. But even the Japanese word *tsunami* is not a perfect choice. *Tsunami* actually means "harbor wave," and the name came from Japanese fishermen who when working offshore could not tell when a tsunami had passed under them, but found their harbor in ruins when they returned. A tsunami is not high in deep water, and because of its long wavelength the motion of the sea's surface up and down is very slow. Thus, the Japanese thought of a *tsunami* as a type of wave dangerous only in a harbor, as opposed to typical high wind waves from a storm, which are dangerous out at sea, but not in a protected harbor. Of course, tsunamis hit entire coasts, not just harbors. Also, a "harbor wave" could easily be confused with a *seiche,* which is a short-period oscillation of the water level in a harbor usually caused by a changing wind, similar to water sloshing up and down at opposite ends of a bathtub. While not an ideal choice, "tsunami" has now become the accepted term. However, "tidal wave" is still used frequently in the popular media.

As we have already seen with the Lisbon earthquake, a tsunami can travel thousands of miles. This is because it is a very long water wave, and such long waves do not lose energy very quickly. A long wave also travels very fast; the deeper the water depth, the faster it travels.[11] Over the deep ocean (two to three miles deep) a tsunami can travel 400 miles per hour, about the speed of a jet airliner. The other important characteristic of a tsunami is that, because it is a long wave, it produces currents that accompany the rise and fall of the water surface. These currents cover the entire water depth, like a tidal current but generally much faster when a tsunami is in shallow water. This is the reason a tsunami wave coming into a beach is so much more dangerous than a large wind wave breaking on the beach, the kind that surfers ride. A twenty-foot wind wave breaking on a beach can certainly be dangerous to a person standing in shallow water, but it crashes down on you from above. Swimmers caught by such a wave avoid danger by diving into the lower part of the wave, which is moving much slower than the crest, and can actually be moving slightly seaward while the crest is moving landward. Not so for the twenty-foot tsunami wave. Because it is a very long wave, all of its water, from the sea bottom to the crest, is moving forward with tremendous power. It is literally a violently moving wall of water that bulldozes everything in its path when it reaches land. A swimmer's only hope is to go over the tsunami wave, usually a very difficult undertaking, although boats have sometimes managed it if the tsunami is not too high. Also, whereas most of the wind wave's energy is lost by its breaking onto the beach, a large tsunami wave does not stop at the beach. It continues inland until friction (and gravity, if it is going up a hill) eventually stop it—sometimes not before it has flooded inland a mile or more. Often the trough of a tsunami wave arrives first at the shore, in which case the sea recedes, leaving large areas of bare sea bottom exposed—the best warning that the destructive crest will arrive in a few minutes. The seaward return current of a tsunami also covers the entire water depth, the water draining off the flooded land with great power.[12]

The 1755 Lisbon tsunami was not the first tsunami to hit the Portuguese, Spanish, and Moroccan coasts, nor was it even the largest. Marine geologists have determined that eight large tsunamis struck this coastal area over the last 12,000 years—roughly one huge tsunami every 1,500 years.[13] They found eight layers in a sediment core from the sea bottom that were produced by those eight large tsunamis, the most recent of which radiocarbon dating showed was created around 1755. Though the 1755 tsunami was one of the largest tsunamis in recorded history, it apparently pales in comparison to a tsunami 12,050 years ago, whose sediment layer, the deepest of the eight, was five times larger than the 1755 layer. What is most intriguing, though, is that this tsunami occurred around the time when Plato said his story of the destruction of Atlantis took place, and not far from bathymetric features west of the Strait of Gibraltar that appear to match details in his story.[14]

Whether Plato's story of Atlantis has any truth, it is an indication of an ancient awareness of destructive tsunamis, although as one might expect, they had no understanding at all of what caused earthquakes and tsunamis (or even that one caused the other). Another, better-documented tsunami (by archaeological and geological studies) from ancient times (around 1600 BC) was produced by the eruption of a volcano on the island of Thera (now called Santorini) in the eastern Mediterranean Sea. It was one of the largest eruptions in recorded history, and its tsunami struck the island of Crete (70 miles to the south), home of the great Minoan civilization that predated the civilization in Greece.[15] Although the Minoan civilization appears not to have been completely wiped out by the Thera tsunami, it never really recovered from the death and destruction, especially the destruction of its sailing fleet, so important to its trade. It declined greatly over the century following the eruption of Thera.[16] Thucydides in *History of the Peloponnesian War* was the first person to write an eyewitness account about a tsunami flooding the land. He wrote that during the summer of 425 BC in the sea at Orobiae there was an earthquake, after which the water dramatically receded from the coast but then returned as a huge wave that inundated the town, killing all who did not run to higher ground. The tsunami wave also washed away an Athenian fort on an island. After describing the destruction, Thucydides states, "The cause, in my opinion, of this phenomenon must be sought in the earthquake. At the point where its shock has been the most violent, the sea is driven back and, suddenly recoiling with redoubled force, causes the inundation."[17]

Plato's student Aristotle also wrote about earthquakes and tsunamis and came up with his own theory about how they were produced. He suggested that both earthquakes and the huge waves that inundate the shores were caused by powerful explosions of air (winds) that had been trapped within the Earth (this "air" being the same in Aristotle's mind as water vapor produced by heating water inside the Earth). Rather than have these wind explosions cause an earthquake and then have the earthquake cause the tsunami, Aristotle believed the winds caused them both. To produce a tsunami he thought the wind from the earth pushed back the sea and then released it, allowing it to violently return. Aristotle's theory was repeated in various forms by many ancient writers, such as Strabo and Pliny, and by many others for two thousand years, in spite of it being completely wrong.[18]

Without an understanding of how earthquakes and tsunamis were produced, there was no way to predict when they would occur. Only Ælianus had a suggestion for predicting earthquakes and the accompanying great waves, a suggestion that would be repeated many times over the next two thousand years—and a suggestion that had a surprising element of truth, as we shall see later in this book. Ælianus said to watch the animals.[19] He wrote of animals leaving Helike, in Greece, in 373 BC before an earthquake occurred and a great wave swept away the city. Over the next two millennia no scientific or technologic progress would be made in earthquake or tsunami prediction or even in understanding the causes of these phenomena. In the meantime, on the other side of the world , earthquakes and tsunamis struck Japan often. The earliest written documentation of them, found in the *Chronicles of Japan,* describes the Hakuho earthquake of AD 684 and the sea swallowing up five hundred thousand *shiro*[20] of cultivated land in the province of Tosa. The governor reported that many ships were sunk owing to that high-rising great tide (called *oshio,* this being the word used before *tsunami*).[21]

By 1755 it was commonly believed that an earthquake under or next to the sea was responsible for producing a tsunami—in a manner similar to the way in which a large rock dropped in a pond causes waves that spread outward from the point of impact. But there was no understanding about what caused an earthquake. The leading theories of the day, with one exception, were just variations on theories that had been around for many centuries. Some thought an earthquake was an explosion caused by compressed air trapped beneath and pushing up on the surface of the Earth. Some thought the pressure of this air was made greater by being heated by the Earth's internal heat, or that water was heated to produce steam, its pressure lifting the earth. The only new theory was the one that suggested an earthquake was somehow caused by electricity, which had been discovered by Benjamin Franklin in 1750 and so was very much the in-vogue subject for scientific investigation. An earthquake was proposed to be the earth equivalent of lightning in the air—the logic being that seismic vibrations from an earthquake travel very fast through the Earth, just as electricity travels very fast through the air, as lightning. A few scientists even proposed that comets caused earthquakes, in spite of the fact that most earthquakes occurred without the arrival of a comet.[22]

Not knowing what caused the Lisbon earthquake and tsunami, which killed between 30,000 and 60,000 people, had serious psychological and philosophical effects all across Europe. In 1755 religion still dominated all thinking, but scientific enlightenment at the end of the Renaissance was beginning to have an impact. The question asked everywhere was, why did the earthquake and the great water waves happen? The fact that this disaster occurred during Mass on the most important religious holiday of the year, killing thousands of the faithful in Lisbon, the center of the Portuguese Inquisition, appeared to be punishment from God, which was very demoralizing to the devout. But some began to question whether the earthquake and tsunami was really an act of God to punish sinners. They asked whether this was instead a violent catastrophe produced by nature. Famous philosophers were on both sides of the debate. Rousseau, on the side of religion, wrote optimistically of a divine Nature, while Voltaire, on the side of reason, wrote of a physical and natural evil in the world. Kant went beyond Voltaire and carried out a scientific investigation into the cause of earthquakes, coming up with an erroneous theory, but one no worse than any of the others.[23]

No one, of course, could predict when such disasters would occur. For the devout it was more a question of whether they could act righteously enough to prevent them, rather than predicting them. For those who did not believe that God used nature to produce disasters, it was perhaps even more disconcerting, for they realized that they had very little understanding of how nature worked. Their only sensible path seemed to be to use science, and to study such disasters in hopes of finding clues to the natural forces behind them. And the Lisbon earthquake and tsunami were studied more thoroughly than any earthquake or tsunami or other natural event before it. The Marquis of Pombal, who began running the Portuguese government for the king after the Lisbon disaster, had sympathies with the enlightened scientific view and sent out detailed questionnaires to all parishes in Portugal to gather information related to the earthquake and tsunami.[24] Other eyewitness accounts were collected and published by the relatively new Royal Society in London. This gave an unprecedented view of how the earthquake tremors had spread over Europe and how the tsunami had propagated along European coasts and to North Africa and North America.

Five years later this relentless search for a scientific explanation for the Lisbon disaster finally produced some significant results. Reverend John Michell at Queen's College in Cambridge published a landmark scientific paper based on hundreds of eyewitness accounts. He showed that the Lisbon earthquake had generated very long waves in the ocean that traveled great distances. From looking at the arrival times of these long waves at different locations around the Atlantic Ocean, he proved that they traveled faster in deep water than in shallow water (as was also true for the long wave motion of the tide, he pointed out, although the cause of these waves was an earthquake, not the moon).[25] Thus, he explained, these long waves arrived at Barbados in the Caribbean before they arrived in Swansea in Wales, even though Barbados was four times as far from Lisbon as Swansea (3,500 miles compared with 900 miles), because the waves to Barbados traveled over much deeper waters.

With this fact in mind, Michell found that if he compared his theoretically predicted arrival times of the tsunamis with the actual reported arrival times, they matched only if the center of the earthquake was west of Lisbon. He calculated that the earthquake's *epicenter*, as it later came to be called, was between 30 and 45 miles west of the Portuguese coast under the Atlantic seafloor.[26] Michell provided the first understanding of a tsunami, correctly proposing that it was generated by the vertical movement of the sea bottom, and learning some characteristics about how it propagated through the ocean once it was generated.[27] Michell's work also represented the beginning of seismology, by proposing that earthquake vibrations could propagate as waves through the solid Earth. However, his theory on how those seismic waves were produced was wrong, since he based it on his own version of the hot-air-under-the-earth theory of earthquakes.

In 1755 there were no scientific instruments to detect and record tsunamis, and that was still true a century later. In 1854 there were also no instruments to detect and record the earthquakes that generated tsunamis—practical operational seismographs were yet to be invented.[28] But by 1854 the U.S. Coast Survey began using for the first time the *self-registering tide gauge* (see Chapter 2). This instrument consisted of a float on the sea surface (protected by a long well) attached by a wire running over gears to a pencil that touched a slowly rotating drum of paper. As the tide slowly rose and then fell and the float with it, the pencil moved up and down and drew a tide curve on the moving paper. Not only did this gauge automate tide measurement, but it also for the first time produced a continuously drawn tide curve, which was a huge improvement over manual measurements of only high water and low water (and occasionally some additional measurements in between). Three of these new gauges had been put into operation in 1851 at San Diego and San Francisco in California and at Astoria in Oregon.[29]

Another of the gauge's benefits was that the tide curve also showed any shorter-period oscillations of the water level that might occur. On December 23, 1854, some unusual extra wiggles showed up on the curve drawn at the gauge in San Diego. These small wiggles represented half-foot oscillations, and their crests were about thirty minutes apart. There had been no nearby storms, and even if there had been, the wind waves and sea swell generated would have had periods much shorter than thirty minutes. Similar but slightly larger oscillations were seen on the tide curve from San Francisco. Lieutenant William Trowbridge, who maintained the gauges, sent a letter to the Coast Survey Superintendent, Alexander Dallas Bache (grandson of Benjamin Franklin), describing the unusual oscillations and suggesting they might have been caused by a submarine earthquake somewhere offshore. Considering that no one had experience with continuous water level curves from tide gauges, this was very insightful. The surprise would be just how far offshore this submarine earthquake had been.[30] Months later Bache learned that an earthquake had occurred on December 23 near Japan's port of Simoda, on the southeastern coast of the island of Niphon. Bache realized that the small oscillations on the tide curves at San Diego and San Francisco were due to what he called seismic sea waves (tsunamis) that had made the 5,000-mile trip across the Pacific Ocean to California.[31]

In Japan on the morning of December 23 the first of five earthquake shocks had been felt around 9 A.M. The waters in Yeddo Bay, from which ships entered Simoda Harbor, had become violently agitated. Then the sea rapidly retreated, revealing the bottom of the harbor in some places and leaving only four feet of water in places where it had been thirty feet deep. The sea then rushed back in as a long wave at least thirty feet high, overflowing Simoda and reaching the tops of houses. When the water again retreated, it swept houses and temples into the sea, leaving only sixteen buildings that had been on higher ground. The townspeople ran for their lives up the hills, but hundreds were overtaken by the wave and pulled back into the sea to drown. The sea withdrew and returned five times, tearing down buildings and trees and covering the shores with the ruins of houses and the wrecks of ships torn from their moorings. Large American frigates, which had been anchored in thirty-six feet of water and had managed to sail over the advancing wave, were laid over on their sides after the retreating wave left them in only four feet of water. Simoda was one of two Japanese ports recently opened to foreign ships as result of a treaty signed by Commodore Matthew Perry earlier that same year.[32]

Two hundred miles to the west of Simoda, the city of Osaka was swept away by the tsunami, as were many other towns in between. Fifty miles south of Osaka, the seaside village of Hiro (now called Hirogawa) was also destroyed by the tsunami, but only a few lives were lost because of the selfless action of an elderly town leader. The village was edged by hills, and his house was on a plateau overlooking the village and bay below. He had felt the mild earthquake tremors and then seen the sea recede. From the stories his grandfather had told him, he knew a tsunami was coming. There was not enough time to get down to the village to warn everyone, nor was there enough time to go farther up the hill to the temple and ring the bell. The only thing he could do was set fire to rice sheafs (*inanura*), the stacks of rice that had just been harvested and in which he had much money invested. When the villagers saw the burning rice harvest, they ran up the hill to put out the fires and were thus saved from the tsunami. After the tsunami the village was rebuilt, and under the guidance of this same town leader an embankment was built to protect the village from future tsunamis. Almost a hundred years later, in 1946, it did just that.[33] The woodblock print in figure 7.2, by the artist Furuta Eisho (Furuta Shöemon), shows the village of Hiro being overwhelmed by the tsunami of 1854 and the villagers running up the hill having seen the burning rice sheafs.[34]

The tsunami from this 1854 earthquake must have traveled to all parts of the Pacific Ocean, but only the self-registering tide gauges of the U.S. Coast Survey on the California coast were able to detect it. After traveling almost

Figure 7.2 Japanese woodblock print by Furuta Eisho of the 1854 tsunami striking the village of Hiro. In it the villagers are running up the hill to save the burning rice harvest, which had been purposely set on fire by the village leader so that the villagers would run up the hill and escape being drowned by the tsunami. (*U.S. Geological Survey*)

5,000 miles, its crests and troughs had become much smaller, showing up as only half-foot oscillations superimposed on the tide curves. Those small wiggles on the tide curves from San Diego and San Francisco were the first measurements of a tsunami ever made and the first remote sensing of a distant earthquake (via its tsunami). Bache published information about these detected waves in hopes of stimulating others to set up similar instruments all over the Pacific, an idea that a century later would lead to the first tsunami warning system. Bache also used this tsunami wave data, and some basic long-wave hydrodynamic theory, to make the most accurate calculation at that time of the average depth of the Pacific Ocean between Japan and California.[35]

In later years other tsunamis would be detected by the self-registering tide gauges of the U.S. Coast Survey. Typically, the Coast Survey would announce the detection of a tsunami and perhaps guess the general area from which it came, but then the Survey would have to wait a week or more to find out through general news sources where an earthquake or a volcanic eruption had occurred. By 1868 there had been some progress on the construction of practical seismographs and also in laying submarine cables to connect the telegraph networks of different continents and islands. But that progress had not been enough to allow detection of the earthquake near Arica, Peru (later

Chile), on August 13, 1868. Coast Survey gauges at San Diego, San Francisco, and Astoria showed thirty-minute fluctuations with amplitudes of 2.6, 1.8, and 1.1 feet, respectively. Later it was learned that the tsunami had come from an earthquake near the Peru-Chile Trench, probably with an 8.5 magnitude. Twelve hours before the tsunamis reached the U.S. West Coast, they killed an estimated 25,000 in Arica and perhaps 70,000 overall along the Pacific coast of South America. The wave was apparently about fifty feet high when it destroyed Callao, 600 miles northeast of Arica. Newspapers in the United States ran stories on the disaster for a week, but the first article did not appear until three weeks after the Coast Survey detected the tsunami.[36]

When the volcanic eruption of Krakatoa in Indonesia occurred in 1883, it was a different story, at least from the standpoint of worldwide news delivery. By then submarine cables connected telegraph networks on different continents, and for the first time the entire world was linked. News organizations, such as Reuters, sprang up, with reporters around the world using this global telegraph network to send news quickly to Europe and the United States, sometimes hours after it happened, instead of weeks. So when tide observers in the renamed U.S. Coast and Geodetic Survey (C&GS) saw wiggles on the tide curve for August 27 from their Sausalito tide gauge in San Francisco Bay, indicating a tsunami, they did not have to wait days to find that its source was the volcanic eruption of Krakatoa.[37] They were, however, probably the first to realize how large the volcanic eruption was, because the outside world was cut off from Krakatoa before the extent of the catastrophe could be reported. An initial telegraph message about the eruption had been received in England and Boston from the port of Anjer, on the coast of Java, 26 miles across the Sunda Strait from the island of Krakatoa. But then the telegraph cable from Anjer to Batavia (later called Jakarta) was broken by a small tsunami. A short while later Anjer was destroyed by a 50-foot tsunami that caused *onshore run-up heights* of 115 feet.[38] Batavia was also hit by a tsunami, but being twelve miles east of Anjer and not in the Sunda Strait, the wave height was only about six feet. Reuters agents, telegraph operators, and government officials in Batavia did not know the catastrophe that had befallen the towns on the islands Java and Sumatra across the Sunda Strait from the island of Krakatoa, and so neither did the rest of the world.[39] But the C&GS scientists had seen on their tide gauges tsunami oscillations of one-foot amplitude, even after the wave had traveled almost 9,000 miles across the Pacific Ocean from Indonesia—this must have been a big eruption!

Once the underwater telegraph cable was fixed and news of the 36,000 deaths reached the rest of the world, Krakatoa became the first true global news story, with new details and new eyewitness stories available in hours via telegraph to newspapers around the world. The stories enthralled everyone, as did the idea that Krakatoa, although on the other side of the world, had touched them physically, albeit in subtle ways. For example, the sound waves produced by the eruptions traveled around the world seven times, as seen from the changing pressures on weather barometers. People on Rodriguez Island, an incredible 3,000 miles to the west of Krakatoa, actually heard the eruptions. A few miles from the volcano the booming explosions were so loud they burst the eardrums of sailors. Beautiful sunsets were seen around the world due to the millions of tons of ash that Krakatoa had spewed into the Earth's atmosphere. And the following year, the world's weather would be cooler, because of the millions of tons of sulfur dioxide thrown into the upper atmosphere, where it combined with water vapor to produce sulfates that blocked some of the sun's radiation. Krakatoa's effects were heard and seen around the world. But it was the tsunami that killed. Over 35,000 people drowned, mostly along the coasts around the Sunda Strait. The other 1,000 deaths, in southern Sumatra, were people burned alive by hot ash blown there by the wind.[40]

After many small explosions, Krakatoa had four major explosions, each generating a tsunami. It was the third of these tsunami waves, at least fifty feet high when it struck the coasts along the Sunda Strait, that destroyed 165 villages on the Java and Sumatra coasts. The horror of fifty-foot walls of water striking the coast was made worse by the frightening darkness caused by the thick clouds of smoke that blocked the sun, so the tsunami waves could be heard but not seen. The ships at sea in the Sunda Strait, although not threatened by the tsunamis like those on land, were rocked mercilessly by large shorter-wavelength waves and ran into enormous floating masses of pumice. But several ships survived to give detailed accounts from the closest vantage point to the volcano.[41]

Once the tsunamis traveled beyond the northeastern end of the strait and spread out into the Java Sea, their heights decreased dramatically. This is the reason Batavia's tsunami was so much smaller than the one that hit Anjer. In the other direction tsunamis headed out into the Indian Ocean. More than 1,800 miles away, on the southwest coast of Ceylon (today Sri Lanka), the tsunami heights were still fairly large. At the city of Galle the sea withdrew hundreds of feet from the shore, revealing a sea bottom with flopping fish and old steamship wrecks. The people ran for high ground, resisting the temptation of free fish. Luckily when the sea came back in, it did so slowly, rising seven feet without a turbulent wave crest, but the people had demonstrated that they knew what to do to remain safe when a tsunami is coming—knowledge that unfortunately their children's children did not posses 121 years later.[42] At most other locations, including the coast of India, the tsunamis were not large enough to make a strong impression on the people, and thus they did not raise their awareness of the dangers of a tsunami. When the tsunamis from Krakatoa reached Europe and the United States, their heights were too small to be noticed except on tide gauges such as the one in San Francisco.[43]

One might wonder now whether no one had foreseen the possibility of a major eruption of Krakatoa since there were modest eruptions back in May of that year, and since there was increasing volcanic activity days before August 27—smoke clouds, deep loud rumblings, and the smell of sulfur. Or perhaps the question should be, even if they foresaw that possibility, did they worry about it? Krakatoa seemed a safe distance away, on an island with twenty to forty miles of water separating it from villages along the coasts of Java and Sumatra. So the real question is, did anyone know that volcanic eruptions could generate tsunamis that could cross that expanse of water and destroy lives onshore? Unless they had been told of the tsunamis from the eruption of Tambora in 1815 (on Sumbawa Island, about 850 miles east of Krakatoa), the answer would be no.[44] The only one worried before August 27 was a circus elephant in Batavia, which became very agitated and finally went berserk just before the eruption and tsunami. But no one saw that as a warning sign, nor would people a century later when other elephants would sense the arrival of the next major Indian Ocean tsunami.[45]

Krakatoa's 1833 volcanic eruption and tsunami became the most scientifically studied natural disaster since Lisbon's 1755 earthquake and tsunami.[46] Telegraph networks quickly provided information that the scientists could use in their hunt for clues on what had caused the eruption and the tsunamis. Krakatoa's eruptions generated seismic waves (not to be confused with seismic sea waves, or tsunamis) that traveled around the globe through the solid earth, although they were not measured on any seismographs, because seismographs were not yet capable of measuring distant earthquakes.[47] That changed only six years after Krakatoa, when an earthquake near Tokyo on April 18, 1889, was detected halfway around the world in Germany.[48] Over the next couple of decades improvements were made in seismographs, and they would become critical for tsunami prediction.

Seismographs would be used to recognize several types of seismic waves, and their times of arrival would be used to locate the epicenter of an earthquake. By the turn of the century it had been determined that the *primary* (*P*) *waves* arrived first from an earthquake, followed some minutes later by *secondary* (*S*) *waves.* Primary waves are longitudinal compression waves, secondary waves are transverse shear waves, and both travel through the body of the Earth. These are then followed by *surface waves,* which travel along the surface of the solid Earth and which later were determined to also have longitudinal and transverse components, Rayleigh waves and Love waves, respectively. The time differences between the arrival of these waves at three or more seismographs allow scientists to determine the location of an earthquake's epicenter, along with the time of its occurrence and an estimate of its size. These seismic waves also began to help scientists understand the structure of the solid Earth (the crust, mantle, and core), which would eventually lead to a correct understanding of what caused the earthquakes and the volcanic eruptions that caused tsunamis.[49]

Whatever tsunami awareness had been raised by Krakatoa in 1883 was long forgotten by 1946, since there were no major tsunami disasters over that time period to serve as reminders. Hawaiians were perhaps the only ones besides the Japanese who maintained some degree of tsunami awareness. Sitting on the tops of underwater volcanoes, the Hawaiian Islands have experienced volcanic eruptions and numerous submarine earthquakes. Even more important, they are situated in the middle of the Pacific Ocean, whose coastal boundaries make up the *Ring of Fire,* where more than three-quarters of the world's largest earthquakes and volcanic eruptions take place. If a tsunami is generated by an earthquake in the Ring of Fire, chances are good that it will hit the Hawaiian Islands. At least forty-two tsunamis hit the Hawaiian Islands during the forty-six years preceding 1946, of which fourteen were generated locally and twenty-eight came from earthquakes in the Ring of Fire.[50] A locally generated tsunami strikes only minutes after an earthquake, but a tsunami arriving from the Ring of Fire takes hours to reach Hawaii—and thus there is time for a warning if the earthquake can be detected quickly enough and if it can be determined that it has generated a tsunami (often not the case). But it was going to take a large tsunami hitting Hawaii to stimulate any government action toward establishing a tsunami warning center.

On April 1, 1946, a submarine earthquake of 7.8 magnitude south of Unimak Island in the Aleutian Islands of Alaska triggered a huge underwater landslide, which produced a large tsunami (larger than would have been produced by the earthquake alone).[51] Tsunami run-up heights reached 130 feet on Unimak Island, wiping out a village and a Coast Guard lighthouse and killing five people. The main path of the tsunami, however, was to the south, eventually reaching all the way across the Pacific Ocean to Antarctica, along the way being observed at tide gauges on the coasts of North and South America and on South Pacific islands. About four and a half hours after the earthquake, the tsunami hit the city of Hilo on the island of Hawaii and produced run-up heights of over fifty feet. At least 159 people were killed, among them at least 44 who were dragged out to sea and never seen again. Some were children and teachers from a school next to a beach north of Hilo. More lives would have been lost if Hilo had not had a breakwater, half of which was destroyed by the tsunami. Even with the breakwater some buildings were "crushed like eggshells and swept from foundations."[52]

After the devastation people protested that they should have received a warning from the U.S. C&GS, since the submarine earthquake that had occurred four and a half hours before the tsunami hit Hawaii had been detected by C&GS seismographs (although the data were not seen until hours later).[53] The Director of C&GS tried to explain that few earthquakes actually generate tsunamis. Reporting every earthquake, without knowing whether it had generated a destructive tsunami, would lead to hundreds of false alarms, and then no one would pay attention to the warnings. While the Director's comments were true, the protest did stimulate C&GS to establish the world's first tsunami warning center, initially called a seismic sea wave warning center.

With both seismographs and tide gauges in operation, C&GS was the logical agency to take up the challenge of providing tsunami predictions.[54] But the problem was that someone would have to stand by each gauge around the clock and watch the tide curves being drawn in the (rare) event that tsunami oscillations showed up. To get around this problem the scientists at C&GS invented a *seismic sea wave detector*, which would sound an alarm when a tsunami occurred.[55] A system was also invented so a seismograph could sound an alarm when seismic waves from a large earthquake arrived, which for this purpose was a big improvement over the typical seismograph, which used a photographic technique that required time to develop the film.[56] Next, a high-speed communications network was needed between C&GS's Honolulu Observatory at Ewa Beach and the dozens of tide and seismological stations around the Pacific. This was accomplished with the help of the Defense Department and the Civil Aeronautics Administration (later called the FAA) but without any funding from Congress.[57] Finally, using basic long-wave hydrodynamic theory not much more complicated than what Bache had used a hundred years before, calculations were made on how long it would take a tsunami wave to travel from an earthquake's epicenter to various locations in the Pacific. These calculations were presented as a series of *tsunami travel-time charts* for many locations around the Pacific, including Hawaii, each being a chart of the Pacific with contour lines showing travel times to that location from many locations around the Pacific (see figure 7.3).[58]

The Seismic Sea Wave Warning System for Hawaii was operational by 1949. If an earthquake occurred, the following steps were taken to determine whether a tsunami had been generated and if so, when it would arrive at Hawaii and how large it would be. First, C&GS would begin receiving seismic wave data from seismographs that detected the first-arriving P waves. A few minutes later the S waves would begin arriving. The distance of a seismograph from the earthquake's epicenter would be calculated from the difference in the arrival times of the P and S waves. Using the calculated distances to three seismographs, a calculation would be made of the location of the earthquake's epicenter. Knowing the location of the epicenter, data from the

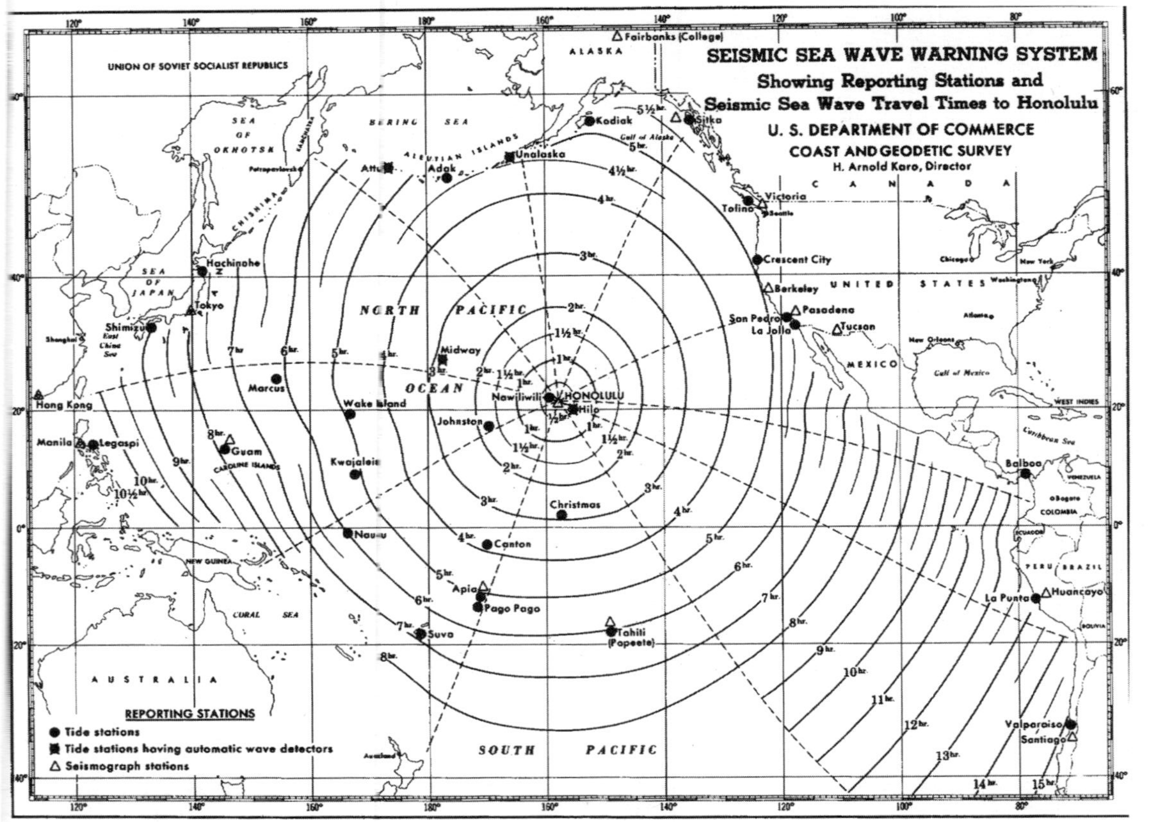

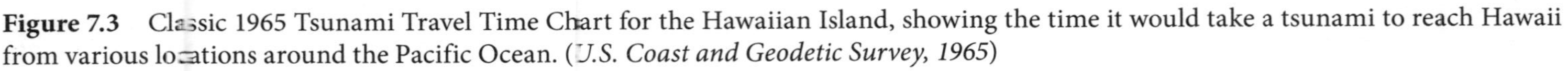

Figure 7.3 Classic 1965 Tsunami Travel Time Chart for the Hawaiian Island, showing the time it would take a tsunami to reach Hawaii from various locations around the Pacific Ocean. (*U.S. Coast and Geodetic Survey, 1965*)

nearest tide gauge stations would be examined to see if there were tsunami oscillations, unless data had already been received from gauges equipped with tsunami detectors. If a tsunami had been generated, time-travel charts (see figure 7.3) would be consulted to find the time it would take the tsunami to reach Hawaii. Finally, on the basis of the size of the tsunami at the tide gauges and the amplitudes of the seismic waves, the expected size of the tsunami when it hit Hawaii would be estimated. This last prediction was difficult because many factors affected tsunami wave heights, for example, earthquake location, the possibility of submarine landslides, oceanwide bathymetry, and local bathymetry and shoreline shape. Information about the predicted tsunami would be turned over to Hawaii civil defense authorities, who handled the warnings to the people. Programs were set up to educate people about what they should do and not do if they hear a warning and to warn them that the first wave is often not the largest. The most important guidance was for people to run inland as fast as possible. Ships were advised to head out to sea, where the tsunami wave would be small.[59]

Over the next fifteen years four large tsunamis would test the new Seismic Sea Wave Warning System, including tsunamis generated by the three largest earthquakes in modern history up to that time (in 1960, 1964, and 1952; the 1957 earthquake was the seventh largest).[60] In all four cases the system was accurate in predicting tsunami arrival times at Hawaii. Accurately predicting the size of the tsunami proved to be more difficult, but the warnings saved lives in every case. In 1952 an earthquake (8.6 magnitude) off the coast of the Kamchatka peninsula in Russia produced a sixty-five-foot tsunami that destroyed Serevo Kurilsk in the Kuril Islands, only about a hundred miles southwest of the earthquake. It also sent a tsunami to Hawaii, where wave heights were similar in size to those in 1946. But this time there were no deaths, because the Seismic Sea Wave Warning Center had issued first a watch and then a warning for Hawaii.[61] On March 9, 1957, an earthquake occurred one hundred miles southeast of Adak Island in the Aleutian chain (about 550 miles west of where the 1946 earthquake occurred). Again the Seismic Sea Wave Warning System went into action, and again there were no deaths.[62]

Three years later the largest earthquake in modern history (9.5 magnitude) occurred off the coast of Chile on May 22, 1960, at 9:11 A.M. Hawaiian time. A tsunami (or *maremoto* in Spanish) as high as eighty feet crushed the Chilean city of Valdivia, killing perhaps as many as six thousand along the Chilean coast. Fifteen hours later a thirty-five-foot tsunami hit Hilo, with ten-foot waves hitting the rest of the Hawaiian Islands. In this case, even with an accurate warning from the Seismic Sea Wave Warning Center, 61 people died and another 282 were injured. The warning had been provided more than five hours before the correctly predicted midnight arrival, and it had been widely communicated by police, civil defense, and others. But the warning was ignored by many people for a variety of reasons—confusion over a new siren system, false alarms in the years since 1952, and a lack of concern because the last tsunami at Hilo had been small. There was also misleading information from the media, for example, a radio station broadcast that the tsunami had been only three feet high at Tahiti, four hours before it arrived in Hawaii, which convinced some people to stay home. But Tahiti did not have a continental shelf to amplify the tsunami; instead it had steep bottom slopes to reflect the tsunami wave. It also had coral barrier reefs around it to wear down the tsunami wave.[63] Adding to the confusion, the first tsunami wave to arrive at Hilo was only about three feet high, and that was also broadcast on the radio, which led many evacuated people to return to their homes, thinking the danger was over. But it was the third wave that was deadly, thirty-five feet high and shaped like a steep tidal bore. Entire city blocks in Hilo were swept bare, and the city was devastated. The tsunami picked up twenty-two-ton boulders from the bay-front seawall and carried them inland six hundred feet. The force of the water on two-inch-thick pipes holding parking meters bent them parallel to the ground. It swept away an eleven-ton tractor. Aside from Hilo, the tsunami heights on the rest of Hawaii's Big Island were on average around nine feet high, similar to those that hit the other Hawaiian Islands.[64] Hilo again was the location with the greatest tsunami heights, as it had been in 1946, 1952, and 1957. Its funnel-shaped bay was partly responsible for amplifying the wave, as was the submarine ridge that focused energy (by wave refraction) onto a shallow continental shelf outside the bay.[65]

The tsunami from the Chilean earthquake continued past Hawaii and seven hours later reached Japan (twenty-two hours after the earthquake), where it was about ten feet high, although at a few locations it was as great as twenty-five feet. At least 142 people were killed, and damage was estimated at fifty million dollars. The Japanese Meteorological Agency, responsible for tsunami predictions in Japan, had no experience with large tsunamis from distant sources (the numerous local earthquakes and tsunamis kept them busy). They thought that any tsunami from Chile would be small, not a surprising opinion, since the tsunami would by that time have traveled 10,000 miles, halfway around the world.[66] But hydrodynamics that they did not understand at the time had increased the size of the tsunami.[67] After these tsunami-caused deaths, Japan joined an international effort to improve tsunami warning, as did Russia, which was also struck by the 1960 tsunami. C&GS offered the Seismic Sea Wave Warning Center in Hawaii as the operations center for the entire Pacific Basin, and changed its name to the Pacific Tsunami Warning Center. More than a dozen nations around the Pacific joined; there are now twenty-six member nations.[68]

Only four years later the second-largest earthquake in modern history and the largest U.S. earthquake in history (9.2 magnitude) occurred on March 27,

1964, in Alaska, near the northern end of Prince William Sound. The predictions from the Pacific Tsunami Warning Center for a tsunami crossing the Pacific were again correct, and the highest tsunami again hit Hilo, but this time it was only about twelve and a half feet high. Communication procedures were greatly improved, the evacuation was successful, and there were no casualties. Significant damage and casualties, however, occurred in Alaska. A local tsunami within Prince William Sound killed 106 people, 32 of whom were in Valdez, an oil port where ruptured oil tanks burst into flames. Others died when the Pacific tsunami propagated southward along the Alaskan and Canadian coasts to reach the U.S. mainland. Eleven died in California at Crescent City, and four children died on the Oregon coast.[69] The tsunamis in Alaskan waters occurred much too quickly after the earthquake for a warning to be possible from the Warning Center in Hawaii. It was therefore decided in 1967 to establish a second warning center, the Alaska Regional Tsunami Warning System at Palmer, Alaska (forty miles north of Anchorage). The emphasis of this center was on a better understanding of Alaskan seismic data and rapidly obtaining local tide data for faster and more accurate warnings.[70] In 1996 it would be renamed the West Coast/Alaska Tsunami Warning Center and given the added responsibility for British Columbia, Washington, Oregon, and California.[71]

The tsunami that did the great damage in Prince William Sound in 1964 was called a *local tsunami* because, although it was destructive, it did not travel very far. Local tsunamis are often produced by landslides. Though a tsunami is local, it can be enormous. In fact, the largest tsunami ever recorded was a local tsunami in Alaska. It was caused by an earthquake-triggered rock slide in Lituya Bay, Alaska, on July 9, 1958. Forty million cubic yards of rock fell 3,000 feet from the steep northeast wall of the fjord-like inner part of the bay. It displaced so much water from the bay that it surged over the 1,700-foot-high ridge on the opposite side of Gilbert Inlet. Run-up from that 1,700-foot wave stripped the land bare of its forests and eroded the ground down to bedrock. But that run-up was technically not a tsunami. The tsunami was the wave that then barreled down the bay toward the entrance to the Pacific Ocean, at a speed of over 100 miles per hour. It was an incredible 500 feet high (the height of a forty-story building). After traveling seven miles, the tsunami wore down but still was over 100 feet high as it reached the bay's entrance. About a mile from the entrance the tsunami picked up a 38-foot fishing boat and carried it toward the sea, its two petrified passengers looking over the side at trees 85 feet below on a land spit that the tsunami was crossing. After reaching the deeper ocean, the tsunami rapidly decreased in height and set the boat down, its two passengers unhurt. Two other fishing boats were also picked up by the tsunami, one with a father and son who survived to tell their own astounding story. The two passengers of the third boat were never seen again.[72]

Although some landslides like the one in Lituya Bay are above-the-water landslides, most landslides occur underwater. These *submarine landslides* are usually caused by collapses of sediment slumps that have built up over hundreds or even thousands of years. Submarine landslides are the second-leading cause of tsunamis after submarine earthquakes, although a distant second. In many cases an underwater slide is triggered by an earthquake or a volcanic eruption and causes a tsunami wave larger than would have been produced by the earthquake or volcano alone. In some cases, the size of the growing underwater slump reaches a critical unstable point, and a submarine landslide occurs without an earthquake. For example, large wind waves from a severe storm can trigger a submarine landslide.[73] Submarine landslides can cause dangerous tsunamis, as we saw with the 1946 tsunami from the Aleutian Islands earthquake. Another occurred November 18, 1929, in the Atlantic Ocean south of Newfoundland near the Grand Banks. An earthquake triggered a sediment slide down the continental slope that lasted for hours and broke twelve undersea transatlantic telegraph cables. It produced a thirty-foot tsunami that killed twenty-eight people along the sparsely populated Canadian coast. Tsunami oscillations showed up on tide gauges as far south along the U.S. coast as Charleston, South Carolina, and across the Atlantic as far away as the Portugal coast and the Azores.[74] Unlike this tsunami and the 1946 Pacific tsunami, most submarine-landslide-generated tsunamis have only local major effects, the slides typically not having enough volume or lasting long enough to generate very long waves. A submarine landslide off the northern coast of Papua, New Guinea, which was triggered by a 7.0 magnitude earthquake, generated three fifty-foot tsunamis that killed more than 2,200 people on July 17, 1998. But beyond fifteen miles from the landslide, wave heights decreased rapidly.

Probably the granddaddy of all submarine landslides, the Storega Slide, took place about 7,900 years ago off the coast of Norway. Offshore sediments had built up for tens of thousands of years during the previous ice age and became unstable once the climate changed and a warmer interglacial period was under way. The quantity of sediment that slid down the continental slope was truly huge, enough to cover the entire continental United States with a foot and a half of sediment. At least one slide—it is not known how many slides might have occurred around this period—happened fast enough to generate tsunamis that inundated all the lands relatively near it, especially present-day Norway and Scotland. Measured tsunami run-up heights from studies of sediment deposits indicate twenty-foot heights in Scotland, forty-foot heights in Norway, and sixty-five-foot heights in the Shetland Islands (between Norway and Scotland). Along with a fine sand, scientists also found marine diatoms at these inland study sites.[75] A couple of the study sites in Scotland were just "downstream" from Loch Ness, but so far no one has suggested that any marine organism larger than a diatom could have been carried into Loch Ness by the tsunami.

A tsunami can also be generated by an above-the-water landslide due to the lateral collapse of a volcano's flanks. Multibeam sonar images of the bathymetry around La Palma Island in the western Canary Islands show apparent scars of prehistoric landslides into the sea from the flanks of the Cumbre Vieja volcano. Such landslides would have theoretically caused huge tsunamis, at least locally. A study in 2001 concluded that these landslides were large enough to have generated tsunamis that would have caused great destruction along the European coast and even along the present coast of the United States. That study received a lot of media attention because it also said that sometime in the future there would be another huge landslide at the Cumbre Vieja volcano that might cause gigantic tsunamis that would ravage European coasts and cross the Atlantic Ocean to inundate the entire East Coast of the United States many miles inland.[76] Most other oceanographers and geologists, however, disagreed with those conclusions. Even though the tsunamis from such landslides would be large and dangerous in the Canary Islands, there are many reasons why their impact would not be catastrophic beyond the islands, as the study had suggested.[77] When a future landslide might occur at La Palma is unknown. In the prehistoric past, landslides there have occurred on average every hundred thousand years but with no pattern. For coasts close to the landslide, however, the impact could be devastating, so being aware of the possibility of a landslide is important, and there are ways to monitor the activity of a volcano and the stability of its flanks.

It was not until the mid-1960s that earth scientists finally gained a real understanding of what caused earthquakes, and thus how tsunamis were generated.[78] After twenty centuries of theories based on hot-air winds, water vapor heated by subterranean fires, underground and undersea caverns, electricity, and comets, there was now finally a scientific theory backed up by real data that explained how earthquakes occurred and why volcanic eruptions occurred. But it took a theory that revolutionized earth science to finally come up with this correct understanding of earthquakes—the theory of *plate tectonics*.[79] The journey to this understanding had begun centuries earlier with the recognition that the shapes of the east coast of South America and the west coast of Africa were similar. This was even more evident when looking at the shapes of the edges of their continental shelves, after accurate depth data accumulated from the bathymetric surveys of the late nineteenth and early twentieth centuries. It looked like the two continents could fit together like two puzzle pieces. In 1912 Alfred Wegener proposed the theory of *continental drift* and searched for evidence to prove that Africa and South America had at one time been next to each other.[80] Three years earlier Andrija

Mohorovičić had discovered a sharp transition (the Moho discontinuity) between the outer edge of the Earth (the *crust*) and a denser, larger layer below it (the *mantle*), which is indicated by a sudden increase in the speed of seismic P waves. Aware of both these results, Arthur Holmes a decade later proposed an explanation for *why* the continents of South America and Africa had drifted apart (and also why the other continents had also drifted apart; at some time in the past they had all been together as one giant continent).[81] His theory, however, would not be recognized as correct for another thirty years.

In Holmes' theory, which set the stage for plate tectonics, he proposed that the continents floated on and were moved by extremely slow-moving currents near the top of the Earth's mantle. These currents were, he said, due to convection in the plastic mantle caused by the heat from radioactive decay in the Earth's interior.[82] Holmes suggested that new hot material slowly rose through the mantle to the Earth's crust on one side of a huge convection cell, and that cooler material sank from the crust back into the mantle on the other side of the cell. In the 1960s it was proved that rising hot material emerged from the seafloor at the center of the Atlantic Ocean, where it cooled and created the Mid-Atlantic Ridge, an enormous underwater mountain chain in the middle of the Atlantic Ocean. Holmes said that the seafloor spread apart on both sides of the ridge, and that deep ocean trenches formed where much-older, cooled crustal material sank back into the mantle. It would take many millions of years for oceanic crust formed at a ridge to move horizontally to a trench and sink into the mantle.

Holmes' theory of continental drift was viewed skeptically by most geologists until 1966, when an amazing discovery was made after surveying the seafloor on both sides of the Mid-Atlantic Ridge. As one moved away from the Ridge, on either side, a pattern was seen in the magnetic characteristics of the seafloor rock. The pattern was one of alternating stripes of north-oriented and south-oriented magnetism that exactly matched the reversals of the north and south magnetic poles (which switch polarity every few million years). Magma had emerged from the seafloor and cooled, the particles of iron in the magma had oriented themselves with the magnetic pole, and then that orientation was locked in when the magma cooled. This alternating pattern as one moved away from the Ridge, on both sides, could exist only if the seafloor was spreading away from the Mid-Atlantic Ridge on both sides, which meant that the continents were drifting apart.[83]

By 1967 it was shown that the moving masses floating on top of the mantle are enormous chunks of crust, or *plates* (eight major plates and several minor ones), some of which have continents on them, and some of them ocean floor.[84] The boundaries (*faults*) between tectonic plates tend to be where earthquakes occur, especially those locations where plates push against each other, such as around the edges of the Pacific Ocean, the Ring of Fire. At these *subduction zones* an oceanic plate (with dense oceanic crust) sinks into the mantle by pushing under a lighter continental plate. But these two massive plates grab on to each other, and the pressure builds to tremendous levels as the oceanic plate pulls the seaward edge of the continental plate down with it. At some point the pressure grows too large, and the plates release each other (the fault *ruptures*), causing an earthquake.[85] These subduction zones are underwater, so when the released continental plate lifts quickly upward, it raises the seafloor with it, displacing a huge quantity of water vertically and thus generating a tsunami.

Over the forty years between the 1964 Alaskan earthquake and December 2004, there were 362 confirmed tsunamis of known origin worldwide. Even though any forty-year period may not be typical, we can still learn some useful facts about tsunamis by looking at this data set (it being a time period with good data coverage).[86] The great majority of tsunamis were caused by submarine earthquakes (86 percent). The stronger the earthquake, the better the chance a tsunami was generated. Earthquakes with magnitudes of 8.0 to 8.9 had a 58 percent chance of producing a tsunami. This dropped to 30 percent for earthquakes with magnitude 7.0 to 7.9. The drop was dramatic for earthquakes of magnitude 6.0 to 6.9—only a 2 percent chance. Below 6.0 magnitude, tsunamis rarely happened. In the 5.0-to-5.9-magnitude range, there were fifty-seven thousand earthquakes and only twenty-four tsunamis. However, the Pacific Tsunami Warning Center still had to analyze seismic data from all these earthquakes, to determine size and whether the epicenter's location was on land or under the sea. Over this forty-year period this included analyzing eight hundred thousand earthquakes with magnitude less than 5.0. There were also a few tsunamis generated by volcanic eruptions and landslides. But for this forty-year period that amounted to only eleven tsunamis generated by volcanic eruptions (just 3 percent) and forty landslide-generated tsunamis (11 percent). Of these landslide-generated tsunamis, seventeen were triggered by an earthquake, two by a volcanic eruption, and twenty-one by neither.

A huge majority of the 362 tsunamis from March 1964 through December 2004 (84 percent) were generated in the Pacific Ocean. This explains why in 2004 there was only a Pacific Tsunami Warning System. Over those forty years only three other areas of the world had tsunamis. There were thirty-one tsunamis in the Mediterranean Sea (9 percent), twenty-three in the Atlantic Ocean (6 percent, mostly in the Caribbean, near Norway, and near Portugal and Spain), and only three in the Indian Ocean (1 percent). But this is somewhat misleading, because about 10 percent of those "Pacific" tsunamis occurred in Indonesia, and almost a third of them caused 4,400 deaths, including approximately 2,500 casualties in 1992 at Flores Island. They were all essentially local tsunamis that occurred near islands east of Krakatoa, but they were counted as Pacific tsunamis, because Indonesia is sitting on the Ring of Fire and is a member of the Pacific Tsunami Warning System. None of these Indonesian tsunamis traveled into the Indian Ocean (as the Krakatoa tsunami had done in 1883), but they were frequent enough and deadly enough that one might have hoped that the Indonesian government would have developed an awareness of tsunamis' danger, which possibly could have saved lives in 2004.

In the meantime, NOAA's Pacific Tsunami Warning Center, though operating with a minimal budget, continued to make improvements in its operations and tried to improve its capabilities to accurately predict the next big tsunami.[87] Though over those forty years, as we have seen, only a couple of tsunamis caused a large loss of life, there was still a huge work load (automated as much as possible) to check out the twenty thousand earthquakes that occur each year and to follow up on those that were determined to be under the sea and large enough to potentially cause a tsunami. Improvements in hydrodynamic tsunami models, seismic data analysis, and real-time instrumentation all played a role. The ultimate key to successful prediction, however, was the capability to confirm by measurement that a tsunami had been generated and to do this fast enough to be able to warn others. Real-time water level stations, still called tide stations by most, were critical to this prediction. NOAA's transition to the rapid-sampling acoustic water level gauges of the Next Generation Water Level Measurement System greatly improved the detection of tsunamis at many more locations, and a special tsunami detection sensor was no longer required.[88]

But water level gauges were often installed in protected harbors or other locations where local bathymetry could greatly affect the tsunami signals they measured. Most important, there were large gaps in their coverage around the Pacific. Many places had too few real-time water level gauges; for example, in the South Pacific between Chile and Hawaii, where only Tahiti and Karibati (formerly Christmas Island) had tide gauges. Ultimately, the Warning Center solved this problem with the installation of DART buoy systems at key open-ocean locations around the Pacific.[89] DART, which stands for Deep-Ocean Assessment and Reporting of Tsunamis, was developed by NOAA's Pacific Marine Environmental Laboratory (PMEL).[90] Its key component is a pressure sensor sitting on the ocean floor that can detect a tsunami if one passes over it. And that is its first function, when a significant earthquake has occurred to tell the Pacific Tsunami Warning Center whether a tsunami from that earthquake has propagated through that part of the ocean. The size of the tsunami measured at these open-ocean locations is very small, but combined with accurate hydrodynamic tsunami models and an ever-increasing database of past measurements, the DART measurements can be used to estimate the size of the tsunami when it hits a coast. The first prototype, then known as a *tsunameter*, was installed in 1995, and by 2004 there were six DART buoys in operation in the Pacific Ocean.

In 2004 the Pacific Tsunami Warning System operated only in the Pacific Ocean. The statistics we've just covered justified that emphasis. But given that 10 percent of the so-called Pacific tsunamis were generated in Indonesia, members of the Warning Center and PMEL suggested that a tsunami warning system in the Indian Ocean would be valuable. Funding, however, could never seem to be found, nor did the national governments around the Indian Ocean seem to listen to their own scientists who pushed for such a system.[91] Sumatra has thirty-three volcanoes along its length, although the island had not had a major eruption on or near it since Krakatoa's eruption in 1883. The destructive tsunamis due to Krakatoa's eruptions were just distant memories, as were the even earlier earthquakes and tsunamis that occurred along Sumatra in 1797, 1818, 1833, 1843, and 1861.[92] Since then underground pressure had been building again along the subduction fault west of Sumatra, and although major earthquakes can never be predicted, the stage was being set for a tragic surprise in the Indian Ocean in December 2004, as we will see in the next chapter.

CHAPTER 8

December 26, 2004 (Part 1)

Tragic Surprise in the Indian Ocean

For a million years a gargantuan subterranean battle had been under way eighteen miles beneath the deepest part of the Indian Ocean. East of the five-mile-deep Sunda Trench and only fifty miles west of the Indonesian island of Sumatra, two massive tectonic plates had been butting up against each other along a battle line known as the Sunda fault.[1] On the west side of the fault was the Indian tectonic plate, made up of oceanic crust, and on the east side was the Burma microplate, made up of continental crust.[2] The Indian plate was trying to push under the Burma plate, but for the last thousand years these two humongous crustal masses had been holding tightly on to each other, so that as the Indian plate pressed downward, it was very slowly taking the Burma plate down with it. At only two inches a year the pace of this battle was excruciatingly slow, but after a thousand years the pressure had built to such an incredible level that at some moment in time these two massive plates would finally have to release each other. When that happened, the result would be violence on a grand scale.

No one knew when that huge earthquake would occur. Although by now scientists understood much about how and why tectonic plates moved, they could not predict the year or even the decade when the Indian plate and the Burma plate would finally let go of each other—much less the day or the hour. That day and hour of sudden violent release came on December 26, 2004, at 7:59 A.M. (local Indonesian time).[3] This date and time have gone down in history, not because of the earthquake, but because of the power that the earthquake released to the sea and the destruction and death that the sea brought to 300,000 people around the Indian Ocean in less than two hours.

The tranquil sunny weather and calm waters that Sunday morning were in stark contrast to the violence that was brewing eighteen miles below. When the two plates finally released and slid along each other, the edge of the Burma microplate rose twenty feet in only seconds, violently lifting the seafloor with it.[4] The location above where this submarine earthquake first unleashed its power (the *epicenter*) was fifty miles west of the coast of Aceh Province, at the northern end of Sumatra, and only thirty miles north of the island of Simeulue. But this location would not be the limit of crustal violence, or even its high point.[5] It was merely its beginning, for the rupturing of the Sunda fault and the vertical movement of the seafloor moved northward for over 900 miles, moving past the northernmost Andaman Islands in less than ten minutes.[6] The upward movement of 80,000 square miles of seafloor (almost the length of California and Oregon and about half their width) displaced hundreds of billions of tons of seawater. On both sides of this north–south fault rupture, long tsunami waves began to move across the sea, primarily eastward and westward. To the east the tsunami headed toward Indonesia, Thailand, Myanmar (Burma), and Malaysia. To the west it crossed the Bay of Bengal to Sri Lanka and India and then continued across the Indian Ocean to East Africa. Although the westward edge of this piece of seafloor had moved upward, lifting tons of water, the eastward edge actually moved slightly downward and the water with it. Thus, the eastward tsunami moved with a leading depression (a wave trough), so that when it reached the coasts of Sumatra and Thailand, the sea at first receded. The suddenly dry sea bottom enticed the people out to look at the revealed coral heads and the fish flapping around in the foreign air. But then after a few minutes the first wave crest came, bringing with it horrific destruction. The westward tsunami moving toward Sri Lanka and India led with a wave crest, but it was small enough at some coastal locations that it was not noticed. So again the first thing people observed was the sea receding, as the trough between the first and second crests arrived, with the same terrible results when the larger, second wave crest struck the coast.

However, long before the tsunamis reached any coast, seismic waves from the earthquake traveled through the earth's crust and mantle, reaching those coasts first. These seismic waves caused earthquake tremors that were felt in countries all around the Indian Ocean. Along the Aceh coast (50 miles from the epicenter) the ground began shaking less than a minute after the initial quake. In Thailand (350 miles northeast) the shaking began in about a minute and a half. In Sri Lanka (950 miles northwest) the ground began shaking in less than four minutes. At greater distances the seismic waves, though much smaller in amplitude and too low in frequency to be felt by humans, were detected by seismographs. In roughly twelve minutes seismic waves reached Europe. After twenty-one minutes seismic waves had been detected around the world, at government scientific agencies, universities, and tsunami warning centers. All these seismic data were sent immediately to NOAA's Pacific Tsunami Warning Center in Hawaii, as well as other tsunami and seismic centers around the world.

The width of the fault rupture was many times greater than the water depth. In less than ten seconds a large enough section of ocean floor was lifted to generate a wave with wavelength long enough to be considered a tsunami. Because they are very long waves, tsunamis can travel great distances without much energy loss (as we saw with the Lisbon and Krakatoa tsunamis). It also travels through the ocean with a speed that depends on water depth.[7] As we will see, this characteristic, the fact that a tsunami travels faster in deep water than in shallow water, explains much of the December 26 tsunami's behavior, including its size when it hits a coast.

Indonesian fishermen from Aceh out on the ocean that Sunday morning were the first to experience the effects of the earthquake. The tsunami waves were too long in the deep water to have a perceivable effect near the epicenter, but the fishermen witnessed other strange happenings. They felt a sudden shock, and some thought they had collided with something or run aground, even though the water was hundreds of feet deep. Others checked their engines, which they thought had broken down. Typical was one fishing vessel about forty miles east of the ruptured fault. It felt a shock and then was shaken up and down for about five minutes.[8] The agitated sea around the boats was described by many as bubbling or looking like it was boiling.[9] These effects were due to compression (primary, or P) seismic waves that had traveled from the Earth's crust into the seawater and up to the ships at the surface.

There were also large water waves, but with shorter wavelengths than a tsunami, that were caused by seafloor motions of lesser horizontal extent but still large vertical extent, perhaps augmented by small submarine landslides. They did not travel far but were tall enough to threaten fishing boats for the short time they lasted. When the captain of a shaking ship turned off his engine, his vessel was carried back and forth three times some distance by a strong current. Six miles closer to the Aceh coast, a smaller fishing boat, after also feeling five minutes of strong shaking, was lifted thirteen feet by a large wave and then lowered again. A second, approximately sixteen-foot wave did the same, and then a third wave broadsided the vessel and capsized it. Fishermen on another fishing vessel saw the bubbling water but also heard sounds like bombs exploding just before a wave the height of a coconut tree (about forty feet) passed by them. A second wave passed under them, lifting them up and down in an instant.[10]

When after ten minutes the "boiling" water and shaking of the sea subsided, the crews on the fishing boats outstretched their arms in prayerful submission, thanking Allah for sparing their lives. Some fishermen spoke of a ghost in the sea. Only one or two guessed that there had been an earthquake, or at least that's what they claimed later. As they got closer to the coast and the water became shallower, they unknowingly began to see and feel the effects of the tsunami. A few fishing boats a short distance east of the Breueh Islands (north of the northern end of Aceh) saw water rapidly pulling away from the shore, exposing a half-mile of ocean floor. They saw people from the shore running onto the dry sea bottom to pick up fish. But then a black wave passed under their boats and headed for the people on the exposed sea bottom, growing to the height of one and a half coconut trees (approximately sixty feet) by the time it reached them. The fishermen said the wave looked like the head of a cobra ready to attack. Some captains aimed their boats into the wave and rode over it. But once the wave had passed, half the boats were gone, as were all the people who had been on the shore. The fishermen who survived suddenly had a terrible feeling about their families back on the Aceh coast. As they headed for home, their feelings of dread grew as they passed floating dead fish and then thousands of floating coconuts. That was before they began to see floating bodies.[11]

The closest land to the earthquake's epicenter was the Indonesian island of Simeulue, only thirty miles to the south. The whole island shook from the earthquake for about five minutes. Eight minutes after the shaking stopped, a thirty-foot-high tsunami wave struck the village of Langi at the northern end of Simeulue. This wall of water bulldozed everything in its path. The entire village of Langi was leveled, leaving nothing to show that families had once lived there except concrete foundations of houses (which were only inches above the ground). But in spite of this total destruction, not a single person among the 8,000 villagers died. For as soon as the shaking began, everyone ran as fast as they could for the high ground of the mountains. Because of the long duration of the shaking, some villagers ran even without checking to see if the sea had receded from the shore (which it had, revealing a half mile of exposed sea bottom, again with fish helplessly thrashing about). That was the unequivocal sign that a tsunami was coming, for the villagers knew that when the sea pulled out, it would return, and when it did, it would come back with much more force and height than when it left. Along the east coast of Simeulue at the town of Air Pang, where a fifteen-foot tsunami washed away most of the houses and damaged a hundred fishing boats, the people there too had run for the mountains, some even going together in groups led by community leaders. All over the island it was the same, with the people who lived along the coast (95 percent of the population) running for high places. Many had prearranged places for family members to meet once they were safe, with prebuilt bamboo frames over which they could throw a tarp to give them shelter for the hours or even days they would wait there until it was safe to return to the coast. Only seven people out of 78,000 who lived on Simeulue died in the tsunami, even though most coastal villages were destroyed by at least three successive tsunami waves.[12]

The people of Simeulue knew to run because many years before their ancestors had experienced another tsunami, which they called a *smong*, which translates roughly as "ocean coming onto the land." Smongs had happened more than once before, but the last time was almost a hundred years ago, in 1907, when a tsunami devastated the west coast of Simeulue and killed possibly half the population. A century might have been enough time for that disaster to fade from memory, but here it was not forgotten. *Smong* was still in the local vocabulary because it had become part of the culture of the Simeuluese people. Orally passed-down stories of the 1907 tsunami told of bodies found in the tops of tall coconut trees and on inland hills and of large pieces of coral transported inland to rice fields, where they can still be seen today. In 2002 the island felt another strong earthquake, but this one did not produce a tsunami. The villagers had stayed in the hills for a day before they decided it was safe to return to their homes. The fact that it turned out to be a false alarm did not lessen their resolve to head for the hills on the morning of December 26, 2004, showing how strong an impression had been made by the smong stories told by their grandparents. All the islanders knew that there were three steps in a smong: first, the earth shaking for a fairly long time; second, the sea receding from the land; and third, the sea rapidly returning and flooding the land. In 2002 the second step had not occurred, and thus step three, the tsunami, did not occur.[13] Unfortunately, only a few others around the Indian Ocean on December 26 possessed a similar awareness of the natural signs of an approaching tsunami.

If most local peoples had little knowledge of the natural signs preceding a tsunami, there was unfortunately even less in the way of technological assistance available to warn them. On December 26, 2004, there were no DART buoys in the Indian Ocean as there were in the Pacific Ocean, and there were no real-time tide gauges within a thousand miles of the epicenter of the earthquake. This was understandable from one perspective, for as we saw in Chapter 7, most tsunamis occurred in the Pacific Ocean. Because of the rarity of recent damaging tsunamis in the Indian Ocean, there were no real-time oceanographic sensors that could have warned the people that a tsunami was heading for their coast (although afterward the tsunami signal was recognized in many satellite and research data sets). Without real-time oceanographic data, an indication that a tsunami had been generated by a massive earthquake could come only from two sources, one scientific and one not. The only scientific source of information was from real-time seismographs around the Indian Ocean, which detected the seismic waves that began propagating from the earthquake at 7:59 that Sunday morning. The other source of information was simply human observation of arriving tsunami waves and the reporting of those sightings by news media to warn people at other locations. Neither of these sources of information turned out to be timely enough to warn the Aceh coast of Indonesia or even for warning the more distant shores of Thailand, Sri Lanka, and India.

The first seismic waves from the epicenter took just under three and a half minutes to travel through the Earth to the two closest seismographs in the region, at Pallekele in Sri Lanka (950 miles northwest of the epicenter) and at the Cocos Islands, an Australian territory (1,000 miles south). Less than two minutes later the seismic waves reached seismographs on the British island of Diego Garcia (2,000 miles southwest) and in Tibet, China (1,800 miles north). These primary (P) waves alone could not provide enough information to determine the location or size of the earthquake. The slower secondary (S) waves took three more minutes to arrive at Sri Lanka and the Cocos Islands and four and a half more minutes to arrive at Diego Garcia and Tibet. These were followed soon after by the surface seismic waves (Love waves and Rayleigh waves), which caused the largest tremors and brought down some buildings in Aceh. Since the difference in the speeds of P and S waves is known, the difference in their arrival times at a seismograph allows calculation of its distance from the epicenter. Knowing distances from three seismographs allows one to calculate the epicenter's location. As more seismic wave information arrives from these and additional seismographs, the location can be made more precise, and the earthquake's magnitude can be estimated.[14]

All these seismic data were automatically sent in real time to NOAA's Pacific Tsunami Warning Center in Hawaii,[15] as well as to the other tsunami and seismic centers around the world.[16] Eight minutes after the initial rupture (8:07 A.M. Indonesian time[17]) these seismic data signals triggered an alarm at the Warning Center. Three minutes later the Warning Center sent a message to other observatories in the Pacific region with preliminary parameters for a submarine earthquake that they determined to be fifty miles southwest of northern Sumatra. From these early data the Warning Center made a preliminary estimate that the earthquake had a magnitude of 8.0 (a value also calculated by some other geophysical observatories). Although this was a large earthquake, this did not guarantee that a tsunami had been generated. In fact, just three days earlier, on December 23, there had been an 8.1 earthquake south of New Zealand near Macquarie Island, the largest earthquake in four years, but it only produced a tiny tsunami.[18] When in doubt the Warning Center preferred to err on the side of caution and put out a warning, but those warnings had led to a high rate of false alarms, which can undermine public confidence and cost money since evacuations disrupt businesses.[19]

The Pacific Tsunami Warning Center had not yet received enough seismic data to determine whether the December 26 earthquake off Sumatra was larger than 8.0 magnitude, a rare occurrence, but given that it was below the seafloor, this earthquake had the potential to produce a tsunami of some size, if most of the energy released went into vertical movement of the seafloor. Fifteen minutes after the initial earthquake (8:14 A.M. Indonesian time) the Warning Center sent out its first bulletin on the earthquake to its twenty-six member nations around the Pacific Ocean (the Warning Center's area of jurisdiction). This included Indonesia and Thailand, because they had coasts whose waters were connected to the Pacific Ocean and they were considered to be along the Pacific's Ring of Fire (which, as we saw in Chapter 7, is the tectonically active area around the edges of the entire Pacific Ocean). The bulletin gave the location as 3.4° north, 95.7° east, the initial size estimate of 8.0 magnitude, and it stated that "this earthquake is located outside the Pacific. No destructive tsunami threat exists."[20]

By this time, unknown to the scientists at the Pacific Tsunami Warning Center or anyone else, a tsunami had already struck the island of Simeulue and was only a couple of minutes away from striking the coast of Aceh Province on northern Sumatra, the next closest land to the ruptured fault. The Aceh coast was lined with fishing villages and towns along a hundred-mile stretch from Lhok Nga Bay (ten miles southwest of Banda Aceh, Aceh's largest city) south to Meulaboh (Aceh's second-largest city). On Aceh some people heard bird calls and noticed flocks of birds flying from the open sea toward the interior of Sumatra. Others saw large numbers of cranes fly out of mangroves and head toward hills inland. Though many considered this an ominous warning, they did not see it as foreshadowing a tsunami.

The Acehnese fishermen at sea that morning were not sure what was happening, but having experienced the strange shocks on their boats, the bubbling waters, and the large waves, the boat captains had headed home fearing for their families back on shore. Those fears had grown as they passed dead floating fish and thousands of floating coconuts, and then the dead floating bodies. Since it was their moral obligation, they picked up as many of the bodies as they could fit on their boats. Occasionally they would rescue someone alive, floating by on a tree or on a mattress or on a piece of rooftop, often with their clothes partially ripped off. One ship came across fifty people still alive. By now they were hearing stories of monstrous waves hitting villages along the coast and even the city of Banda Aceh. The fishermen were confused, for all the big waves they had seen in the past had been accompanied by rain and high winds, and even the biggest of those wind waves had not killed people in the numbers they were finding in the sea. As they got closer to shore, they had to go slowly because there was so much debris in the water. They strained their eyes to see the shore in the distance because it looked different. For some reason they could not make out their villages. Eventually they got close enough to see why. The villages were no longer there. Their homes were gone. The coast was barren except for a few mosques and a few trees. Were their families gone also?[21]

The villages along the shore of Lhok Nga Bay were the first Acehnese settlements struck by the tsunami. This shore was less than sixty miles from the location on the rupturing fault that produced the largest tsunamis (about two hundred miles northwest of the epicenter). As at Simeulue it began quietly. Calm, clear turquoise waters covered beautiful coral reefs next to pearly white beaches that edged the entire bay. Lhok Nga Bay was the perfect spot for diving, fishing, surfing, and enjoying the beach, and it typically drew crowds on Sunday mornings from Banda Aceh and nearby villages. Luckily this early in the morning the beaches were not yet filled. Around 8:00 A.M. the ground began shaking and rumbling. After five minutes it stopped, but then five minutes later the sea began to retreat, pulling back more than half a mile and revealing the large coral heads that had attracted so many divers to the bay. But, as would be the case at so many beaches that morning, few people knew they should immediately run. Instead they were drawn toward the amazing sight of the dry sea bottom and fish flailing about easy for the taking. Then three loud muffled booms were heard from the ocean as though it were announcing that it was now returning—but returning with a vengeance, with a roaring sound that more than one person later compared to the sound of a jet plane. The first wave, green with a lip of foam, hit the shore thirteen minutes after the earthquake had ended, and though "only" sixteen feet high, it was fast, powerful, and deadly. The people finally ran for their lives, but too late.[22]

The first wave, however, was only the prelude to the second wave, which was an incredible 115 feet high (as high as a ten-story building). It destroyed everything in its path. The town of Lhok Nga was completely leveled, the water washing away every piece of it. Of its 7,500 inhabitants only 400 survived, either by running immediately upon seeing the first wave or by being away from home that day. Other seaside villages were also swept out of existence. After the waves retreated, nothing was left but barren land. The very few remaining coconut and casuarina trees were completely stripped of foliage. Five-ton chunks of coral had been ripped from the reefs and carried onto the land at Lhok Nga Point.[23]

As the tsunami moved over the land, friction reduced its height considerably, but its power was still overwhelming. At areas of the shore backed by hills the water ran up the slopes to great heights. Between Labuhan and Leupung at the southern end of the bay the water ran up a cliff to a height of 167 feet, overtaking even those few who had managed to reach high ground there. A few hilltops farther inland became sanctuaries for those who had started running soon enough, from the first wave. Those who made it to the tops of the hills looked down at a harrowing sight as the now fifty-foot-high black wall of water bulldozed trees, buildings, and everything else in its path. The water that whirled around hilltops was filled with debris from the destruction. The people called out prayers to Allah as bodies floated by, sometimes with their now blank faces staring up at them. This second wave surged four miles inland in some directions, while in other directions it was stopped short by steep tree-covered hillsides. A mile inland, the Lhok Nga mosque survived the tsunami, some said miraculously. But being a stronger building and, most important, having an open design on the first floor, the rushing waters mostly flowed through it, rather than against it. About thirteen minutes after the first earthquake tremor people had come running to the mosque shouting, "Air laut naik!" ("The sea is coming!").[24]

Moving south along the coast of Aceh, village after village would suffer the same fate, each tsunami wave striking just a little later, because the shallow continental shelf widened to the south and it took the tsunami a little longer to reach the next town. One by one, towns were wiped off the face of the Earth along with their families of mothers, fathers, children, and grandparents. Just south of Lhok Nga Bay, the tsunami obliterated the town of Leupeng, its power evidenced by the shattered steel bridge over the Kreung Raba River. As elsewhere, ordinary houses were scraped out of existence. Nothing was left to show that a town had been there, no landmarks remaining to show a relative where a family member's home had once stood. Only the painted outline of a pedestrian crosswalk remained as proof that a school had once stood nearby. Ninety percent of Leupeng's 10,000 residents disappeared with their town, in spite of having nearby hills to which they might have escaped. What had been a group of homes became an open field of gray mud, with scraps of wood, rags, and corpses, as far as the eye could see. Many corpses were without clothes, which had been scraped off by rapidly flowing sand-filled water; some were without faces.[25]

High on top of a jungle-covered hill inland from the town of Lhoong, south of Leupung, rebels from the separatist Free Aceh Movement were witnesses as the tsunami wiped out the town. The rebels had felt the earthquake tremors, which made the hills shudder and dislodged rocks. Then they had heard sounds from the sea, so loud even that far away that the rebels had thought the government was carrying out aerial bombing against them. But then in the distance they saw the sea turn black and rush inland. It frightened them even more than government troops. Only Allah could send such a destructive force.[26] The rebels had been fighting government forces for twenty-eight years, seeking independence for Aceh. Their reaction to the booming noises of the tsunami was not unusual. Many residents did not run immediately from the tsunami, because they thought the sounds they heard were the sounds of bomb blasts and gunfire from another skirmish. Many residents were also relatively new to the Aceh area, and they had no older generation to tell them stories of past great waves.

Still farther south, the town of Calang, situated on a peninsula, was smashed from two directions. Again only its main mosque survived. At least

6,500 people died, 70 percent of its population.[27] Fifty miles farther down the coast, the city of Meulaboh was closer to the epicenter than towns north of it but farther from the strongest tsunamigenic (or tsunami generating) point of the rupture (north of the epicenter), and it had a wider continental shelf across which the tsunami had to travel. Thus, the tsunami arrived there later than at the villages farther north and was somewhat smaller, estimates ranging from sixteen to thirty feet. Having the second-largest population (120,000), Meulaboh had the second-highest number of deaths on Sumatra. An incredible 40,000 people died, a third of its population, with another 50,000 left homeless. The sea had first withdrawn here as well, at least a third of a mile, and again the people had rushed to pick up stranded fish. The Minangkabau people who made up the majority of Meulaboh's population did not see this as a warning sign. They were fairly recent immigrants to the area, most arriving since 1970, and they had no cultural knowledge of large waves from the sea. Most had not even experienced an earthquake. However, one cultural tradition might have saved some lives, for their shopkeepers built sturdy two-story *rukohs,* to which many Minangkabaus escaped. Meulaboh also had an army camp which was destroyed along with 500 rebel prisoners. Most rebels were safely hidden in the mountains and so survived the tsunami, but many ran down to destroyed villages to help families and neighbors.[28]

As in any coastal disaster, the greatest factor in determining the number of casualties is the size of the population living close to the sea. At the north end of Aceh Province, at the northernmost end of Sumatra, the capital city of Banda Aceh had at least 250,000 people, with still more living in its suburbs. A large number of them lived along the coast, which faces north, as well as along rivers and streams connected to the ocean, and near rice fields, lagoons, and other low-lying areas. Although there were middle- and upper-class neighborhoods, most of those living near the sea were poor, and their houses could not withstand a tsunami. The people had strong religious beliefs, but they had long since lost a cultural memory of tsunamis.[29]

In Lambada, a fishing village two miles northeast of Banda Aceh, they had felt the first tremors at 8 A.M. and heard three deep, booming sounds. But when a short while later the sea withdrew a half mile, no one saw this as a warning sign. And no one thought to leave, until the sea returned as a towering sixty-foot wall of water. Then they ran.[30] Masses of people flowed through the streets, with the wall of water flowing after them. The few trying to escape by car were surrounded by a sea of people and so were limited to haphazard lurches forward whenever a space in the crowd opened up. The motorbikes fared somewhat better in this frantic exodus, but even they had to frequently stop to avoid hitting mothers holding children with both hands, the young ones already tiring and slowing down, even as the raging waters behind them came ever closer. Time and time again hands lost their grip, and screams were emitted as small children fell back and were trampled by the running crowd. But no one stopped, because the roar of the water behind them kept getting louder. When the front edge of the wave reached them, it pushed their feet out from under them and then engulfed them. They gasped for air but instead swallowed filthy water. Pieces of debris carried by the wave—broken wooden boards and glass, metal scraps of roofing, bicycles, pieces of cars, and other hard objects—gashed and gouged their bodies, opening wounds for the dirty water to infect. Once swept over by the wave, those who did not drown were pulled along by the current, grasping for anything to hold onto—large floating objects that could keep them above water, or if very lucky, a tree that might still be standing. But most of those who managed such small successes were overcome shortly thereafter by a second wave, or by a third. Within twenty minutes Lambada had been washed away, leaving only its mosque along with a minaret and a municipal water tower. Only 636 survived out of 2,200 villagers, and of these only 12 were children.[31]

Boats from Lambada, Ulee Lheue, other coastal villages, and the Aceh River were carried miles inland, some ending up on top of houses. Even a 2,600-ton floating power plant docked in Ulee Lheue was picked up by the tsunami and carried two miles to the Punge village of Blang Cut, too far to ever bring back to the coast. Later, still in working order, it produced power for its new village. Many people who survived were carried away to other

Figure 8.1 Picture showing the destruction caused by the December 26, 2004, tsunami along the northern Aceh coast of Sumatra, leaving a lone surviving mosque. (*U.S. Geological Survey*)

villages far away, such as the villagers of Bitai, who were carried a mile southeast to the village of Lamteumen, arriving with clothes in shreds or gone.[32]

Lamjabat, another suburb of Banda Aceh, was a little farther from the seashore than Lambada, about a half mile. Being a Sunday morning, the rocking ground jolted many of its residents out of bed. Dazed, but with enough presence of mind to know their houses might collapse, they ran outside. But in a few minutes they began hearing the cry that was all over Aceh. "Air laut naik!" People were running up Pendidikan Road and other streets, with something dark and high and ominous following them. The dark wall of water obliterated all the houses, even those on six-foot stilts, in which a few residents stayed, thinking stilts would save their houses and them from the sea. Mothers and fathers held on to their children, some in their arms, some by their hands, but when the water caught them, they were all ripped apart. Those still alive were swept along in the violent current and battered by rubble. They stared in stunned confusion as they were carried by the water past tall city buildings or even through stands of bamboo in the open areas. Sometimes people being pushed toward a violent collision with a house were saved when something big hit the house first, destroying it. Logjams of rubble were everywhere, some of them growing into piles high above water, becoming moving islands onto which those with still enough energy climbed for safety, often half dressed or even naked, bleeding from scrapes and wounds.

Those who managed to climb onto floating rooftops or mattresses rode the water inland. But when the flow reversed as the sea retreated between successive tsunami wave crests, these temporary survivors were carried out to sea, most never to be seen again. Later there was the occasional story of someone miraculously rescued at sea on a floating mattress.[33] Too exhausted and too fearful even to watch where they were being taken, most were spared looking at the bodies staring up at them from the dark water or at the almost-dead people staring down at them from the trees where they hung, some still chanting prayers. But those in trees or on rubble mounds or floating on mattresses were a very small percentage. Slowly, most of the people who had begun running joined those unable to run, and they were now floating among the debris, being swept along with refrigerators, bicycles, mattresses, and rubble piles. Farther inland videos would be taken of the floating islands of junk, some with people still on them, but by then there were no longer any violent waves, just strong currents moving down the streets between the buildings. Within twenty minutes Lamjabat had been destroyed. Of 3,600 original residents, only 210 survived, of which only 15 were children and only 40 were women.[34]

Blang Padang field was a twenty-acre park near the center of downtown Banda Aceh, about a mile and a half from the sea (which was to the north) and a half mile from the Aceh River (to the northeast). At the western corner of this area of green was a monument of a plane, a Dakota Seulawah RI-001, the first plane of Indonesian Airways. Blang Padang was the largest park in Banda Aceh and, like any large park in any large city, the center of many community activities. On Sunday mornings thousands were typically drawn to the park for exercise or to begin a leisurely walk down to the sea. On this Sunday the crowds were even larger because a minimarathon, the 10K Aceh Open 2004, was to start and end at the park. The mayor of Aceh, along with other dignitaries, was there to hand out trophies. The race began around 7:30 A.M., and by 8:00 A.M. some runners were beginning to cross the finish line at Blang Padang when the ground began to shake. The runners could not keep their balance. Everyone sat or lay down on the ground. In the openness of the park, they seemed safe, far from falling buildings. They sat there and looked around as some of the buildings on the edge of the park began to break apart. The first floor of the large, curved Kuala Tripa Hotel collapsed. When the tremors stopped, people got up and walked around to look at the damage. But then slowly some began to notice a rumbling sound coming from the direction of the sea. When the twenty-foot wall of black water came into view north of the park, the people ran, screaming. Water swept between buildings, over the park, and between more buildings, in some places being joined by water coming from the Aceh River, up which the tsunami had forged and then flooded its banks. Hundreds of runners and thousands of onlookers died, as did the mayor.[35] When the waters pulled back, the park had become a debris-filled bog with only the Dakota Seulawah RI-001 still standing, perhaps due to its aerodynamic design.

Approximately a third of the city of Banda Aceh was leveled by the tsunami. In the two miles closest to the sea there was essentially nothing but gray mud covering a denuded landscape. A mile farther inland the destruction was more haphazard, leveled areas next to still-standing structures. The remainder of the city was left untouched. For some, at least during the turmoil, it was a sign that the end of the world was at hand, and prayers were heard everywhere. But as it became clear that it was often a quirk of geography or of topography that some lived and others died, and that some still had homes while others were left with nothing, it was less clear whether this was Allah's punishment. And how could the end of the world take place on such a sunny day under such a blue sky? Some Acehnese continued to feel that they were Allah's chosen people, so it made sense to them that Allah had aimed these destructive waves at them, because they had not lived up to his standards. One fact was clear though: across Aceh tens of thousands had died, and tens of thousands still alive had lost everything—their families, their homes, their means of livelihood. Instant homelessness.[36] In Aceh Province an estimated 242,347 people died or were missing and presumed dead, of which probably more than half were from Banda Aceh, about a fifth were from Meulaboh, and the rest were from the dozens of coastal villages wiped out, from Banda Aceh to Meulaboh.

That Sunday morning the emails sent by the Pacific Tsunami Warning Center had arrived at the appropriate Indonesian government office in Jakarta, but the official in charge was not at work, and no one read them until Monday.[37] Even if someone had seen them, there was no infrastructure in place ready to act on the information. For some years scientists from Indonesia had expressed a need to develop an effective tsunami warning system. There had even been a special tsunami workshop held in Jakarta in 2003 on the 120th anniversary of the tsunami caused by the eruption of Krakatoa in 1883. It was sponsored by the Intergovernmental Oceanographic Commission (IOC) of UNESCO (United Nations Educational, Scientific, and Cultural Organization) and attended by one hundred Indonesians and many international experts.[38] The workshop was intended to raise awareness of the tsunami threat to the region and to facilitate establishment of a tsunami warning system in the Indian Ocean. There had been other deadly tsunamis in Indonesia, though not on the scale of the December 26 tsunami. A tsunami in 1992 killed 2,080 people at Flores Island, another in 1994 killed 223 people along the south coast of Java, and another in 1996 killed 108 people on the island of Irian Jaya.[39] These were at the other end of Indonesia from Aceh (more than a thousand miles away), so they would not have helped local Acehnese awareness, but they should have helped the Indonesian government's awareness. There was an earthquake in November 1998 to which the government did respond, evacuating Mangole Island, but it turned out to be a false alarm, with no tsunami hitting the island.[40] By 2004, lack of funds or lack of concern or both had apparently discouraged the implementation of a tsunami warning system for Indonesia.

Minutes after northern Aceh was hit, the tsunami hit the Nicobar Islands, north of Aceh, and the Andaman Islands, still farther north, both part of a territory owned by India. Being a relatively short distance from the northern half of the ruptured fault, and much of that distance being over fairly deep ocean waters, these islands were hit by the tsunami only minutes after Banda Aceh, and with large wave heights (which decreased in size near the northernmost islands). On South Andaman Island, Port Blair, the largest city and the capital, was hit by a tsunami wave at 7:15 A.M. Indian Standard Time, which is 8:45 Indonesian time and forty-six minutes after the earthquake. For the population of roughly 350,000, spread out over thirty-eight inhabited islands, the tsunami's effect varied considerably depending on the topography and location of the island.[41] The higher forest-covered islands suffered damage to their beautiful beaches and the villages next to them, while the low-lying islands were completely washed over, the erosive power of the tsunamis carving some of them into groups of smaller islets. From three to five tsunami waves, thirty to fifty feet high, brought utter devastation, washing away entire villages, as well as valuable pigs and coconut trees.[42]

About 10 percent of the population of the Andaman and Nicobar islands is made up of six indigenous tribes, five of which (Jarawa, Sentinelese, Shompen, Onge, and Great Andamanese) have remained essentially isolated from the larger Indian population. DNA testing has revealed that two of these tribes, the Onge and the Great Andamanese, have avoided intermixing with other populations for fifty thousand years. They are one of the last Stone Age cultures left on Earth. There are perhaps only 850 of these ancient people living on a few islands that were declared Aboriginal Tribal Reserves by India, their numbers having decreased in the past whenever they came in contact with the germs of modern society. The sixth and largest tribe, the Nicobarese, numbering almost 30,000, had assimilated with the Indian population. After the tsunami almost 10,000 Nicobarese were dead or missing and presumed dead. But the aboriginal tribes escaped the tsunami, apparently because of cultural wisdom passed down orally from generation to generation about shaking ground followed by large waves. Their actions on December 26 were all similar. An elder of the Jarawa tribe led his people to a hilltop when the earthquake was felt. The Onge tribe fled into the hills after seeing the level of a creek by their village suddenly drop as the sea receded. Down the coast, in Hut Bay, where Onge lands had been taken by settlers, the waves killed forty-eight people. The Great Andamanese tribe on Straight Island ran up a hill, outracing the oncoming water. No one knows how the Sentinelese escaped, because they are hostile to outsiders. When an Indian helicopter flew over their island to see if they were okay, arrows were fired up at it.[43]

There was, however, one incident in which modern information helped save lives. About 1,500 Nicobarese on Tarasa Dwip (Teressa Island) were saved because an employee of India's Port Management Board liked to watch the National Geographic Channel, where he had learned that earthquakes caused tsunamis. He and a colleague warned the nearby villages, and the people scrambled up the hills to safety, just before the first of three large waves demolished all five villages.[44]

The island of Car Nicobar, about sixty miles to the northwest, and only about fifty miles east of the ruptured fault, was also hit very hard. An Indian Air Force base on the southeast coast was flattened by the tsunami, killing 100 people. News of this disaster was sent to mainland India, about 800 miles to the east. The Air Force base at Chennai received a message from the Car Nicobar base at 7:30 A.M. Indian time (one hour after the earthquake) that a massive earthquake had occurred. Twenty minutes later, just before communication links went down, Chennai received a last message from Car Nicobar over a high-frequency channel. The message said: "The island is sinking and there is water all over." The calamity at the Air Force base in Car Nicobar was an early warning for mainland India, the only one they would receive. The question was, would it be recognized as a warning and acted on?[45]

One hour and five minutes after the earthquake (at 9:04 A.M. Indonesian time), after analyzing additional seismic data that had arrived, the Pacific Tsunami Warning Center sent out a second bulletin with a revised earthquake magnitude. The new magnitude estimate of 8.5 meant that the earthquake was at least 5.6 times more powerful than originally thought, although that still did not necessarily mean that the earthquake had produced a tsunami.[46] This most recent analysis also did not preclude the earthquake from being even larger than 8.5 magnitude, but there was not yet enough seismic data to determine that. It had been forty years since there had been an earthquake greater than 8.5 magnitude. This second Warning Center bulletin again stated that "no destructive tsunami threat exists for the Pacific basin," but "there is the possibility of a tsunami near the epicenter." Of course, by this time the tsunami had already struck Aceh Province, and most of the more than 240,000 Indonesian victims were already dead, killed within thirty minutes of the submarine earthquake. But no one outside of Aceh knew, and probably no one inside Aceh knew the extent of the losses.

And no one would know for another hour and a half. News did not come out of Aceh Province rapidly, since access to the province by outsiders was restricted by the Indonesian government because of the ongoing rebellion. There were no Western reporters there and few Indonesian reporters. The news agencies in the capital city of Jakarta on the island of Java, twelve hundred miles away, had little reason to cover news in one of the poorest provinces of Indonesia. And even inside Aceh there was little awareness of more than each local catastrophe. With villages spread down the coast, miles from each other, those who survived each village's obliteration suffered Allah's displeasure in isolation somewhere inland, alone, perhaps on a hilltop. Only from the city of Banda Aceh would word begin to leak out. About the same time that the Warning Center's second bulletin came out, the first news report of the earthquake was put out by Reuters news service, but there was no mention of a tsunami. Thirty minutes later, at 9:34 A.M. (Indonesian time), one hour and thirty-five minutes after the quake, an Associated Press newswire report said an earthquake had caused dozens of buildings to collapse in Banda Aceh, but again there was no mention of a tsunami. Agence France-Presse reported the earthquake but was also unaware of the tsunami.[47]

About this time the Pacific Tsunami Warning Center called Emergency Management Australia and found that they were aware of the earthquake and had gotten the Warning Center's email. Thailand had also received the email. In Thailand earthquake tremors had been felt as far away as Bangkok, about 800 miles northeast of the epicenter, and were strong enough that three duty officers at the Seismic Monitoring and Statistic Center in Bangkok were overwhelmed with phone calls. They had known by 9:00 A.M. (Indonesian and Thailand time) that the epicenter was off Sumatra, but to them that seemed far away, and a tsunami had never hit Thailand before as far as they knew. So it was without a sense of urgency that they faxed details of the earthquake to local radio and TV stations.[48]

As a portion of the tsunami brought widespread death and destruction to Aceh Province, another portion of the wave moved westward, between Aceh Province and the Nicobar Islands and across the southern end of the Andaman Sea to Thailand. This entire region had felt tremors within one and a half minutes of the initial fault rupture. In Thailand, at approximately 9:55 A.M. local time,[49] almost two hours after the first tremors, the sea began to retreat at Patong Beach, a popular tourist location on the west coast of Phuket Island. Thousands of tourists, many from Europe, flocked every winter to Patong and other Thailand beaches along three hundred miles of the Andaman Sea coast because of the beautiful white sandy beaches and the turquoise waters with coral reefs.[50]

The sea receded hundreds of feet and drew curious tourists and locals like a magnet. They walked out on the dry sea bottom, looking at the coral heads and collecting stranded fish. Some of those who wondered about it thought maybe it was an extreme low tide due to the full moon, though the locals said they had never seen a tide go out so quickly. Eventually, some noticed on the horizon a small band of white. It looked like one long breaking wave perhaps a mile out, but what was so strange was that it was along the *entire* horizon. Slowly the white band of water grew larger, its turbulence more apparent as it came closer, distributed along the entire beach. Yachts and fishing boats were tossed about, and some with their bows pointing at the waves went over it, perhaps keeping the wave's power from appearing as ominous as it should have to those still transfixed by the sight. This was not your typical breaking ocean wave that a surfer could ride or a swimmer could dive under at its base to avoid being smashed. This was a solid, churning, white wall of water. At some point people began to realize the danger and started to run and yell, but for many it was too late. The water rushed over them and up onto the shore, its power still perhaps not apparent to those looking down from higher floors of the many hotels, until they saw houses ripped apart and cars and buses carried inland like toys. But to those unlucky enough to be in hotel rooms on the first floor the tsunami's power was all too evident. Some rooms literally exploded from the pressure. If the occupants were lucky, they were washed out of their rooms by the water that had crashed through the windows and doors, but many drowned as water filled their rooms to the ceilings. And this was only the first wave. Then the second wave came. Then a third.[51]

The height of the tsunami as it reached the shore varied from location to location depending on sandbars, reefs, bathymetry, and beach alignment.

The largest waves to hit the beaches on the west coast of Phuket Island were eighteen to twenty feet high.[52] Beach conditions influenced how much power each wave had as it barreled inland. Karon Beach had a sand dune in front of it, which helped reduce the wave height, whereas Kamila Beach had a large near-shore rock platform and suffered significant loss of life. The penetration of the tsunami was inhibited where there was a greater density of stronger buildings such as the hotels, restaurants, and other commercial businesses in Patong. City streets leading directly inland from the beach received the strongest flow and carried the most lethal debris, including cars and other large objects that piled up at the ends of these roads. City streets running parallel to the beach, and thus perpendicular to the primary water flow, were mostly sheltered by buildings. Some people survived because they went down these side streets instead of running farther inland, trying to outrun the two-story tsunami wave.[53]

Hotels and businesses in Patong and Kamila, and in Phuket on the east coast, were heavily damaged, mainly on the ground floors. Cars floated through hotel lobbies. Floating objects smashed through the first-floor windows and doors of shops, restaurants, bars, and hotels. The water would then rush in and sweep everything out, leaving the rooms empty except for mud. Large buses were carried hundreds of feet and usually demolished whatever got in their path. In some places the first wave was the largest; in some places it was the second wave. In most places, the backwash after the third wave also proved to be lethal, pulling anything not held down out to sea. In between the three main waves, some people who survived but were hurt or in positions that seemed precarious tried to move uphill but were caught by the next wave, to the horror of those watching safely on hotel balconies. After the first three large tsunami waves, many smaller waves continued to come for a couple of hours.

Even harder hit than Patong Beach was Khao Lak, about sixty miles north, another popular tourist location with white sandy beaches lined with hotels, restaurants, bars, and shops in front of rolling green hills. After the initial sea withdrawal (about thirty minutes after the one at Patong Beach), a large wave appeared (perhaps reaching twenty-five feet), its size probably due to the broad, shallow, gently rising sea bottom in front of Khao Lak Beach. Khao Lak had the highest measured tsunami run-ups in Thailand, ranging from thirteen to forty-six feet, because behind the beaches the ground sloped up to hills. At least 25 percent of those at Khao Lak Beach died that morning. In one of the luxury beach-resort hotels half the four hundred guests perished, most in ground-floor rooms. At some places the wave reached the roofs of three-story buildings. More than 80 percent of the hotels were damaged, and many were totally destroyed.[54]

Khao Lak was a popular spot for Scandinavian families, along with Ko Phi Phi, a small island thirty-five miles southeast of Patong, which had the second-highest number of fatalities in Thailand. Ko Phi Phi was popular with young people while Khao Lak was more popular with families. In Khao Lak many people were apparently swept out to sea from the beaches, during morning swims, while at Ko Phi Phi those who died were sleeping in their rooms when the water swept in. About 5,000 people were confirmed dead, half of whom were foreigners. Because so many bodies were unidentifiable, forensic teams from a dozen countries were flown in to carry out DNA testing.[55] The province of Phang Nga, which includes Khao Lak, suffered the most fatalities in Thailand (71 percent of the 7,593 people reported dead or missing in Thailand). In addition to tourists this included many locals, most from fishing villages that were completely destroyed.[56]

The tsunami continued on past Thailand to Myanmar in the north and toward Malaysia and Singapore in the southwest, but the frictional effect of the shallow water was wearing it down. The wave, in fact, never made it all the way down the Malacca Strait to Singapore (600 miles southeast of Phuket Island, Thailand). It decreased significantly as it moved through the strait, along the southeast coast of Malaysia and the northeast coast of Indonesia. The seventy deaths that occurred in Malaysia were near the border with Thailand, where the tsunami waves were still high.[57] The seventy-one reported deaths in Myanmar occurred closer to the (northern) Thai border, the wave heights then decreasing from nine feet to only a foot as they moved northward along the Myanmar coast.[58]

But there was another huge part of the tsunami traveling westward, this wave moving through the deepest part of the Indian Ocean, unimpeded by shallow waters. It would last a lot longer and go much farther than the eastward moving wave—in fact, it would eventually reach most of the world. And, as we will see in the next chapter, it would add tens of thousands of deaths to the already horrible toll. We will also see how the lessons learned and the extensive data set from the December 26 tsunami promise improved tsunami predictions in the future.

CHAPTER 9

December 26, 2004 (Part 2)

Learning from a Tragedy

The people who saw the natural warning signs of the December 26 tsunami and told others, thus saving lives, were few enough that their actions made news. We have already seen the tsunami awareness in the people on Simeulue Island and the ancient tribes on the Andaman and Nicobar islands. Their stories of big waves that came after an earthquake and a receding sea had been passed down orally from one generation to the next. Cultural knowledge again came into play with Thailand's Moken tribe of sea gypsies on South Surin Island, about fifty miles northwest of Khao Lak. The elders of the tribe recognized unusual movements of the sea and told two hundred tribe members to follow them to a hilltop. Minutes later the tsunami destroyed all their houses on stilts in their beachside village.[1]

For the rest of the more-modern population only a very few had an awareness of the warning signs of a tsunami. The most famous example was the ten-year-old English schoolgirl who remembered what she had learned about earthquakes and tsunamis in geography class two weeks before coming with her family on holiday to Mai Khao Beach, about twelve miles north of Patong Beach, Thailand. She was on the beach with her parents and younger sister when she noticed the sea retreating and the water looking bubbly. It took a great deal of effort to convince her parents that they must leave the beach, which they finally did along with others who had heard her frightened pleas. Because of her action no one on Mai Khao Beach died, unlike other beaches in the area.[2] At the same time, a few miles south of there, a Scottish teacher was at Kamala Beach when the water suddenly pulled back, leaving boats on a dry sea bottom. He immediately ran up the beach to his family yelling, "Tsunami!" He then commandeered a hotel shuttle bus and, with the help of a Thai doctor who knew English, convinced the driver to take the bus up to high ground. But the driver made a wrong turn, and the bus ended up traveling south on a road parallel to the shoreline only a half mile from the sea. Its passengers watched in fear as the first tsunami wave got closer and closer to them. It eventually caught them but did not overturn the bus.[3] A mile north of Kamala Beach at the Amanpuri Hotel, the man in charge of the water ski boats called his boss when the water disappeared from the beach. His boss told him to get everybody off the beach immediately and then called friends at other hotels to warn them.[4]

Numerous animals also reacted to the earthquake and tsunami, and saved unsuspecting people in the process. Eight elephants at Khao Lak that were giving rides to tourists suddenly began trumpeting and "crying" around the time the earthquake tremors began. They calmed down, but an hour later they became upset again and could not be comforted by their handlers. The elephants took off for the jungle-covered hills behind the resort beach with tourists aboard, who therefore were not among the 3,800 people who died at Khao Lak, half of whom were foreign tourists. The elephants that were not working broke their chains and also ran to the hills. Their handlers could not stop any of the elephants, but they managed to turn them in the direction of some running tourists. While on the run, the elephants lifted the tourists with their trunks and put them on their backs, carrying them to safety.[5] Nearby at the Khao Lak Elephant Trekking Centre, the same thing happened. Two elephants trumpeted and broke their chains, something they had never done before, and ran for higher ground, ignoring the commands of their trainers. The four Japanese tourists who had just climbed on them were unknowingly carried to safety.[6]

Although reference is often made to a possible sixth sense in animals with regard to their apparent awareness of earthquakes and anticipation of tsunamis, in elephants this ability is due to their hearing and their feet. They can hear sounds (vibrations) with much lower frequencies than humans can, including the type of *infrasonic* waves created by large water waves breaking in the ocean. Such low-frequency acoustic vibrations are less attenuated and travel longer distances through the air than the higher-frequency sounds audible to humans. It could have been this infrasound that upset the elephants around the time of the quake, for the submarine earthquake indirectly produced infrasonic waves when its seismic waves reached the mountains in northern Sumatra and shook them, producing low-frequency sound waves that the elephants in Thailand could have heard. The tsunami itself also produced infrasound when it reached shallow water, its wavelength decreasing to an infrasonic wavelength. Even closer, infrasonic waves would have been created by the tsunami when it first began crashing down on the dry sea bottom. But elephants can also sense surface seismic waves through their feet. Elephants, in fact, stamp the ground to send signals to other elephants miles away. Thus, in addition to sensing infrasonic waves through the air, they may have also sensed both the initial P waves and the surface seismic waves of the earthquake and then later the seismic waves produced by the tsunami itself as it crashed over the shallow waters near the beach. Their trunks are thought to be useful in both sensing the seismic waves (when they lay them on the ground) and in forming the low-frequency rumbling sounds that they use to communicate with other distant elephants.[7] But it was not just the elephants. Water buffalo were grazing along a beach in Ranong Province, Thailand, when suddenly the entire herd lifted their heads in unison and looked out at the sea, their ears perked up. They immediately turned and stampeded up a hill, with the confused villagers chasing after their precious livestock. Minutes later the tsunami came and destroyed their village.[8] Birds are also apparently sensitive to infrasonic waves in the air (and may even use wind waves breaking on beaches as a navigational tool), which might explain why flocks of birds were seen in many places flying inland from the sea.

The westward-moving tsunami wave produced by the ruptured fault crossed the Bay of Bengal and struck the east coast of Sri Lanka about seven hundred miles west of a strong tsunamigenic point along the fault. It arrived at Sri Lanka one hour and forty-six minutes (at 8:45 A.M. Sri Lankan time) after the earthquake.[9] This was ten minutes *before* the eastward-moving tsunami wave arrived in Thailand, even though the westward wave to Sri Lanka traveled almost twice the distance. In the deep waters of the Bay of Bengal (about two and a half miles deep) the westward tsunami wave traveled at roughly 430 miles per hour, comparable to the speed of a jet plane. In addition, the continental shelf along the east coast of Sri Lanka is narrow, in most places less than ten miles wide, so that the tsunami waves were not significantly delayed from hitting the coast by a long stretch of shallow water. In the deep ocean waters on the way to Sri Lanka, the tsunami wave heights were not great, the first crest being only a couple of feet high, according to satellite measurements.[10] Its wavelength was very long, about 270 miles from one crest to the next, so ships at sea would not have noticed it passing under them.

But once the tsunami hit shallow water, it slowed dramatically and its height quickly increased. Whereas the wave that hit Indonesia and Thailand led with a trough, so that the sea initially receded for a few minutes before the first destructive wave crest arrived, the westward-moving wave that hit Sri Lanka led with a crest. But this first tsunami wave crest was not very large, not more than three feet at most coastal areas. In some locations it was not even noticed, so it was the retreat of the sea between the first and second wave that was the first thing people saw. The sea rapidly pulled out so far and stayed dry so long that, just as in Aceh and Thailand, people ran down to see the colorful fish left stranded, some with bags into which to stuff the helpless fish. Most of these people were struck by the second wave crest, which was much larger, ranging from fifteen to forty feet.[11] There had been no warning, even though Indonesia had been hit an hour and a half earlier.

The first town in Sri Lanka hit by the tsunami was the fishing village of Oluvil.[12] Oluvil was struck first not because it was the most easterly point on Sri Lanka, but because it was near the shoreward end of a submarine canyon that cut through the continental shelf. Twenty minutes after Oluvil was hit, the rest of the 250-mile-long Sri Lankan east coast was being flooded. This was not an area with many tourists. They preferred the south coast, away from the fighting between Sinhalese government troops and Tamil rebels, which has been raging for twenty-one years. It was a poor and vulnerable area where fishermen and rice farmers lived in small wooden houses near the shore, so that few survived when the thirty-foot tsunami waves struck. Half of the total deaths in Sri Lanka would be along its east coast—15,000 people dead, 10,000 injured, and 800,000 left homeless.[13]

Yet in the town of Oluvil only two people were killed. Just eight miles north of Oluvil, more than 8,500 died in the town of Kalmunai, making it the most devastated city in all of Sri Lanka. At Oluvil the water simply rose like a fast tide, but at Kalmunai a violent forty-foot tsunami wave struck with tremendous force.[14] Oluvil fishermen called it a blessing, which it was of course, but there was a more scientific reason why Oluvil survived. In Kalmunai, the surviving members of a family wiped out by the tsunami would have found it difficult to comprehend that their tragic loss (while others just down the coast were barely touched) was determined by the shape of the sea bottom below the dark ocean waters to the east. Oluvil was near the shoreward end of a submarine canyon, while Kalmunai was near the shoreward end of a submarine ridge. The submarine ridge acted like a waveguide, the energy of the tsunami being focused onto the ridge by refraction (which we talked about in Chapters 5 and 7), because the speed of a long wave like a tsunami depends on the water depth. The parts of the tsunami in deep water on both sides of the ridge travel faster than the part of the tsunami in shallow water above the ridge, so the wave front bends toward the ridge from both sides, focusing the energy onto the ridge and producing greater tsunami heights and thus greater flooding when the wave hits the shore. On the other hand, wave energy is focused away from the deep waters of a submarine canyon, producing smaller tsunami heights when that wave hits the shore. Off the east coast of Sri Lanka there was an alternating pattern of submarine canyons and submarine ridges, which produced an alternating pattern of death and destruction—great damage, then less damage, then great damage again moving south along the east coast. Near Kalmunai, at least five locations had tsunami heights greater than thirty feet, alternating with locations with heights less than fifteen feet. Some villages were washed away, while others had little damage and many survivors.

Casualties were also affected by natural protection on shore, such as sand dunes or some type of strong vegetation (although, being an area exposed to regular ocean wind waves, mangrove forests did not usually do well here). Offshore reefs also absorbed tsunami wave energy, although they were damaged in the process. And barrier islands reduced the height of the tsunami that would hit the mainland, as was heartwarmingly demonstrated by the story of a group of orphans from Navalady, about fifteen miles north of Kalmunai. Someone had yelled, "The sea is coming!" as an ashen-colored thirty-foot wave approached the barrier island. Thirty orphans and staff jumped into a motorboat, which luckily started on the first pull, and headed across the lagoon toward the mainland. The roaring wall of water demolished their orphanage, crossed the island to the lagoon, and chased after them. But the tsunami wave lost a great deal of energy when it crossed the island, and some more energy traveling through the shallow water of the lagoon, and the boat of orphans managed to stay ahead of it and reach the mainland safely.[15]

The tsunami wave front had hit the east coast of Sri Lanka head-on, but at the southern end of the island nation, it hit the coast at an angle. The part of the wave front that was in deeper water traveled faster than the part in shallower water, and so the refracted wave front bent around, eventually coming directly at the southern coast of Sri Lanka. Along the south coast, tsunami wave heights generally decreased from east to west, in most places being greater than ten feet but in some places in the east reaching over twenty feet. The greatest tsunami heights were at the Yala National Park, at the eastern end of the southern coast. Offshore bathymetry focused tsunami wave energy toward the thirty-mile-long coast of this wildlife reserve, with wave heights over forty feet in places, the tsunami causing much damage and death at locations not protected by sand dunes. However, there seemed to be almost no dead animals. In and around the park there were reports of elephants and buffalos running inland, as well as birds flying away from the sea, apparently yet again having sensed the infrasound from the tsunami's arrival.[16] About 10,000 people died along the southern coast of Sri Lanka.[17]

The refracted tsunami waves followed the bending coastline to the west and then north. The tsunami waves on the west coast of Sri Lanka were generally smaller, varying from about twelve feet in the south near the city of Galle to less than seven feet in the north near the capital city of Colombo.[18] However, a thirty-foot tsunami struck the village of Peraliya (nine miles northwest of Galle), a location where a railroad track was close to the coast. A passenger train called the Samudradevi (which means "Queen of the Sea") left Colombo a little after 9:00 that morning heading south to Galle with over 1,700 passengers packed to overcapacity in its eight cars, because it was a holiday weekend and a Full-Moon Day, a day when Buddhists offer special monthly prayers and visit relatives. With the worst luck possible the

Samudradevi reached Peraliya at exactly the same time as the first tsunami wave, around 9:15 A.M. That first wave was small, but large enough to wash over the tracks to about waist high. The train stopped, not far from the village's high school. The tsunami had caused panic in the village, and many people hoisted their children up on the roofs of the train cars, which looked like sturdy areas of safety. The water retreated, but the train did not move on, perhaps because of the people on its roofs. Not leaving proved to be a fatal mistake, because about ten minutes later a second wave came, thirty feet in height and with so much power that it twisted the steel train tracks and hurled the train cars into houses and trees so forcibly that their steel wheels were broken off. A third wave followed, not as large, but still powerful. Almost all of the 1,700 passengers died, plus over 200 villagers. It was the worst train disaster in history, with twice as many deaths as the second worst, in 1981 in India.[19]

In Hawaii, two and a half hours after the initial epicenter quake (9:30 A.M. Sri Lankan time), the scientists at the Pacific Tsunami Warning Center saw reports on the internet about casualties in Sri Lanka.[20] Now it was clear to them that a tsunami had indeed been generated—a large earthquake off Sumatra, death and destruction 700 miles across the ocean at Sri Lanka—only a tsunami could have caused it. But now was there time to warn anyone else who might be in the tsunami's path?

The scientists at the Warning Center may have realized that there had been a huge tsunami generated by the Sumatra earthquake, but that fact had not yet shown up in the various news reports that had finally begun to come out. One Reuters report, for example, mentioned the earthquake off Aceh, large waves at sea, and rising sea levels, but it also spoke of a "flash flood" hitting part of Banda Aceh, and saying that "it was not immediately known where the water came from."[21] Even an hour later some news reports still wondered if the giant wave that hit Sri Lanka was related to the earthquake. The news media was running way behind the tsunami wave and was in no position to provide warnings to nations still in its path.

India was about to be next. Here at least there had been one piece of information that could have potentially provided a warning—the messages from the Car Nicobar Air Force base that they had been rocked by an earthquake. The base's last message, at 7:50 A.M. Indian time, had said that "the island is sinking and there is water all over." This was received one hour and fifteen minutes before the tsunami would hit Chennai and the rest of the east coast of mainland India. However, it was not until 8:15 A.M. (fifty minutes before the tsunami would hit) that the Indian Air Force Chief asked the Assistant Chief of Air Staff Operations to alert the Defense Ministry.[22] What happened after that, and whether anyone in Chennai or higher in the government even understood that Car Nicobar had been hit by a tsunami, is unclear. From the

Air Force message sent from Car Nicobar the sender could easily have believed that the earthquake had caused the island to sink and that was why "there is water all over." So it may have been less a matter of government incompetence than simply a lack of education on the subject of tsunamis. Even if the messages had been passed up the chain of command more quickly, would anyone have realized that a tsunami had hit Car Nicobar and was about to hit the Indian mainland? Whatever the reason, no warning was given to the public. Another part of the Indian government, the Indian Meteorological Department, also received information about the earthquake from its office in the Andaman and Nicobar islands. It received and analyzed seismic data from numerous seismographs, but no one thought that a tsunami would be heading for India.[23]

A portion of the westward-moving tsunami wave passed north of Sri Lanka and slammed into the east coast of India, 200 miles north of the northeastern tip of Sri Lanka. It struck the city of Chennai at 9:05 A.M., two hours and thirty-six minutes after the initial earthquake and about fifty minutes after Sri Lanka was first hit.[24] It also hit the city of Vishakapatnam, 400 miles farther up the Indian coast to the northeast, at almost exactly the same time as it hit Chennai. At both cities the tsunami wave heights were smaller than elsewhere along the Indian coast (around eight feet, with run-up heights up to fifteen feet in some places), but they were still powerful and deadly to those in a vulnerable position. For Hindus it was Full-Moon Day, and at Manginapudi Beach, halfway between Chennai and Vishakapatnam, many women and children were in the sea for ritual bathing. Then the waves came, and lifeless bodies were thrown onto the shore or dragged out to sea.[25]

The largest tsunami heights and the greatest damage and loss of life in India occurred south of Chennai. When the huge wall of dark water moved toward the fishing village of Tharangambadi at 9:17 A.M., most of the men were at sea fishing.[26] A few miles offshore in their boats they felt a strange slow rising and even noticed that the sea had turned a dark inky color, but it was not until they headed home and saw the floating bodies that they realized that something frightening had happened. In the meantime onshore, being a Sunday, the children were in their seaside wooden homes rather than farther inland at (better built) schools. Their mothers were at home doing chores or at the fish markets or walking on the beaches with their children. So when the screams began and the huge wave was sighted bearing down on the shore, they were the ones who ran for the children and then ran away from the shore, carrying or pulling them along desperately. Having only two hands often meant saving only two children unless an older child could run alongside. The first wave to hit the village was only a little taller than a man, and some mothers and children had made it to high enough ground, but then many mothers went back to get children left behind. Then a second, larger wave came, taller than the trees. When the water washed over them, handholds were broken and children were lost, and so were many mothers. The few men who were onshore generally could run faster and, if caught by the water, could better grasp onto trees or other lifesaving structures. In Tharangambadi, of the first 206 bodies found after the waters drew back, 84 were children, 96 were women, and only 26 were men. There were still at least 200 missing, mostly children.[27] In the days following the tsunami, mass graves would contain a preponderance of little bodies. With few priests to chant last prayers, women simply threw marigold petals and burned incense. A sobbing mother or father, or occasionally both, would stand at the edge of the pit and stare at one of the tiny corpses laid in neat rows, say something, and then move on, their place taken by another. The large numbers of parents waiting to say good-bye allowed no more time than that.[28]

In the district of Nagapattinam, more than 6,000 people died. A small number of villages managed to escape with few casualties, because they had mangroves in front of them, or they were on higher elevations, or they were behind a rare seawall. Trees saved the village of Naluvedapathy.[29] About twenty-five miles south of Tharangambadi, another 2,000 people died in Velankanni, including many who had come on Christmas to see the beautiful Basilica of the Virgin Mary and its holy beach.[30] Much farther to the south the Indian coast was in the shadow of and protected somewhat by Sri Lanka. The portion of the tsunami that passed south of Sri Lanka would have missed southern India altogether if it had continued strictly westward, but the wave front refracted around to the north and hit the southern tip of India.[31] The tsunami waves were smaller in southern India, but 400 fishermen from villages on the coast were killed.[32] And once again, the animals seemed to know when to move away from the sea. Along India's Tamil Nadu coast, where thousands of people died, the Indo-Asian News service reported that many animals were found unharmed. Flamingoes that normally bred this time of year at the Point Calimere wildlife sanctuary were seen flying to higher ground before the tsunami came.[33]

An hour after the tsunami hit Sri Lanka, the American ambassador to that nation had called the Pacific Tsunami Warning Center to set up a notification system in case there were aftershocks, in which case he would pass on the information to Sri Lanka's prime minister. Except for Indonesia and Thailand, which were members of the Pacific Tsunami Warning System, the Warning Center had no contact phone numbers for countries in the Indian Ocean region that they could call to warn of the tsunami coming their way. But they kept trying to reach whomever they could in nations in the western half of the Indian Ocean. Being a Sunday did not make it any easier, nor did the fact that it was a holiday weekend in many of the countries.[34] The Warning Center spoke with the U.S. State Department Operations Center about the threat to the western Indian Ocean and the coast of east Africa.

The State Department eventually set up a conference call with U.S. embassies at Madagascar and Mauritius. Besides warning those two governments, they promised to warn Kenya, Somalia, the Seychelles, and whomever else they could.[35] Those at the Warning Center knew that if this earthquake and tsunami had occurred in the Pacific region, they could have saved thousands of lives. Now there would be some warnings for east Africa, but that did not help their sense of frustration.

About an hour after the tsunami hit Sri Lanka, news reports began to come out about India feeling mild earthquake tremors, but there was not yet word in the media about deaths or damage. Some news reports continued to wonder if the giant waves that hit Sri Lanka were related to the earthquake. Three hours and forty-one minutes after the earthquake, Reuters reported that the Indonesian government had put out an official statement that Aceh Province had been hit by a tsunami.

By this time, the tsunami had done most of its damage and killed 99.9 percent of the approximately 297,800 people who lost their lives. Over the next nine hours the tsunami would hit a dozen more countries in the Indian Ocean, killing another 313 people. Reuters finally put out a report that 100 people had been killed in Phuket, Thailand, two hours after it had happened and four hours and ten minutes after the earthquake. This was immediately followed by a report of 100 killed by a tsunami in Sri Lanka, and then by a report that rising waters were causing damage along India's coast. Thirty minutes later deaths were reported from tsunamis in Indonesia, Thailand, Sri Lanka, and India, but only very small numbers were reported. The first indications of a much larger disaster finally began to come out, but these reports emphasized Thailand and Sri Lanka, because of the many tourists there, rather than the much-harder-hit Aceh Province on Sumatra in Indonesia.[36]

The many hotels along the Thai and Sri Lankan beaches had internet and cell-phone access, which thousands of tourists from dozens of countries around the world used to send emails with pictures and videos. Some supplied information to the many blogs that were being set up all around the world. Videos from Thailand showed the receded sea with hundreds of feet of revealed sea bottom, and families with children walking out to look at it, overcome with curiosity. In the videos the sea did not look that dangerous at first. When the ominous-looking white strip of turbulent water along the entire horizon was in the distance, it might have been simply wind waves breaking over a long barrier reef. But its strangeness grew as it came closer and grew larger. And of course, by the time we saw these videos, we knew what had happened, and we knew that many of those curious tourists standing on the dry sea bottom would die a few minutes later. There were no videos of the 115-foot tsunami wave that struck the Aceh coast, a sight that would have looked much more frightening. Later videos of debris-filled water flowing through the streets of Banda Aceh were a somber sight but lacked the sheer terror that had been experienced at the Aceh coast.

Four hours and twenty-seven minutes after the earthquake, the Pacific Tsunami Warning Center received a revised estimate of the size of the earthquake from Harvard University's seismology center, which used a different seismic model. The new estimate of 8.9 magnitude meant that the earthquake had almost thirty times the destructive energy of the initially estimated magnitude of 8.0. The greater magnitude explained the great destruction caused by the tsunami throughout the Indian Ocean—this was a large quake. With more data the magnitude would later be raised to at least 9.0 and days later, by some calculations, to 9.2 or 9.3. A magnitude 9.3 would make the Sumatra earthquake the second-largest in modern history, after the 1960 earthquake along the coast of Chile, and 150 times stronger than the originally estimated 8.0. Whether second-largest or not, it had produced the most deadly tsunami in recorded history.[37]

The tsunami moved on, spreading out into the Indian Ocean. Originally, because of the north–south orientation of the ruptured fault, the tsunami waves had moved primarily westward across the Indian Ocean. But then ocean bathymetry began to play a more important role. The tsunami waves tended to follow submarine ridges, as we saw off Sri Lanka, but on a larger scale. Some of the wave energy still took the deep-water route, those waves arriving first at distant coasts, but the later-arriving waves that followed the ridges had greater heights.[38] The tsunami hit one island after another in the western Indian Ocean—the Maldives (about three hours after the earthquake), where eighty-two died; Diego Garcia (almost five hours);[39] Rodrigues (almost six hours); and Mauritius (more than six hours). The last three were protected by coral reefs, and no one died.[40] And then the tsunami reached the Seychelles (more than seven hours after the earthquake), which had received a warning (and passed it on to Kenya), but two people still died.[41] About the same time, the tsunami arrived at the coast of Oman in the Middle East (2,800 miles from the epicenter), where its strong currents caused a good deal of turmoil with many ships in the port of Salalah.[42] At Yemen's Socotra Island (seven and a half hours) there were twenty-foot run-ups and one death.[43] To the south on the coast of Somalia (seven hours), the highest tsunami run-ups in east Africa occurred, thirty feet.[44] Although now more than 3,000 miles from the epicenter, the tsunami swept away several Somalian fishing villages and killed at least 298 people. Tsunami waves had also gone up the west coast of India, finally reaching Okha near the Pakistan border (eight hours), but they were barely noticeable, having been worn down by a thousand miles of shallow water.[45]

Next the tsunami struck Kenya (nine hours), but the Kenyan government had received a warning. At the port of Mombasa an emergency plan was put into action, with the police, navy, and port authority coordinating with each other. It was actually an emergency plan for oil spills and fires, but it was all they had. The news media spread the word, local authorities were mobilized along the Kenyan coast, radio messages were sent to ships, and the police began evacuating the beaches. Since it was the Boxing Day holiday in Kenya, thousands were at Jomo Kenyatta Beach in Mombasa.[46] When many tourists did not respond to the evacuation order, armed riot squads were sent in to move the people away from the sea. When the tsunami hit, the first wave was not large, so the first thing that most people noticed was the receding sea and boats sitting on dry seabed. The currents were strong. Hippopotamuses in rivers were dragged five miles out to sea. So were three swimmers who had not heard the warnings and became the only deaths in Kenya. The water came in, in eight to ten rapid cycles, each at least ten minutes long, rose about five feet, and then receded.[47]

The tsunami hit Tanzania (almost ten hours after the earthquake), killing at least ten teenagers swimming near beaches in Dar es Salaam.[48] From there the tsunami moved on to Mozambique, the Comoros Islands, and South Africa, reaching Port Elizabeth (about twelve and a half hours) with wave heights of less than a foot. The tsunami traveled south of its epicenter to the coast of Australia and even farther away to the coast of Antarctica (eleven hours), but was barely noticed there.

Guided by deep-ocean ridges and continental shelves, the tsunami then left the Indian Ocean and headed up the Atlantic Ocean. Refraction bent the wave front around the Cape of Good Hope, and the tsunami traveled up the west coast of Africa along the continental shelf. By the time it reached Takoradi in Ghana (almost twenty-five hours after the earthquake), having traveled 9,000 miles, the waves still had a trough-to-crest height of one and a quarter feet.[49] Refraction focused the tsunami wave energy so that the Southwest Indian Ridge acted as a waveguide to the Mid-Atlantic Ridge, which acted as a waveguide up the center of the South Atlantic Ocean. About halfway up, due to a bend in the ridge, some of the tsunami waves propagated over to South America, reaching Brazil near São Paulo (twenty-one hours) with a height of four feet and Argentina near Buenos Aires (twenty-four hours) with a height of about a foot.[50] The tsunami waves continued north into the North Atlantic, registering on numerous tide gauges. After traveling over 14,000 miles from the epicenter in about twenty-nine hours, the highest tsunami wave heights in the North Atlantic occurred at Halifax, Canada (1.3 feet), and at Atlantic City, New Jersey (0.7 foot). Most likely the waves followed the northern part of the Mid-Atlantic Ridge until it curved to the east, at which point a portion of the wave headed straight for Halifax and Atlantic City.[51] Tsunami waves reached the Pacific coasts of North America by a different route. Most tide gauges along the coasts of Mexico, the United States, and Canada showed tsunami heights of only about 0.3 foot, after

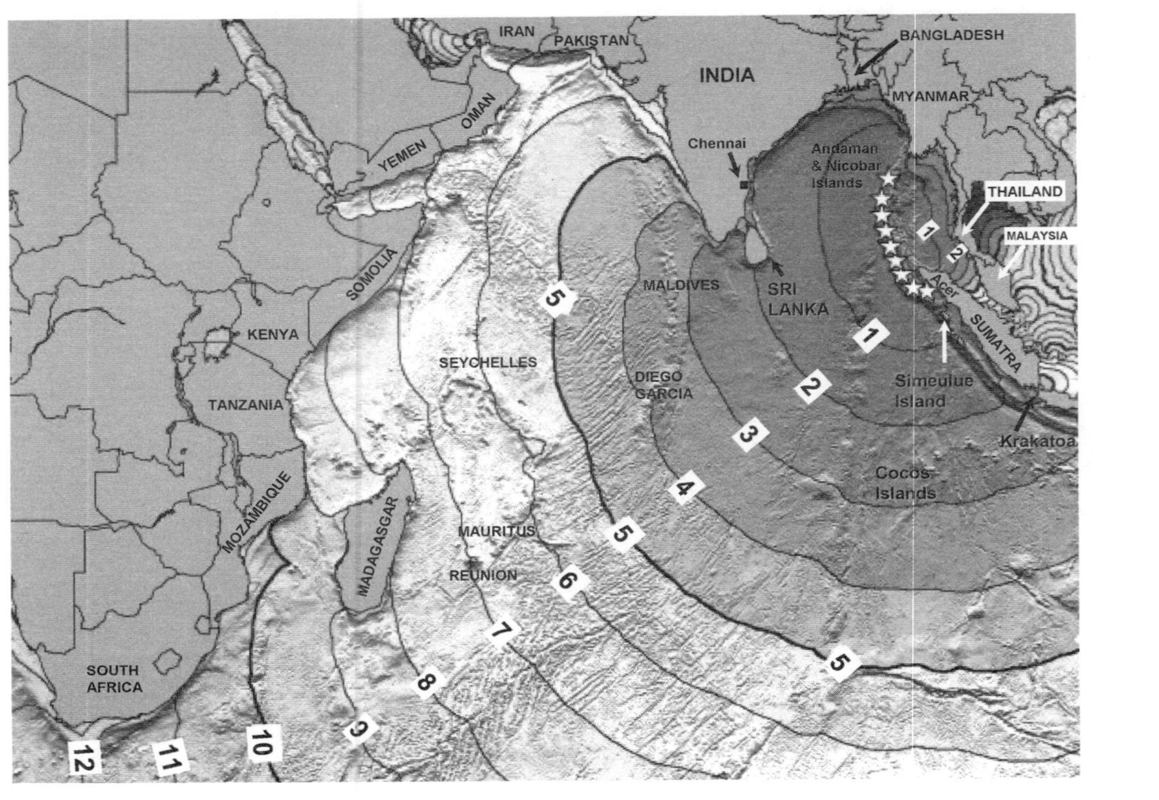

Figure 9.1 Map showing the arrival times of the leading wave of the December 26, 2004 tsunami in the Indian Ocean. The stars indicate the ruptured fault of the earthquake that generated the tsunamis. (*Modified version of model-produced map from NOAA's National Geophysical Data Center*)

traveling for twenty-eight hours over 15,000 miles from the epicenter. But several locations had larger heights, for example, 2.9 feet at Manzanillo, Mexico, and 2.0 feet at Crescent City, California.[52] These larger values were because those bays or harbors had natural periods of oscillation that closely matched the periods of the tsunami waves when they arrived.

The ten deaths in Tanzania were the last deaths caused by the tsunami, not including those still suffering in Aceh in Indonesia from infected debris-inflicted wounds or from having dirty water forced down their lungs and stomachs as the tsunami waves had overwhelmed them. The exact death toll will never be known. The numbers below are best estimates and are at least correct with respect to their relative sizes. The total number who died from the tsunami is probably at least 297,000. The best estimate for those who died in Aceh Province on Sumatra in Indonesia is 242,347, or 82 percent of the total death toll. Sri Lanka probably lost at least 30,957, or 10 percent (some estimates are 13,000 higher than that). India lost 16,389 (6 percent), and Thailand lost at least 7,593 (3 percent), many of the dead and presumed dead being foreigners. The remaining countries in the Indian Ocean that suffered casualties were Somalia (298), the Maldives (82), Malaysia (68), Myanmar (61), Tanzania (10), Kenya (3), the Seychelles (2), and Bangladesh (2). Because of global tourism, in this case primarily in Thailand and Sri Lanka, almost 4,000 tourists from forty countries died, most from Sweden, Germany, and the United Kingdom. The total number of deaths is staggering, especially when one is moved by the tragic stories that were reported and then realizes that there were 297,000 stories that were never heard.

Short of a climatic catastrophe, only a tsunami could produce such worldwide trauma. People around a large ocean had died almost simultaneously. Parents grieved for drowned children in Indonesia and Thailand and Sri Lanka and India and Somalia. Because of global tourism, the grieving spread to forty other countries around the world. Because of modern telecommunications, stories of the horrors and the sadness caused by the tsunami flashed around the world—not just about the nearly 300,000 who died but also about the millions left homeless, many sick or injured, all needing help. The world responded with an outpouring of aid and volunteers—food and medicine and clothes provided first, and later homes built by the thousands. It made a difference and was appreciated. On February 23, 2009, a Tsunami Museum was opened in Banda Aceh in the Aceh Thanks the World International Park, not far from the Dakota Seulawah RI-001 airplane that survived the tsunami.[53] The tsunami, and the cooperative spirit that followed during reconstruction, also seemed to have affected people's view of life and political conflict, and eight months after the tsunami a peace agreement was signed between the Acehnese rebels and the Indonesian government. In December 2006 regional Acehnese political parties won provincial elections, and a former rebel became the first democratically elected governor.[54] Unfortunately, in Sri Lanka the tsunami, and the way disaster relief was distributed afterward, reduced tension in its civil war only temporarily, and ended up intensifying ethnic differences. The civil war ended only after a complete military victory by the Sinhalese government in May 2009.[55]

As one might expect, tsunami awareness was now at a high level in all the countries that had been hit on December 26. People had been educated about tsunamis the hard way. This was demonstrated just three months later when another earthquake occurred, on March 28, 2005, at 11:09 P.M. local time, its epicenter only seventy miles southeast of the December 2004 epicenter, halfway between Simeulue and Nias Islands. The Pacific Tsunami Warning Center sent out a bulletin twenty minutes after the 8.6-magnitude earthquake, including a warning that "this earthquake has the potential to generate a widely destructive tsunami in the ocean or seas near the earthquake." It suggested that coasts within 600 miles of the epicenter should be evacuated. Governments around the Indian Ocean responded immediately, as did the people, some as soon as they felt the tremors, even though it was late at night. In Aceh, in Thailand and Malaysia, in Sri Lanka and India, hundreds of thousands of people moved inland, sometimes causing large traffic jams. A tsunami was generated, but a much smaller one than anticipated from the earthquake's magnitude. In spite of the false alarm the newspapers quoted most evacuees as saying, "Better safe than sorry."[56] Since there were no DART buoys or real-time tide gauges near the epicenter, there was no way to confirm whether a tsunami had occurred until one hit Simeulue. In this case, everyone was lucky, but it again pointed out that a real-time tsunami warning system needed to be installed in the Indian Ocean.

For decades there had been no serious earthquakes and no significant tsunamis, and now there had been two tsunamis within three months. A little more than a year later there was a third. On July 17, 2006, at 3:19 P.M. local time, an earthquake occurred about 135 miles off the south coast of the Indonesian island of Java (about 250 miles southeast of the remnants of Krakatoa). Though it had a smaller magnitude (7.7) than the 2005 earthquake, it produced a more destructive local tsunami that killed over 700 people and left thousands injured, leading some to guess that a submarine landslide may have been triggered. The Pacific Tsunami Warning Center had put out a bulletin seventeen minutes after the beginning of the earthquake stating that there was a possibility of a local tsunami, specifically mentioning Java and Australia's Christmas Island. This was about twenty minutes before the tsunami struck the Java coast, but officials in the Indonesian government did not pass on the warning to the people. The casualties might have been higher, but many people, now tsunami aware, responded to the earthquake tremors and the receding sea by running inland.[57]

A year later there was a fourth earthquake (four earthquakes in that area in less than three years). On September 12, 2007, at 6:10 P.M., there was a large earthquake (8.5 magnitude) off Sumatra, this one about 700 miles southeast of the 2004 epicenter and 270 miles northwest of Krakatoa. By this time there was an Indian Ocean Tsunami Warning System that had been implemented through international funding and cooperation. When the Pacific Tsunami Warning Center put out its first bulletin fourteen minutes after the earthquake, it looked quite different than bulletins for the earlier Indian Ocean earthquakes. It not only said that a tsunami watch was in effect for twenty-seven countries around the Indian Ocean but also gave predicted arrival times for a tsunami, if one was generated, for all twenty-seven countries. Now there were also instruments in the water to better detect a tsunami if one occurred, including two DART buoys and dozens of real-time tide gauges. The tsunami turned out to be small and did not cause much destruction, but the system had worked, even though it was not yet complete and there were other improvements to be made.[58]

By the fall of 2009 there were four DART buoys in the Indian Ocean, two contributed by the United States but operated by Thailand and Indonesia, and two operated by Australia. The Indian Ocean also had thirty-two more real-time tide gauges than there were in 2004 and one hundred more seismographs. Temporarily NOAA's Pacific Tsunami Warning Center and the Japanese Meteorological Agency have been providing bulletins with tsunami watches and warnings. However, twenty-six national warning centers have been established, and by 2011 there should be five Regional Tsunami Watch Providers for the Indian Ocean (Australia, India, Indonesia, Malaysia, and the Asian Disaster Preparedness Center in Bangkok, Thailand) that will take over the duties that U.S. and Japanese warning centers have been handling.[59]

The Pacific Tsunami Warning System has also been enhanced. As of March 28, 2008, there were thirty-nine U.S. DART buoys, most around the Pacific Ring of Fire but some in the Caribbean and off the Atlantic Coast of the United States, these last two sets of gauges used by the West Coast/Alaska Tsunami Warning Center. All the DART buoys are maintained and operated by NOAA's National Data Buoy Center. The Pacific Tsunami Warning Center put into operation in 2009 a new tsunami forecast system that was designed by NOAA's Pacific Marine Environmental Laboratory, which had designed the DART buoy. The forecast system makes use of improvements in hydrodynamic tsunami models based on lessons learned from the 2004 tsunami.[60] This improved system got its first test with a small tsunami generated by an earthquake near the Andaman Islands, north of the 2004 epicenter, on August 10, 2009, the DART buoys and tide gauges confirming the model's forecasts, although the tsunami was only four inches high. All these real-time tide gauges, DART buoys, and supporting instrumentation and the international cooperation on tsunami warnings are part of the Global Ocean Observing System (GOOS).

The Lisbon tsunami in 1755 resulted in extensive data that enabled the first scientific investigation of a tsunami and the earthquake that caused it. The Krakatoa tsunami in 1883, with the help of the new worldwide telegraphic network and newly developed scientific instruments, enabled the first global investigation of a tsunami and the volcanic eruption that caused it. The Indian Ocean tsunami in 2004 provided the most extensive scientific data set ever accumulated for a tsunami—from satellite and other global data-gathering instruments that had never before been in operation for such a large tsunami. These data enabled a much greater understanding of tsunamis and the earthquakes that cause them, and better understanding leads to better prediction. By December 2004 there had been numerous advances in satellite altimetry, the global positioning system (GPS), digital seismology, infrasound, gravity, and magnetic measurement from satellites in addition to direct tsunami measurement by tide gauges and DART buoys. And all had proved valuable in either detecting the tsunami or acquiring better information about the earthquake that generated it. This led to research efforts to see how these data sources, if received in real time, might improve tsunami prediction.

To better understand modern advances stimulated by the 2004 tsunami, we must first categorize tsunami prediction into two general types. The first type is prediction *after* a tsunami has been generated by an earthquake, volcanic eruption, or landslide. This involves predicting where the tsunami waves will travel, at what times they will hit particular coasts, and how large they will be when they hit those coasts. The second type is prediction *before* a tsunami is generated, which essentially means we are predicting that an earthquake, volcanic eruption, or landslide will occur and that it will generate a tsunami.

The first type of prediction begins as soon as there has been an earthquake, volcanic eruption, or landslide. We need first to determine whether a tsunami has been generated, and if so, then to determine how large the tsunami is and what directions it will be heading. There are two general approaches to doing this. The first approach is to detect the tsunami itself as rapidly as possible so as to give the earliest warning possible. A couple of minutes can be the difference between life and death for coasts close to the epicenter. At least seven different methods were used to detect the 2004 Sumatra tsunami (in data looked at after the fact). We have already seen that the first two methods, using DART buoys and tide gauges, can do this quite accurately, but even after installing hundreds of them, there might not be one near the location where a tsunami was generated. Thus researchers have looked at other instruments that might be used to detect a tsunami. We have already seen the importance of earthquake-produced seismic wave data in calculating the location and size of the earthquake. But tsunamis can also produce seismic waves when they reach the shallow water near a coast. Seismic stations installed on a coast would see low-frequency signals on their seismographs at the same time that a tide gauge would detect the tsunami. For a coastal location with a seismograph but no tide gauge, a low-frequency seismic signal arriving at the time predicted by a hydrodynamic tsunami model would indicate the likelihood of a tsunami.[61] We also saw earlier that infrasound (low-frequency acoustic waves) sensed by elephants in Thailand, Sri Lanka, and Indonesia was the reason they ran from the tsunami. Such sound waves can easily be sensed by real-time acoustic instruments.[62] GPS also appears to be able to detect a tsunami by detecting the tsunami's effects on the atmosphere above it.[63] Tsunamis propagating through the deep ocean also produce underwater sound waves, which can be detected with underwater hydrophones.[64] As we saw earlier, altimeters on satellites directly detected the 2004 tsunami but only when the satellites happened to be over the Indian Ocean.[65] Although very useful for validating a hydrodynamic tsunami model, altimeters on satellites would probably be useful for rapid verification of a tsunami only if future satellites could provide better temporal and spatial coverage.

The second approach to determining that a tsunami has been generated does not involve detection of the actual tsunami, but instead involves measurement of phenomena that usually occur when a tsunami is generated. There are two methods. The first uses records of the seismic waves produced by the earthquake, those data being received at seismographs at various locations. As we have seen, the location of the epicenter of the earthquake can be determined by seismic data from three seismographs. The closer those three seismographs are to the epicenter, the faster the Pacific Tsunami Warning Center can determine its location. So the more seismographs installed the better. Because of the increased number of seismographs installed in Indonesia since 2004, it only took four minutes to determine the location of the 7.3 submarine earthquake that occurred off northwestern Sumatra on May 9, 2010. The Warning Center sent out a Tsunami Bulletin only eight minutes after the earthquake.[66] But scientists have been trying to go beyond just calculating the location of the earthquake with these data. They are trying to see if they can use seismic data to quickly determine whether an earthquake is likely to have generated a tsunami (namely, been tsunamigenic). We have mentioned four types of seismic waves (P, S, Rayleigh, and Love waves), but there are others (T, W, etc.), and these are being analyzed in different ways to see which ones might indicate a high chance that a tsunami was generated.[67] The second method uses GPS data. The primary use of GPS data is to determine the accurate position of a GPS receiver. Using installed high-precision GPS stations, one can measure how the land moved, vertically and horizontally, immediately after an earthquake and even how the seafloor moved after a submarine earthquake. Not only does this allow one to determine whether there was enough vertical movement to generate a tsunami, but the results of the analysis can also be used to drive a hydrodynamic tsunami model.[68] To be useful, of course, these methods must make the determination very rapidly that a tsunami was generated.

Once we know (or believe there is a high probability) that a tsunami has been generated, the next step is to determine where the tsunami waves will propagate. Data analyses can be used to produce initial conditions for a hydrodynamic tsunami model. The model will then quite accurately predict where the tsunami waves will travel, what coasts they will hit, and how large they will be when they get there.[69] A further step in this tsunami modeling is to use high-resolution bathymetry along coasts to determine which locations are likely to amplify the tsunami, for example, near the shoreward end of a submarine ridge or onshore next to a wide continental shelf.

The second general type of tsunami prediction involves predicting a tsunami before it is generated—in other words, predicting the earthquake, volcanic eruption, or landslide that will generate the tsunami. This is extremely difficult and, in fact, presently seems impossible in most cases. But it is something worth pursuing, because the greatest number of casualties are often along coasts so close to the earthquake, volcanic eruption, or landslide that the tsunamis hit them within a few minutes after being generated—well before any of the above-mentioned approaches could provide a warning. There is actually some hope for predicting two of these three tsunami-generating mechanisms. A major volcanic eruption is often preceded by smaller precursory volcanic activity, which can serve as a warning for people to stay away from the shore. Before landslides occur, above or below water, one can measure the slow accumulation of material and thus recognize an increasing instability (as sediment is added to a submarine slump, or as the steepness of the slopes of a volcano increases as it grows), but there is presently no way to predict when it will reach a critical point of instability so that a landslide will take place. However, real-time instruments installed on the accumulated material, such as tiltmeters, can detect a sudden movement and warn that a landslide is taking place.[70] Earthquakes are much more difficult to predict. Rarely is there a small precursory earthquake that might provide a warning of a big one to come. Geologists have been looking at a variety of data types to see if there might be any precursory signs that would indicate that an earthquake was imminent (such as changes in the speed of certain seismic waves, changes in pressure, or changes in electromagnetic properties), so far with no luck, but some of the researchers are optimistic.[71]

No warning system presently available would have been fast enough to warn the people along the northwest coast of Aceh, where the tsunami hit eighteen minutes after the ground began shaking and thirteen minutes after the sea began to recede. Only tsunami awareness through education would have helped. If the people along the beaches of Lhok Nga Bay had started running inland the moment they saw the sea recede, or even better, when they felt the first earth tremors lasting longer than thirty seconds, they might have survived, even considering the incredible size of the ten-story second wave that demolished their villages. They certainly would have survived in other countries. We have seen several cases in which people recognized the natural warning signs of a tsunami, ran inland or to higher floors of strong buildings, and survived. If there had been more tsunami awareness, tens of thousands of lives in the rest of Indonesia and in eleven other countries might not have been lost.

Tsunami awareness has, of course, been high in the Indian Ocean nations since December 26, 2004. There have been educational materials for schools, and signs have been put up along beaches (like the tsunami-alert signs that have been in Hawaii and Japan for decades). But the real source of tsunami awareness has been the memory of the almost 300,000 people who died. Unfortunately, we have seen examples where such awareness slowly diminishes with time. It diminishes even faster if there have been a few tsunami warnings that turned out to be false alarms. Not only do people tend to ignore tsunami warnings if there are too many false alarms, but they can also become quite angry about them. On June 7, 2007, in the hardest-hit part of Aceh Province, villagers near Lhok Nga threw rocks at a tsunami siren and cut its power after it accidentally went off and caused panic. This came three days after several sirens went off by accident in Banda Aceh, sending thousands of people fleeing to high ground and causing huge traffic jams. Both these false alarms were due to mechanical problems and not to inaccurate predictions, but the effect on the people's confidence is the same.[72]

Tsunami awareness does not help, of course, if the tsunami surprises people at night when they are sleeping. There is no doubt as to the critical need for a worldwide tsunami warning system, but a great many lives might also be saved by a small, much more modest technological device, one installed locally by a coastal community worried about being hit someday by a tsunami. For lack of a better name, we might call it a *dry-sea-bottom sensor.* In its simplest form it would be a pressure sensor placed on the sea bottom far enough from the shore so that the sea bottom would never be dry except when the sea recedes prior to the arrival of a tsunami. This pressure sensor would send a signal to a siren onshore when the bottom becomes dry (or when the water level quickly drops lower than the lowest tide ever recorded, whether the bottom becomes dry or not). Essentially we are using an instrument to recognize a natural warning sign of a tsunami, in case the people don't see it (it could be at night) or in case not enough people are warned fast enough by the few people who do see it. Such a system would provide a warning for a locally generated tsunami not yet detected or not predicted by an international warning system. It does not rely on tsunami-aware people to get everyone to move inland. It provides a widespread warning with the inherent drama of the siren to make people respond. If there is the possibility of a crest-first tsunami, a slightly more sophisticated device could be programmed to recognize a rapid rise in water level (after averaging out any wind waves), much like a DART buoy does. A DART buoy offshore would, in fact, be the next option, albeit a very expensive one. One could also install a seismic device on the land that would sound a siren if earthquake tremors lasted longer than thirty seconds. With respect to minimum expense and minimum maintenance, a dry-sea-bottom sensor attached to a heavy concrete slab on the sea bottom, with acoustic and radio communication to a siren nearby onshore, would seem to be a simple device with which concerned coastal communities could do something to protect themselves.

If dry-sea-bottom sensors had been operating off the beaches of Samoa, American Samoa, and Tonga on September 29, 2009, when a tsunami struck each of those islands only minutes after a nearby submarine earthquake occurred, most of the approximately 190 lives lost might have been saved. Entire villages were swept away. The warning from the Pacific Tsunami Warning Center was issued sixteen minutes after the earthquake—as fast as humanly possible in terms of analyzing the incoming seismic data and running prediction models. But the earthquake was too close to the Samoan Islands, and by the time the warning was sent out, the tsunami had already swept in and destroyed people's lives. The 8.3 earthquake occurred at 6:48 A.M. local time, only 100 miles south of Samoa, 120 miles southwest of American Samoa, and 430 miles north of Tonga. The residents of Samoa felt the earthquake tremors. Many realized that these tremors lasted longer than usual, and they immediately ran inland and up the hills. But when the tremors stopped, many others simply went on about their business. Some waited to hear if there would be a tsunami warning, but not hearing one, they thought everything was fine. About five minutes later some people noticed the sea receding, and many of them also immediately ran inland. But others were still in bed and knew none of this until the first of four or five huge waves, estimated at fifteen to twenty feet, swept through their homes. The southern coast of Samoa was damaged the most. On the northern coast near Apia, sirens sounded after the warning and had some benefit. On American Samoa, because of the length of the earthquake tremors, the local NOAA National Weather Service office did not wait for an official warning from the Pacific Tsunami Warning Center but put out its own warning alert. The alert's effectiveness was, however, limited by the lack of an island-wide siren system, but those who heard the alert from the Pago Pago radio station responded well, in part because of earlier training and education programs run there by the National Weather Service with the Pacific Tsunami Warning

Center and because of signs put up and evacuation drills conducted by the government. Later, a survey of the island measured tsunami run-up heights ranging from six to thirty-nine feet. The casualties in the Samoan Islands in 2009 may have been on a much smaller scale than the casualties in Indonesia in 2004, but the individual stories were very similar and just as tragic—children walking along the beach to school swept away by the sea, a husband and wife clinging to a palm tree torn apart by the rushing water and the wife drowning, children ripped from their mothers' hands and never seen again—stories that should not still be happening.[73] An awareness of the signs of a tsunami saved some lives, but for those who had recognized the signs, there had been no easy way to warn everyone. Even a loud bell can make a difference, as was shown by a twelve-year-old girl on Robinson Crusoe Island off Chile who saved 700 people when a large tsunami generated by the Chilean earthquake of February 27, 2010, struck her village.[74]

When a submarine earthquake occurs, the sea responds immediately. Tsunamis can strike a nearby coast within minutes and can cross an ocean to strike another nation's coast in a few hours. Being unable to make a prediction until after the earthquake has occurred means that the prediction must be produced extremely rapidly to be of any use in saving lives. As we saw in earlier chapters, other lifesaving predictions, for example, predictions of storm surges or large wind waves, have a slightly longer time frame (up to a couple of days). Other marine predictions have even longer time frames (months to years or even longer), because they deal with global climate phenomena such as El Niño and global warming, which the sea plays a major role in controlling. Though perhaps lacking the drama of more imminent dangers (such as tsunamis), such longer-term predictions on how the sea will affect our weather and our climate can mean saving even more lives, as we will see in the next chapter.

CHAPTER 10

Predicting the Future—and Saving Lives

El Niño, Climate Change, and the Global Ocean Observing System

In previous chapters we have seen the importance of marine predictions in saving lives. Our emphasis has been on predicting an impending danger such as a tsunami, a storm surge, or a rogue wave—a danger that could arrive in a day, or in a few hours, or in the case of a tsunami generated by a nearby submarine earthquake, in less than thirty minutes. There are, however, other predictions that can be for much further into the future—predictions of catastrophes that result from environmental changes that develop over years, decades, centuries, or even longer, changes also produced or greatly influenced by the power of the sea. While perhaps lacking the drama of a huge tsunami wiping an entire town out of existence, or a storm surge flooding a city, or a rogue wave sinking an oil tanker, these more distant calamities can also lead to loss of life and cause serious economic impacts. The two most important examples are El Niño, which occurs every two to seven years, and long-term climate change, especially with respect to the effects of global warming where the most serious effects could possibly show up in only a few decades. Both are global phenomena with worldwide impacts. Both demonstrate the immense power of the sea on a planetary scale and its critical role in driving weather patterns around the world and in modifying our climate. Being able to predict the onset and effects of El Niño and being able to predict the effects of global warming can be of great economic benefit and can save lives by allowing us to prepare for future dangerous conditions. In the past, strong El Niños and dramatic changes in climate have caused millions of deaths.

Two of the largest El Niños in recorded history took place at the end of the nineteenth century, before the name *El Niño* was known to anyone other than Peruvian fishermen. Those fishermen first used that name for the annual warming of the ocean waters near Peru that happened every year near Christmas, *El Niño* meaning "the little boy" and referring to the Christ child. But every few years that ocean warming would be especially severe, and eventually the name came to be used only for those occasional extreme warming events. This was long before the El Niño off Peru was recognized as being an early stage of a global climate phenomenon that involves both the ocean and the atmosphere and whose effects are felt all over the world. This phenomenon was later called El Niño–Southern Oscillation (ENSO). Southern Oscillation was the name given to the very slow seesawing of atmospheric pressure over the Pacific Ocean (meaning that pressure is high over South America when it is low over Indonesia, and vice versa). Much later it was discovered that the Southern Oscillation was connected to the very slow rising and falling of ocean water temperatures across the tropical Pacific (El Niño being the phase when there are warm waters in the eastern Pacific near Peru).

The El Niño that began with much warmer waters off Peru in December of 1876 lasted for two years and had deadly impacts around the world.[1] The primary commercial fish in Peruvian waters was the cold-water anchovy, which was so plentiful that fishermen could literally scoop them up with their hands and fishing boats could forgo nets and use suction hoses. Hundreds of thousands of tons of anchovy were exported to Europe. When the warm waters of El Niño came, the anchovy disappeared and the Peruvian fishing industry collapsed. The seemingly limitless anchovy population had also supported a seabird population so huge that their droppings (guano) grew into small mountains on islands off the coast of Peru. At that time guano was the most prized fertilizer in the world, and hundreds of thousands of tons were exported to France, Great Britain, and other nations. When the anchovy disappeared, the so-called guano birds, primarily the Guanay Cormorant, died, and the droppings stopped. The mountains of guano shrank, not only because guano exports continued while not being replenished but also because they were washed away by the heavy rains brought by El Niño. One reason the mountains of guano had built up over the years was that Peru is normally one of the driest places on Earth, a desert by the sea, with the only freshwater coming down rivers from the Andes. But during an El Niño, torrential rains came, causing floods that killed thousands of people in Peru and also washing the guano from the islands into the sea. So not only did the fishing industry collapse, but the guano fertilizer business collapsed as well. The Peruvian government, which depended on revenues from these two industries, faltered (and two years later lost a war to Chile).[2]

But much more terrible effects of this El Niño were still to come. Whereas El Niño brought heavy rains to a normally dry Peru, farther to the west, across the Pacific and Indian Oceans, it brought drought to lands that were normally wet and green. El Niño stopped the rain-bringing monsoons, and in 1877–1879 the resulting droughts caused famine and disease on an incredible scale. The death tolls were staggering—ten million dead in India, between ten million and twenty million dead in China, another million in Brazil, and more dead in southern Africa, the Dutch East Indies (now Indonesia), Vietnam, the Philippines, and Korea.[3] Twenty years later, in 1899–1900, a second very strong El Niño again stopped the monsoons and caused droughts in India and China and Africa, killing another eight million in India, another ten million in China, and additional deaths in Brazil and Africa.[4]

When these two catastrophic famines occurred, no one had ever heard of El Niño, other than Peruvians for their own local event. British scientists in India began trying to understand why the monsoon had failed, and they searched for ways to predict when that would happen again. In the process the Southern Oscillation was discovered, but it would be another sixty-five years before anyone would realize the critical role that the ocean plays in this global phenomenon. Once ENSO was understood, however, historians and paleoclimatologists discovered evidence of numerous other El Niño impacts in the past. Even ancient Egypt had been affected by El Niños. Historians discovered that the pharoahs had their engineers regularly measure the flow of the Nile with instruments now referred to as *Nilometers,* because the annual flooding of the Nile was so important to Egypt, providing the water necessary for irrigation and the fertile sediments that caused crops to flourish. The flow of the Nile depended on rain falling in Ethiopia, which varied dramatically with the monsoons. When a significant El Niño occurred, the Nile would have a low flow, as shown by Nilometer records, and Egypt would experience drought and deaths and political unrest.[5] Even in the twentieth century, political events in Africa were affected by El Niño. The 1972–1973 El Niño caused a severe drought in Ethiopia, followed by famine and turmoil, which contributed to the overthrow of Emperor Haile Selassie. Ten years later, another drought caused by the 1982–1983 El Niño led to the overthrow of the man who had deposed Selassie. During this period over a million people starved to death or were killed during the resulting political chaos. These are but a few historical examples of past effects of El Niño. Scientists have tracked El Niños back 130,000 years using paleoclimate records such as tree rings and corals and cores from the sea bottom, even being able to tell that there were fewer El Niños during the Little Ice Age (approximately AD 1300–1850) and more during the Medieval Warm Period (approximately AD 900–1300).

El Niño is the most famous part of the ENSO cycle, but there is also a period with colder than average water temperatures along the Pacific Coast of South America, called *La Niña* ("the little girl"). La Niña's effects on weather around the world are generally the opposite of the effects of El Niño. The strong La Niña that began in December 2007 and continued well into 2008 contributed to the 2007–2008 winter being the coldest winter in the United States since 2001, when the last La Niña occurred.[6] This same winter was the worst in fifty years in China's eastern, southern, and central regions and included a heavy snowfall that stranded 5.8 million passengers throughout China's railway network a week before the Chinese New Year on February 7.

An accurate prediction of drought and famine due to El Niño could have saved countless lives in the past because stockpiles of food could have been made available to help the starving.[7] Today, with global relief organizations and countries willing to help when there is a natural catastrophe (as in the worldwide response to the 2004 tsunami), a similar El Niño–caused drought would probably not lead to millions of deaths, and so the prediction of El Niño would not have as dramatic a benefit as it would have had at the end of the nineteenth century. But knowing that a drought is coming can lead to preparations that would be economically beneficial, including farmers changing to more drought-tolerant crops or increasing crops in other regions not affected by the drought. If El Niño predictions are good enough to predict the locations of future floods, preparation for those floods can save lives. The largest El Niño in the twentieth century, in 1997–1998, had many effects around the world, such as torrential rains in California that caused widely reported mudslides, in which homes slid into the sea. The running joke on late-night TV in 1998 was to blame everything on El Niño. Effects of this El Niño, including the heavy rains across California, were correctly predicted by NOAA's National Centers for Environmental Prediction six months in advance.[8] As a result, overall property losses were a billion dollars less than what they had been for the previous large El Niño in 1982–1983.[9] Today fairly accurate prediction of an El Niño six months to a year in advance has become possible using computer models that ingest millions of gigabytes of real-time data from instruments on buoys deployed across the Pacific Ocean.[10] But even with this system we still cannot always correctly predict the specific effects of an El Niño for particular regions.

Climate, of course, has been changing on time scales much longer than that of ENSO, with the world going into and out of ice ages every 80,000 to 120,000 years, and with other climatic variations on shorter time scales, all of which involved changes in the ocean and its interaction with the atmosphere. The world has been warming since the last glacial maximum, about 20,000 years ago. Along the way there have been notable climatic oscillations in some regions, such as in Europe, where the Medieval Warm Period had temperatures almost as high as they are today and the Little Ice Age had temperatures generally lower, although they fluctuated wildly. Near the end of the last major ice age, when mile-thick ice sheets covered more than a third of the Earth's landmass, sea level was 300 feet lower than it is today, exposing major areas of continental shelf and allowing humans to walk from Asia to North America across what is now the Bering Strait. As the climate warmed, melting glacial ice sheets that had reached to the sea pulled back, exposing a coastal route by which humans were able to migrate down the west coasts of North America and South America. The warming climate also produced changes in the Middle East that led to the beginnings of civilization. Later, during the Medieval Warm Period, centuries of drought caused the Mayan empire in Central America to disintegrate. Farther north during this warm period, the Vikings were able to explore islands and coasts throughout the North Atlantic and colonize Greenland, which at that time was indeed green because of the warmer temperatures. The British successfully cultivated vineyards in England. During the Little Ice Age that followed, however, these vineyards disappeared, and the Vikings had to abandon their North Atlantic colonies. The Baltic Sea froze over, and even Mediterranean countries had very cold winters. With shortened growing seasons there were frequent crop failures, followed by famines and many deaths.

Today, of course, the primary climatic concern is *global warming.* What global warming really means in this case is the *increased* warming of the globe that has occurred in the last century or two, believed by most scientists to be caused by an enhanced greenhouse effect due to the increased carbon dioxide in the atmosphere during that time period.[11] It is critically important to be able to accurately predict the Earth's climatic response to this higher level of carbon dioxide, which has resulted from two actions by humans. First, enormous amounts of carbon dioxide have entered the atmosphere through two centuries of burning fossil fuels. Second, the cutting down of forests over many centuries has reduced nature's ability to take carbon dioxide out of the atmosphere. Since the resulting increased carbon dioxide is a greenhouse gas, which traps heat in the atmosphere before it can escape to space, the result has been additional warming beyond the natural warming that would have occurred since the last ice age without the additional quantity of greenhouse gases added by humankind. How much of the increased warming over the last century is due to this enhanced greenhouse effect has been debated. The Earth also responds (*chaotically,* in mathematical terms) to particular cycles involving orbital motions of the Earth and the sun, so it is also important to be able to predict the Earth's response to these astronomical cycles.[12]

The ocean plays a crucial role in climate change and the effects of global warming. It is the greatest solar collector on Earth (since it covers 70 percent of the Earth's surface), and it stores heat 4,000 times more efficiently than the atmosphere. It has a major role in how carbon dioxide affects climate, since it stores 500 times as much carbon as the atmosphere does and it absorbs up to half the carbon dioxide produced by burning fossil fuels. Its phytoplankton absorbs as much carbon dioxide as trees, grasses, and other plants on land do. And as an important side effect, a gas from these phytoplankton is responsible for a considerable proportion of aerosols in the atmosphere, which become condensation nuclei for cloud formation, an important aspect of climate change that is still not well understood.[13] The oceans contain approximately 97 percent of the Earth's water, although that amount varies as climate changes. During ice ages water from the sea goes into snow that accumulates as mile-high glaciers over vast areas of land, with the result that sea level drops. During the warmer interglacial periods, water that has been locked up in glaciers on land returns to the sea, and sea level rises.

Ocean currents play a major role in transporting heat from the hot equatorial regions of the Earth to the cold polar regions. The Gulf Stream, though produced primarily by winds blowing over the Atlantic, also plays a major role in the density currents involved in the Atlantic's thermohaline circulation. The Gulf Stream moves warm salty water north along the surface of the Atlantic Ocean. When it reaches the North Atlantic, the warm salty water cools, releasing heat into the northern atmosphere, with the resulting colder salty water then sinking to great depths. This deep water then moves very slowly southward, eventually passing through the South Atlantic and the Indian Ocean on the way to the Pacific Ocean, where it finally rises again to the surface and heads back toward the Atlantic. This process is why the thermohaline circulation is also called the conveyor belt. The whole trip can take between six hundred and one thousand years. Large changes in the strength of the thermohaline circulation appear to correspond to beginnings and endings of glacial periods, while smaller changes seem to correspond to other smaller climate oscillations. Past dramatic weakenings of the thermohaline circulation may have been caused by sudden increases in freshwater in the North Atlantic, while various processes that increase salt transport from the south might have strengthened it.

It wasn't until the last fifty years that much progress was made in understanding and predicting El Niños or in understanding some aspects of long-term climate change. In both cases, this was possible only after we were able to acquire extensive oceanographic and meteorological observations from around the world and feed these data into large sophisticated dynamic models run on supercomputers. The first understanding of one aspect of the ENSO cycle goes back to 1904 when Sir Gilbert Walker, the British head of the Indian Meteorological Department, analyzed large amounts of data from weather stations around the world and recognized for the first time the seesawing of atmospheric pressure over the southern Pacific Ocean (which he called the Southern Oscillation).[14] But it was not until 1957 that the Southern Oscillation's connection to El Niño was recognized by Jacob Bjerknes, a Norwegian meteorologist at the University of California at Los Angeles working for the Inter-American Tropical Tuna Commission. Bjerknes analyzed huge amounts of oceanographic and meteorological data from the Pacific obtained by sixty-seven nations working together during the International Geophysical Year 1957–1958. He was able to determine that large water temperature anomalies in the tropical Pacific (El Niño) were connected to large atmospheric pressure anomalies over the Pacific (Southern Oscillation), and he proposed the first theory to explain how this ocean-atmosphere interaction worked.[15] In the 1970s, Klaus Wyrtki, an oceanographer at the University of Hawaii, took the next step in understanding El Niño when he analyzed sea level data from tide gauges around the Pacific.[16] When ocean waters warm, sea level rises because of thermal expansion. When ocean waters cool, sea level falls. Mapping his data, Wyrtki could show that a huge pool of warm water moved from the western Pacific to the eastern Pacific and then moved north to California and south to Chile.[17]

The scientific story of El Niño goes on to involve large oceanographic observation programs, such as TOGA (Tropical Ocean Global Atmosphere), which by 1994 had seventy instrumented buoys deployed across the tropical Pacific.[18] Three years later, data from these buoys allowed the first complete description of an El Niño event and made prediction possible. Eventually, specially developed numerical oceanographic models coupled to weather models would become capable of predicting the onset of an El Niño at least six months in advance. These models can be traced back to the dynamic model first derived by Laplace for use in understanding and predicting the tide (discussed in Chapter 1). The oceanographic models added equations for salinity and heat. The weather models added equations for heat, humidity, compressibility, and other atmospheric effects. To successfully predict El Niño, an ocean model and an atmospheric model had to be coupled, which essentially means they become one model. For other ocean phenomena one could run an ocean model with energy from the atmosphere (in the form of wind stress on the ocean's surface, or heat to warm the ocean's surface, or rain to add freshwater at the surface) being input as what are called *boundary conditions* based on data. Likewise, weather models could be run with boundary conditions that included the heat from the ocean's surface or the amount of water vapor entering the atmosphere from the ocean. But for phenomena where the ocean and the atmosphere greatly affect each other, so that both are changing significantly, then the two models must be coupled. Billions of data points from across the Pacific Ocean and around the world are used to initialize the model and run it for a while, before it is run into the future (without any data to keep it on track) to make a forecast.

It is still not clear what triggers the beginning of an El Niño, or what causes it to end. During non–El Niño conditions, with low atmospheric pressure over the western Pacific and high pressure over the eastern Pacific, the trade winds blow toward the west along the equator. This pushes warmer surface waters into the western Pacific, where they accumulate as an enormous warm pool the size of Canada, causing a great deal of evaporation that turns into heavy rains over Southeast Asia and the Indian Ocean. At the same time, these westward trade winds are pushing water away from the coast of South America, which causes the upwelling of cold nutrient-rich deep waters into the sunlit surface waters of the eastern Pacific, which in turn causes phytoplankton blooms on which the anchovy feed, as well as dry conditions on land. During El Niño, with low pressure over the eastern Pacific and high pressure over the western Pacific (the other half of the Southern Oscillation), the trade winds weaken or reverse, allowing the waters in the western Pacific to shift to the east. This essentially reverses the situation, so that heavy rains and floods now occur in eastern Pacific areas and droughts occur in western Pacific and Indian Ocean areas. In the eastern Pacific the upwelling stops, the waters warm and have less nutrients, the phytoplankton do not bloom, the anchovy are not in abundance, and the guano birds die. For a large enough event, the warm waters move north to California and south to Chile. If, when conditions reverse again and the trade winds start up, the situation becomes very strong in that direction, so that it moves past an average situation, then the waters off South America become extra cold, which is the La Niña situation.

At the start of an El Niño, the ocean affects the atmosphere and the atmosphere affects the ocean, both in the same direction. The trade winds blowing toward the west are slowed by the warm water in the east and reverse their direction toward the east, but this shift in wind direction then further pushes the warm water to the east. This positive feedback mechanism is possible only along the equator when the water movement is toward the east, because here the effect of the Earth's rotation is to keep the water movement in alignment with the winds.[19] But it is not clear whether the warm water or the trade winds make the first move, and why. Even without understanding the initial triggering mechanism, useful short-term predictions have been possible when there are enough real-time data from the buoys across the Pacific.

Long term climate predictions also require coupled ocean and atmosphere models, but the complexity of the models and the difficulty in making accurate predictions are much greater than for ENSO. There are few phenomena more complex than the Earth's climate system, and the greenhouse effect of anthropogenic carbon dioxide is only part of the story. Although the ocean and the atmosphere are most important, all parts of the Earth make a contribution, including the biosphere (the biological effect of all living things on Earth; we have already mentioned forests and phytoplankton), the cryosphere (all land-based ice, such as the ice sheets on Greenland and Antarctica), and the geosphere (the solid earth, including land and tectonic plates, whose effects include volcanic eruptions that put light-reflecting particles into the atmosphere).[20] The biosphere obviously plays a critical role in the carbon cycle and in global warming by carbon dioxide and other greenhouse gases such as methane. Another critical effect is the variation in solar radiation (the ultimate driver of the Earth's climate) over time, especially in terms of the variation in the amount of light that hits particular parts of the Earth. For example, this varies with cycles of 20,000, 40,000, and 100,000 years due to, respectively, changes in the orientation of the Earth relative to the sun, changes in the tilt of the Earth's axis, and changes in the shape of its orbit around the sun. These three cycles appear to affect the timing of ice ages (and can be identified in paleo sea level records), although it is not understood how such seemingly small effects can have such a profound impact. It probably involves positive feedback mechanisms involving the oceans.

We usually test a model and the theories behind it by seeing whether predictions using that model are accurate compared with data. ENSO events occur every two to seven years, and so after each event we can see how well a model did in predicting it. This is not possible when predicting climate change a century or more into the future. The best we can do to test a climate model is to see if it can predict past climate changes, but then we have a problem with having adequate data to compare against the model predictions and adequate data to run the model. Instrument records go back only 50 to 150 years, depending on the meteorological or oceanographic parameters being measured. Beyond that we must develop proxy data sets (using tree rings, corals, ice cores, cores from the bottom of the ocean and lakes, etc.), from which we estimate the air and water temperatures, sea level heights, carbon dioxide, and other indicators of past climate. These can go back thousands of years (and even hundreds of thousands of years), but being proxies, they are indirect measurements based on things like ratios of particular isotopes of an element, with enough uncertainties that sometimes particular questions cannot be answered. Also, because of the long time periods that must be computed, climate models must be run with coarser time and space resolution than ENSO models, and so all physical processes might not be handled as accurately as we would like.

The most sophisticated climate models have been run with and without including the increasing carbon dioxide levels in the global atmosphere over the last century. Only the models that include the increase in carbon dioxide can reproduce the increase in the global air temperatures measured over that same period. Without the increased carbon dioxide levels, global air temperature still increases as it has since the last ice age ended, but not nearly as much. But while all models generally agree about the rise in temperature averaged over the globe, there is less agreement among climate models on the exact changes that will occur at specific regions of the world. Thus, there may be less certainty with respect to some of the predicted regional effects, many of which would be detrimental to human health if they occur. Of course, floods, droughts, and famines have happened without global warming, as we saw at the end of the nineteenth century when two strong El Niños stopped the monsoons and millions died from the resulting famines. While droughts, heat waves, and floods on an unprecedented scale would certainly justify the concern over global warming, until the models agree and can prove that they will take place on a worldwide scale, the skeptics will use such regional uncertainty as an excuse to ignore global warming. But there are other possible calamitous global effects.

A global effect that would certainly be calamitous if it occurs is a significant and rapid rise in global sea level as a result of global warming, since roughly half the world's population lives close to the coast.[21] As we have mentioned, during the last ice age sea level was 300 feet lower than it is today. Continental shelves now underwater were land 20,000 years ago. (The Port of New York and New Jersey, if it had existed then, would have been more than a hundred miles inland from the Atlantic Ocean.) By roughly 8,000 years ago, most of the 300-foot rise in sea level had occurred due to the melting and retreating of the ice sheets. In more recent centuries sea level rise has been primarily due to the thermal expansion of the upper water column of the ocean as it warmed. Over the last century sea level only rose approximately two-thirds of a foot. However, sea level will rise faster if the Greenland or Antarctic ice sheets began to melt more quickly. The ice sheet on Greenland if completely melted would produce roughly a 24-foot rise in global sea level, while completely melted Antarctic ice sheets would produce a 200-foot rise. It would probably take centuries for complete melting, but it is not necessary for entire ice sheets to melt to produce a dramatic rise in sea level. The ice sheets merely need to move faster toward the ocean and break off into huge icebergs, since icebergs raise sea level the same amount as their melted water would. A critical question is how quickly are these ice sheets moving toward the sea, and could this process speed up? Melted water beneath the ice sheets serves as a lubricant for a more rapid glacial flow toward the sea.[22] Based on the latest predictions for the movements of ice sheets on Greenland and Antarctica, estimates for the amount of global sea level rise by the year 2100 range from two and a half to six feet, a rise capable of producing tremendous economic impacts and loss of huge amounts of coastal land for many countries.[23] Satellite gravity data indicate that there was a doubling of ice-mass loss from Greenland and Antarctica from 2002 through 2009.[24] Findings from coral paleoclimate studies of the last interglacial period (about 130,000 years ago), when sea level was higher than it is today but global air temperature was only a little warmer, may indicate that there are mechanisms for sea level to rise higher than was previously thought possible.[25]

Global warming could also possibly affect the one-sixth of the world's population that relies on mountain glaciers or seasonal snow packs for their water supply. As the world gets warmer, the melting of winter snow would occur earlier in the year so that maximum river flows would occur in late winter or early spring rather than in summer or autumn, when the demand for water is highest for agriculture, and unless there are sufficient storage capabilities, that water will be lost to the sea.[26] Another likely global effect of increased carbon dioxide levels would be *ocean acidification,* whose impact on phytoplankton and other global ecosystems is not yet well understood but which has the potential for a variety of serious global effects.[27]

Whether we manage to reduce global warming or not, accurate climate prediction is critically important. The models must be able to reliably predict the local consequences of that warming, namely, how wind patterns and precipitation might change, where there would be droughts and where there would be floods, where water supplies will be threatened by loss of snow storage areas—the kind of thing that governments must know in order to prepare for whatever changes are going to occur. It is vital that climate models continue to improve and be supplied with needed global data sets from modern global observing systems. We also need more data from paleoclimate studies—in this case it is especially true that we have to be able to understand (and predict) the past in order to predict the future.

We also need to be able to accurately predict what would happen if someone actually goes ahead and implements one of the many proposals that have been suggested for artificially cooling the Earth.[28] These proposed geoengineering solutions will have side effects and potentially strong positive feedback mechanisms that we do not understand. Climate prediction models are our only tools to assess the possible dangers. For example, at first look, a controllable method to counteract global warming might appear to be to imitate the way a volcanic eruption cools the Earth. A volcano shoots sulfur dioxide high into the stratosphere, where it forms sulfate particles that reflect sunlight back into space, thus cooling the Earth. But there are also other effects besides the cooling. A year after the volcanic eruption of Tambora in Indonesia (see Chapter 7, note 44) there was significant cooling. The year 1816 was called the "year without a summer"; snow fell in New England in June, and the southern United States had frost on July 4. The cold led to crop failures in the northeastern United States and northern Europe, which resulted in famines and death. The eruption also disrupted the monsoons over India and China. In 1991 the eruption of Mount Pinatubo not only had a global cooling effect but also decreased global precipitation and increased drought.[29] Even if we manage to cool the Earth's temperature using sulfur dioxide injections, the carbon dioxide concentration will still continue to grow higher, and its other effects will not be counteracted—for example, the acidification of the ocean will get much worse. Also, if the injection were suddenly stopped, perhaps sabotaged by a nation that did not like how its climate was being affected, the warming would quickly reach the same high temperature levels, but this rapid change would cause even more damage because there would be less time for ecosystems to adapt.[30] We also do not know enough to rule out the possibility that the cooling might go too far. If a positive feedback mechanism is initiated that keeps cooling the planet even when the injection of sulfur dioxide is stopped, we might move into an ice age, which is another type of devastating climate that could cause death and serious economic problems. We need better climate models to fully understand all these possibilities.[31]

Another proposed geoengineering solution to global warming that has been proposed is to put tons of iron compounds into the ocean to stimulate blooms of phytoplankton and therefore use up more carbon dioxide. The increased phytoplankton also might release an increased amount of dimethyl sulfide particles that would become the condensation nuclei of clouds, thus increasing cloud cover, which could reflect more sunlight back into space. Others have proposed agitating the ocean's surface to spray salt particles into the air, which can also serve as condensation nuclei for cloud formation, or artificially pumping nutrients from the deep parts of the ocean to the surface to create the same kind of phytoplankton blooms as the iron solution. All these geoengineering solutions were once thought to be too risky to even consider. We would, after all, be intentionally playing with the global climate of our only home. But if an extreme climate catastrophe did seem to be happening, such as if ice sheets on Greenland or Antarctica began to rapidly slide toward the sea, such methods might be tried. We must have a better understanding of the climate system and better prediction models to have a chance of accurately predicting what would happen.

But how sure are we that catastrophic climatic change will occur at some time in the future if we continue burning fossil fuels? There is of course some *uncertainty* in the climate model predictions, especially at the regional level. But there is uncertainty in all scientific studies, and critics are holding climate change studies to a much higher standard than other scientific work, for a variety of economic and political reasons. What we are trying to do is simply assess a *risk* (as we would in any risk management application), and then weigh that risk against the costs of trying to minimize that risk. We might start by looking at the possible consequences associated with the positions of the two political extremes in the global warming debate—at the one extreme the *global warming deniers* and at the other extreme the so-called *doomsdayers*. If we put a lot of money into trying to slow down global warming and the deniers are correct, the cost may be serious damage to our economy—something we have managed to do already (the 2008 economic collapse) and so something that is not that unusual. If we do not put any money into trying to slow down global warming and the doomsdayers are correct, then the cost will be unprecedented human suffering. The two sides also differ in the way they view the uncertainty. The deniers say that until we are absolutely certain that global warming is happening and is leading to catastrophe, we should not waste any money on that so-called problem. The doomsdayers say that this uncertainty is actually a reason to worry more about global warming, for if we do not fully understand the nonlinear feedback mechanisms at work in this incredibly complex climate system, the problem could be even worse than we think. The extreme of runaway climate change (like happened on Venus) might not even be out of the question. But perhaps comparing the extreme views may not be the best approach.

What actually is the uncertainty in the model predictions? This is a question that has been hotly debated. The last IPCC report said the results were 90 percent certain, but some critics said that seemed high considering the complexity of the climate system.[32] In any other area of science, even a better than fifty-fifty chance of something bad happening would be enough to stimulate some type of corrective action. But critics have so far successfully blocked any significant action by saying that the cost of reducing carbon dioxide in the atmosphere, or even of keeping it from rising further, would have a very harmful effect on the economies of the United States and other nations. Today that argument should probably carry less weight. First, because of the high price of oil, alternative forms of energy are no longer that much more expensive. And second, there are other reasons besides global warming for needing to break free of our dependence on oil. Before the economic crash in 2008 the price of oil was so outrageous that the economic and political consequences were becoming obvious to almost everyone, and once the economy began to recover, that situation returned. With 70 percent of the oil it needs coming from foreign countries, including the volatile Middle East, the United States is critically vulnerable to the possibility of oil being withheld. The solution to this national security problem—reduce and ultimately eliminate U.S. dependence on foreign oil, especially oil coming from the Middle East—would also be a big step toward solving the global warming problem.

The ultimate solution to this economic/political/national security problem—and to the global warming problem—is renewable energy, such as solar, wind, geothermal, and several types of ocean energy, such as waves, tides, currents, and ocean thermal energy conversion (OTEC). In these economic times, the production of renewable energy systems has the potential to have the same beneficial effect as did the increased wartime production during World War II. Though Franklin Roosevelt had done everything he could think of to end the Depression, it took the incredible production necessary to win World War II to finally transform the weak U.S. economy into a booming economy. Pumping huge amounts of money into real production, in this case building renewable energy systems, not only would provide the stimulus needed to turn the economy around, but also would give the United States energy security, allowing us to escape our dependence on Middle East oil with all its political ramifications. The critical "side effect" would be the reduced carbon dioxide emissions wanted so badly by those worried about global warming.

Throughout this book we have seen that marine prediction has depended on two important activities. The first has been our acquisition of large amounts of data from instruments designed to measure different properties and characteristics of the ocean. The second has been our efforts to understand the physics that describes the movement of the ocean and then to develop models that mathematically represent that physics. Feeding huge quantities of oceanographic and meteorological data into these models has led to accurate descriptions and explanations of many ocean phenomena. With today's technology, there are numerous ways to measure ocean parameters not even conceived of a century ago, with observations now made from satellites and thousands of land- and sea-based platforms. We have seen that for prediction it is important to have real-time data, data delivered as quickly as possible to the models that need them. In today's internet age, such real-time delivery of data is no longer a problem. And with today's ever-increasing computer power, the numerical dynamic models of the oceans, often coupled with atmospheric models, can continue to grow larger, thus allowing the greater spatial and temporal resolution that improves their accuracy.

We have seen numerous examples of how measurement and modeling have evolved over the centuries. The first ocean measurements began with recorded observations of the tide in the Persian Gulf by Seleucus in the second century BC, but it was not until two thousand years later that tide measurement finally progressed from simple tide staffs to self-registering tide gauges. Those gauges allowed a continuous curve of the vertical movement of the sea's surface to be drawn, thus showing other water level oscillations in addition to the tide. We then saw these gauges used to measure the size of storm surges produced by the high winds of a hurricane and to detect tsunamis generated by earthquakes or volcanic eruptions. They were able to measure the slowly changing sea level due to the warm waters caused by El Niño and thus were instrumental in developing an understanding of that phenomenon. Because many of these gauges were in operation for more than a hundred years, they provided data on how fast sea level was rising, a major issue in the discussion of climate change and the effects of global warming. Tide gauges, now more properly called water level gauges, showed that oceanographic instruments could be multipurpose. Many nations around the world installed water level observation networks, making them humankind's first real step toward creating a global ocean observing system.

Wind waves could also be measured by water level gauges, but they were typically treated as noise to be averaged out so as to not bias the measurements of the tides or storm surges or tsunamis. Wind waves ended up being measured more accurately by other methods (for example, with accelerometers on buoys) that could also show the direction the waves were traveling. The sampling was fast enough to allow analyses that produced spectra showing the energy at each frequency. Tsunamis, though measurable by relatively rapid sampling on a water level gauge, also had special instruments devoted to them, such as bottom pressure sensors connected to DART buoys. Other oceanographic instruments were developed to measure parameters such as water temperature, salinity, and the speed and direction of currents. Such data were used to understand El Niño and the ocean's role in climate change, the latter also requiring measurements of carbon, biological compounds, and other ocean parameters. We have seen how such measurement progressed from individual stations to multiple stations in oceanographic studies, such as TOGA, to permanently installed networks of stations providing data in real time for many purposes. And we have seen the importance of the various kinds of oceanographic data now obtainable from satellites.

The earliest hydrodynamic models began with the work of Laplace when he was trying to understand the tides. Yet it was a type of statistical model made possible by Laplace's work that led to the harmonic prediction method, the first tide prediction "model"—a method not usable for any other ocean phenomena because only the tide has energy at specific frequencies, determined by the orbital motions of the moon, the Earth, and the sun. But Laplace's dynamic model grew into other dynamic models that could be used to describe and predict storm surges and tsunamis, and describe and predict El Niño and climate change when coupled to weather and climate models (which also ultimately trace back to Laplace's work). Wind wave models on the one hand had to deal with short length scales when representing wave generation by the wind but on the other hand had to be of global extent when considering sea swells traveling thousands of miles along the ocean surface. And global weather models were needed to predict the storms that would generate the swells—the waves at any particular location being a combination of locally wind-generated waves and the swells that traveled there from all over the ocean. In virtually every situation, accurate predictions required a global approach, both in the modeling and in the data needed for the models.

So it was no wonder that at some point a true global ocean observing system would be needed. It was just a matter of when the technology would develop enough to make such a system possible. By the end of the twentieth century we had satellites for synoptic ocean measurements and for rapid communication, high-powered computers for ocean modeling and for the storage and analysis of trillions of gigabytes of oceanographic data, new technologies for measuring oceanographic properties both in situ and remotely, and the internet along with other communication advances for the real-time delivery of data and predictions. The Global Ocean Observing System (GOOS) was established as an official entity in 1991 by four international agencies,[33] but as we have seen, many of the instrument systems that were integrated under GOOS had been implemented over the previous decades. International programs like TOGA, for predicting El Niño, and WOCE, the World Ocean Circulation Experiment, for understanding the ocean's role in climate change, had moved the scientific community toward real-time global monitoring, creating a foundation for GOOS.[34] A permanent integrated real-time observing system like GOOS (and its U.S. component, the Integrated Ocean Observing System, or IOOS) are now clearly seen as the best way to handle the many interconnected marine phenomena. GOOS and the ocean models that it supports represent the culmination of centuries of marine scientific research and are finally beginning to provide the marine predictions needed around the world.

At the end of the last chapter of *The Sea Around Us,* Rachel Carson's 1951 best seller (ten years before *Silent Spring*), she stated that "even with all our modern instruments for probing and sampling the deep ocean, no one can say that we shall ever resolve the last, the ultimate mysteries of the sea."[35] In the sixty years since she wrote that, our ability to probe and sample the ocean has progressed by orders of magnitude. Some mysteries do remain, including one that many think might threaten life as we know it, but such mysteries no longer appear unsolvable. Perhaps Carson's statement was made not out of pessimism but merely as a way to extol the sea's grandeur, that grandeur keeping the sea from being totally understood. But with our growing Global Ocean Observing System and our ever-improving dynamic computer models, we are on our way to solving the sea's most important remaining mysteries and to being able to predict how the sea will affect us in the future. That does not in any way take away from the grandeur of the sea or decrease our appreciation for the incredible power of the sea.

Acknowledgments

For a book of this scope there are many people who must be acknowledged and thanked. First are the many scientists who have spent entire careers trying to understand the ocean and develop methods for predicting its movements and its catastrophic events. Many of these scientists play key roles in the stories in this book, some mentioned in the main text and others mentioned as part of additional information provided in the endnotes. Many more are also listed in the endnotes as the authors of important scientific works that I referenced because they were relevant to the science discussed in the book. There are many others whom I should have referenced, but could not due to space limitations. Some of these scientists have been my colleagues—at NOAA, at the Center for Maritime Systems at the Stevens Institute of Technology, or at universities and institutions with which I had the privilege of working on various oceanographic projects over the years. Still others have written valuable papers which I have used during my career. A few of these scientists were kind enough to look over portions of this book, and they are acknowledged below.

Thanks must also be given to another large and very important group that is critical to writing a historical scientific book—librarians and archivists. Without their constant efforts to seek out and save important historical documents of all types, a book like this would be impossible. I have seen firsthand how much difference a first-rate library can make, having used for many years the NOAA Federal Library in Silver Spring, Maryland, not only a source of important scientific works on oceanography, meteorology, and other sciences, but also a treasure chest for historical documents, many of them quite rare. I also witnessed how that library, once not much more than a building with dusty shelves of books, was turned around by Carol Watts, a librarian with a vision, who with her staff not only organized those historical treasures but brought the entire library into the electronic age, before passing off leadership to Janice Beatie, during whose tenure it was recognized as the Federal Library of the Year. The NOAA Library has also been dedicated to the history of NOAA, including oceanography, led by Captain Albert "Skip" Theberge, NOAA Corps (ret.). I have seen similar dedication to saving and making available scientific and historical information at other libraries and archives as well, and those librarians and archivists cannot be thanked enough. I also want to thank the S. C. Williams Library at the Stevens Institute of Technology, which gave me electronic access to so many valuable resources, as did the Sheridan Libraries at The Johns Hopkins University. Some of the archives and libraries that I visited and always received valuable help from included: the National Archives at College Park, Maryland; the Library of Congress in Washington, DC; the Eisenhower Presidential Library in Abilene, Kansas; the (British) National Archives (formerly the Public Records Office) in Kew, UK, and the Proudman Oceanographic Laboratory Library in Liverpool, UK. Other libraries and archives are mentioned in the endnotes associated with particular documents that I used. I would also like to thank Google Books, which has digitized so many books, a century or more old, that have proven invaluable to the writing of this book.

When one is writing a historically based book with true stories meant to demonstrate in human terms the impact of the sea (and the value of the science of the sea), one really begins to appreciate the accounts written by reporters and the memoirs written by participants or observers of important historical events. These are, of course, listed as references in the endnotes, but a special mention had to be made of those valuable eyewitness reports of historical marine disasters. Likewise, many libraries and archives have recorded and written down many oral histories, which again are very valuable.

I would also like to especially thank the scientists who were kind enough to review parts of this book for scientific and/or historical accuracy. I am very grateful to Walter Munk (Scripps Institution of Oceanography) for looking over the sections in Chapters 5 and 6 dealing with the wave forecasting for D-Day and the North African campaign, in which he played such a crucial role. He was kind enough to do this while he was getting ready to go to Stockholm to receive the Crafoord Prize from the Royal Swedish Academy of Sciences, the latest in a long list of prestigious awards he has received for his pioneering work in physical oceanography over six decades. I would also like to thank Stuart Weinstein, Deputy Director of the Pacific Tsunami Warning Center for reviewing the three tsunami chapters, and especially the two chapters dealing with the 2004 Indian Ocean tsunami, which Stuart experienced from a unique vantage point at PTWC. I would like to thank Brian Jarvinen, who was the storm surge expert at NOAA's National Hurricane Center during Hurricane Katrina, for reviewing the two storm surge chapters, including the section on the deadly and destructive storm surge due to Katrina. These two storm surge chapters were also reviewed by Will Shaffer (NWS's Techniques Development Laboratory), who played an important role in developing and improving the SLOSH model, and Joannes Westerink (University of Notre Dame), who is a nationally known expert on storm surge modeling. The tsunami chapters were also reviewed by Hal Mofjeld, now with the University of Washington but formerly with NOAA's Pacific Marine Environment Laboratory, which developed the DART buoy and the tsunami prediction models used by the PTWC. The two waves chapters were reviewed by Henrik Tolman (NOAA's National Center for Environmental Prediction), who is the developer of NOAA's global wave model, WAVEWATCH III®. All their reviews were very helpful, but of course the opinions expressed and any remaining errors are solely my own. I also greatly appreciated discussions about the book that I had with Richard Spinrad (Oregon State University, formerly Director, Office of Oceanic and Atmospheric Research, NOAA), Alan Blumberg (Director, Center for Maritime Systems, Stevens Institute of Technology), and Richard Patchen (Coast Survey Development Laboratory, NOS, NOAA). I would also especially like to thank Thomas and Valerie Doodson, son- and daughter-in-law of Arthur Doodson, who took me into their home for a wonderful visit with several former staff members from the Liverpool Tidal Institute who had helped Arthur Doodson produce the tide predictions for D-Day.

I would like to especially thank my agents John Brockman, Katinka Matson, and Max Brockman for immediately seeing the potential of this book, for suggestions that improved the book proposal, for advice along the way, and of course for finding a good publisher. I want to thank all those at Palgrave Macmillan who worked so hard on getting this book out, but especially my editor, Luba Ostashevsky, who was the one who wanted to do this book, supported it totally throughout the process, and did all those things editors have to do to transition a scientist into the world of popular nonfiction book writing.

And I want to thank my family for their support during this venture, including their help in many assorted ways: my son Sean for showing the Brockmans my one-pager about the book; my daughter Courtney and her husband Josue for the trip to the working tide mill in Spain; and my daughter Kimberly for her help in getting me access to a very useful research library. But above all, I thank my wife Diane, the most important person in my life for forty years. She read every word of this book more than once, critiquing it, helping me hear how it sounded, and finding errors that I kept missing. This book is dedicated to her.

About the Author

Bruce Parker first experienced the sea as a young boy, working for his father's scuba diving and water ski schools in the Bahamas. Over the years he mixed his encounters with the sea with academic training and research, to eventually become a world-recognized expert in the oceanographic subjects covered in *The Power of the Sea.* He has a PhD in physical oceanography from The Johns Hopkins University, an MS in physical oceanography from the Massachusetts Institute of Technology, and a BS/BA in biology/physics from Brown University. Before leaving NOAA (National Oceanic and Atmospheric Administration) in October 2004, Dr. Parker was Chief Scientist for the National Ocean Service, and before that Director of the Coast Survey Development Laboratory. He is presently a Visiting Professor at the Center for Maritime Systems at the Stevens Institute of Technology in Hoboken, New Jersey. His awards include the U.S. Department of Commerce *Gold Medal* and *Silver Medal,* the NOAA *Bronze Medal*, and the *Commodore Cooper Medal* from the International Hydrographic Organization. Dr. Parker is also a former Director of the World Data Center for Oceanography, and a former Principal Investigator for the NOAA Global Sea Level Program, and at one time ran the U.S. national tides and currents program. Dr. Parker has published over a hundred papers and articles and written or edited several books.

Notes

INTRODUCTION

1. The official name of Burma since 1989 is Myanmar, but Burma is still the name known by most.
2. In scientific terminology there is a difference between *power* and *energy*. Energy is defined as the capacity to do work. Power is the rate of doing that work, or how quickly the energy is expended.

Chapter 1

1. The primary sources for the story of Napoleon Bonaparte in Egypt and his escape from the Red Sea are: Abbott, John S.C., 1852. "Napoleon Bonaparte," in *Harper's New Monthly Magazine*, vol. 4, no. 21 (Feb.), pp. 310–330 [This is the source of figure 1.1]; Bourrienne, Louis Antoine Fauvelet de, 1891. *The Memoirs of Napoleon*, vol. 3, 1979, Charles Scribner's Sons, New York, 459 pages, chap. 17.
2. Woolf, Stuart Joseph, 1991. *Napoleon's Integration of Europe*, Routledge, 319 pages, p. 190.
3. By 1798 tide prediction had finally become a scientific endeavor, thanks primarily to the work of Pierre-Simon, Marquis de Laplace, whom Napoleon knew very well. As we will see later in this chapter, Laplace had taken Isaac Newton's gravitational theory of the tides and derived dynamic equations with which accurate predictions became possible for the first time. Among the scientists Napoleon brought to Egypt was the great mathematician Jean-Baptiste Fourier, who would later develop techniques that would be used to make tide prediction even more accurate. See: McLynn, Frank, 2002. *Napoleon: A Biography*, Arcade, 739 pages, pp. 26–27, 138, 159–161, 171.
4. More precisely, the average time from low water to high water is approximately 6.21 hours, and the *tidal period* (the time of one complete tidal cycle, e.g., from one high water to the next high water) is 12.42 hours (or 12 hours and 25 minutes). Two such cycles are completed in 24.84 hours, the length of a *lunar day* (one complete rotation of the Earth relative to the moon), which is about 50 minutes longer than a *solar day* (one complete rotation of the Earth relative to the sun).
5. Abbott, op. cit.; Bourrienne, op. cit.
6. Abbott, op. cit.
7. Towers, John Robert, 1959. "The Red Sea," *Journal of Near Eastern Studies*, vol. 18, no. 2, pp. 150–153.
8. Holland, F.W., 1868. "On the Peninsula of Sinai," in *Proceedings of the Royal Geographical Society of London*, vol. 12, no. 3, pp. 190–195; see especially the comments at the end by Sir Samuel Baker. Many other authors have also written about Moses' knowledge

derived by living in the wilderness, some of them referring to the work of Artapanus (80–40 BC) referenced in note 15.

9. Several other points are relevant: (1) No other ocean phenomenon could have been predicted for this purpose except the tide. Some authors have hypothesized a tsunami washing over the Egyptians that was due to an earthquake or volcanic eruption, but that would have been totally unpredictable, and was extremely unlikely to have occurred just at the right time to help the Israelites. Nor could it have been a storm surge, which also would not have been predictable in those ancient times. (2) The Bible says that water was "on their right hand, and on their left," which might mean that the bathymetry of the Gulf of Suez at that point and time was such that the Israelites walked on a wide shoal that was exposed at low tide. Such a shoal, which could be forded at low water, was used by Napoleon, as mentioned in the references in note 1. Other references mention a shoal in the general area of Suez, for example, Holland, op. cit., see especially the comments of Mr. Kennelly. However, at the time of the Exodus, when sea level was higher, a shoal with such exposure would most likely have been farther north. (3) Some Bible experts have pointed out a translation error such that the waters that engulfed Pharaoh's army were not really like great walls on both sides of it (as dramatically portrayed in the movie *The Ten Commandments*) but instead more like great fields or plains of water, which is closer to what Napoleon experienced in 1798 and what we see today at Mont-Saint-Michel. See: Boyle, Marjorie O'Rourke, 2004. "'In the Heart of the Sea': Fathoming the Exodus," *Journal of Near Eastern Studies*, vol. 63, no. 1, pp. 17–27.
10. Parker, Bruce, 2000. "The *Perfect* Storm Surge," *Mariners Weather Log*, vol. 44, no. 2 (Aug.), pp. 4–12.
11. Two relatively recent oceanographic studies have come to the conclusion that the wind could have pushed back the waters and made the sea bottom dry and then stopped or reversed direction, allowing the waters to flood back over the exposed sea bottom: Nof, Doron, and Nathan Paldor, 1992. "Are There Oceanographic Explanations for the Israelites' Crossing of the Red Sea?" *Bulletin of the American Meteorological Society*, vol. 73, no. 3, pp. 305–314; Volzinger, Naum, and Alexei Androsov, 2003. "Modeling of the Hydrodynamic Situation during the Exodus," *Izvestiya, Atmospheric and Oceanic Physics*, vol. 39, no. 4, pp. 482–496. However, in a follow-up study, Nof and Paldor calculated that wind speed and direction needed to push back the waters of the northern Red Sea (at least for the present-day configuration) would likely occur only once every 2,400 years. Nof, Doron, and Nathan Paldor, 1994. "Statistics of Wind over the Red Sea with Application to the Exodus Question," *Journal of Applied Meteorology*, vol. 33, pp. 1017–1024.
12. A century earlier, in 1766, somewhat higher values were found for the tidal range, for example, "At Suez, it flows six foot; the spring tides are nine, and in the variable months, from the beginning of November, to the end of April, sometimes twelve," from "A letter from Edward Wortley Montagu, Esquire, F.R.S. to William Watson, M.D.F.R.S. containing an account of his journey from Cairo, in Egypt, to the Written Mountains, in the desart of Sinai," 1766, *Philosophical Transactions*, vol. 56, pp. 40–57.
13. Because of this higher sea level, the Egyptians had been able to build a canal connecting the Gulf of Suez to the Nile River in the thirteenth century BC. That canal was rebuilt several times over the next two thousand years, but as time went on and sea level dropped, the canal finally became infeasible. This sea level drop over two millennia was probably due to the land rising locally over that period. There are many references in ancient literature to the canal connecting the Nile River to the Gulf of Suez, including Strabo's *Geography* (see note 19), which would have still been in operation when the Exodus was most likely to have taken place, 1500 to 1200 BC. Sea level estimates based on marine deposits (e.g., shell and coral fragments) have led researchers to estimate that sea level during part of this period was higher than it is today; for example, Fairbridge, R.W.,

1961. "Eustatic Changes in Sea Level," in *Physics and Chemistry of the Earth*, ed. L.H. Ahrens, F. Press, K. Raukawa, and S.K. Runcorn, vol. 4, Pergamon, New York, pp. 99–185. Tectonic activity in this area could have raised the land level since the time of the Exodus, thus making the relative sea level look lower.

14. Kirsch, Jonathan, 1998. *Moses—A Life*, Ballantine, New York, 415 pages. This is one of many references dealing with the three biblical authors whose work was combined by a fourth author, in sometimes inconsistent ways, to produce the Book of Exodus.
15. Eusebius of Caesarea, ca. AD 263–339. *Praeparatio Evangelica* (*Preparation for the Gospel*), chap. 27, bk. 9. He quotes from Artapanus, *Concerning the Jews*, 1903, transl. E.H. Gifford.
16. Panikkar, N.K., and T.M. Srinivasan, 1971. "The Concept of Tides in Ancient India," *Indian Journal of History of Science*, vol. 6, no. 1, pp. 36–50.
17. This small tide is because the Mediterranean Sea is too small for a significant tide to be directly generated by the gravitational effects of the moon (explained later in this chapter) and its connection to the Atlantic Ocean is too small to let in a significant tide from that ocean.
18. Harris, Rollin A., 1898. "Manual of Tides," pt. 2, chap. 5, in app. 8, in *Report of the Superintendent, U.S. Coast and Geodetic Survey, 1897*, Government Printing Office, Washington, DC.
19. Pytheas noticed the correlation between the phases of the moon and the half-monthly variation of the tide (neap to spring and back). His writings have been lost, but he was quoted by Strabo in *Geography* (AD 22), the earliest surviving Greek or Roman written work that provided insights about the tide. Strabo, *The Geography of Strabo*, transl. Horace Leonard Jones, 8 vols., Harvard Univ. Press, Cambridge, MA, 1969; originally written by Strabo sometime before AD 24.
20. Strabo, op. cit. The earliest advances in astronomy, including development of the zodiac reference system for the night sky, took place in Babylonia. This difference in height between two successive high waters is now called a *diurnal inequality* (and we will see later in this chapter why this occurs).
21. Harris, op. cit.
22. The two primary sources for the story of Alexander the Great and the tidal bore are: Quintus Curtius Rufus, ca. AD 53, *The History of Alexander*, transl. John Yardley, bk. 9, sec. 8, Penguin, 1984, pp. 232–235; Arrian, ca. AD 140s, *The Campaigns of Alexander*, transl. Aubrey de Sélincourt, Penguin, 1958, pp. 326–333.
23. Rufus, op. cit., p. 232.
24. Ibid., p. 233.
25. Ibid.
26. Ibid., p. 234.
27. The tide being a long wave should not be confused with *tidal wave*, the popular but erroneous name for a tsunami. A tsunami has nothing to do with the astronomical tide and is caused by an earthquake or volcanic eruption (see Chapter 7).
28. Parker, Bruce, 1999. "Tides in Shallow Water," *Mariners Weather Log*, vol. 43, no. 3 (Dec.), pp. 16–24.
29. This quotation from Procopius and the following one from Justin Martyr are found in Gill, Adrian, 1984. "Walter, Aristotle and the Tides of the Euripus," in *A Celebration in Geophysics and Oceanography—1982*, Scripps Institution of Oceanography, Univ. of California, La Jolla, CA.
30. Tsimplis, M.N., 1997. "Tides and Sea-Level Variability at the Strait of Euripus," *Estuarine, Coastal and Shelf Science*, vol. 44, pp. 91–101. Tsimplis took Elias the Cretan's quote from Miaoulis, A.A., 1882. *On the Tides of Euripos*, ed. Koromilas, Athens, 28 pages (in Greek).

31. Parker, Bruce B., 2007. *Tidal Analysis and Prediction*, NOAA Special Publication NOS CO-OPS 3, U.S. Dept. of Commerce, Washington, DC, 378 pages.
32. Defant, Albert, 1942. "Scylla and Charybdis and the Tidal Currents in the Straits of Messina," *International Hydrographic Review*, vol. 19 (Aug.), pp. 30–41. Defant mentions that an earthquake in February 1783 caused the rocks of Scylla to sink into the sea, decreasing the vortices and the roaring noises mentioned by Homer. There may have also been other tectonic motions that widened and deepened the strait, and there is a good possibility that the Charybdis tidal whirlpool was even more violent in Homer's time than it is today.
33. To briefly explain how the tidal whirlpool is generated, we first note that when the water flows from the wide part of the strait into the narrow part, the current follows the shoreline and converges to flow more rapidly in the narrow part. But when the tidal current reverses and flows from the narrow part of the strait into the wide part, the rapidly flowing current cannot suddenly make the sharp left or right turn that would spread the flow over the whole width of the strait. The current's inertia keeps its flow going approximately straight down the middle. This allows water near the edges of the strait to continue moving in the same direction it had been flowing (also due to its inertia), until the bending shoreline forces that edge flow to bend to the middle and meet the flow down the middle, creating a rotating whirlpool.
34. Azzaro, F., F. Decembrini, F. Raffa, and E. Crisafi, 2007. "Seasonal Variability of Phytoplankton Fluorescence in Relation to Straits of Messina (Sicily) Tidal Upwelling," *Ocean Science*, vol. 3, pp. 451–460.
35. Julius Caesar, ca. 52 BC. *The Conquest of Gaul*, transl. S.A. Handford, bk. 3, Penguin, 1982, p. 79. The title of Caesar's *Commentarii de Bello Gallico* has been translated as both *Commentaries about the Gallic War* and *The Conquest of Gaul.*
36. Ibid., bk. 4, p. 101.
37. Cartwright, David Edgar, 1999. *Tides—A Scientific History*, Cambridge Univ. Press, 292 pages, pp. 13–14. A book written by Isidore, the Archbishop of Seville, Spain (AD 560–636), was typical, merely restating Greek and Roman ideas about the possible cause of the tide, mentioning not only the possible role of the moon but also the idea about the tides being the oscillations of the body fluids of an Earth animal. The only original contribution was in a book by the Christian monk Venerable Bede (673–735), who lived in the kingdom of Northumbria in medieval England. In his book Bede included actual tidal observations from the coasts of England, probably based on sailors' observations, from which he could tell that high waters occurred earlier in the north of England than in the south.
38. Cartwright, D.E., and C.P. Conway, 1991. "Maldon and the Tides," *Cambridge Review*, vol. 112, pp. 180–183. The tides would have saved the day if the Anglo-Saxon leader hadn't been talked into letting the Viking army cross a narrow causeway to the island in order to have a fair and honorable fight.
39. Gillingham, John, 1989. "William the Bastard at War," in *Studies in Medieval History Presented to R. Allen Brown*," ed. C. Harper-Bill, J. C. Holdsworth, and J. L. Nelson, Boydell Press, pp. 141–158. He uses the writing of William of Jumièges as his source.
40. The Qiantang River has been referred to by various names, some of them probably through differing translations of the Chinese characters. These names include Ch'ien T'ang, Tsien-tang, the River Chê, and the River Chih.
41. Scidmore, Eliza R., 1900. "The Greatest Wonder in the Chinese World, the Marvelous Bore of Hang-Chau," *Century Magazine*, Apr. 1900, vol. 59, no. 6, pp. 852–859. [This is the source of figure 1.2.]
42. Moore, W. Osborne, 1999. *Report of the Bore of the Tsien-Tang Kiang*, Hydrographic Office, Admiralty, London.

43. The tide tables for the bore at Yanguan were printed using movable type, two centuries before movable type was used in Europe. This tide table is described in some detail by A.C. Moule in "The Bore on the Ch'ien-T'ang River in China," in *T'oung Pao*, 1923, vol. 22, nos. 1–5, pp. 135–188.
44. For example, a reference to this is found in a poem by Mei Sheng from this period, as indicated on page 485 of Joseph Needham's *Mathematics and the Sciences of the Heavens and the Earth*, 1959, vol. 3 of *Science and Civilisation in China*, Cambridge Univ. Press, 926 pages.
45. Ibid.
46. The moon appears north of the equator during the Northern Hemisphere summer because the axis of rotation of the Earth is tilted toward the sun during summer (thus receiving more direct sunlight). The Earth is tilted away from the sun during winter (receiving less direct sunlight).
47. Yang, Zuosheng, K.O. Emmery, and Xui Yui, 1989. "Historical Development and Use of Thousand-Year-Old Tide-Prediction Tables," *Limnology and Oceanography*, vol. 34, no. 5, pp. 953–957.
48. This story is found in Wang Chung's *Lun Heng*, in which he went to great lengths to show its absurdity, as reported by Joseph Needham (see note 44). A couple of versions of the story were written down, and called "nonsense," by Ch'ien Yüeh-yu in AD 1274, that account being reproduced by A.C. Moule (see note 43).
49. Yang, Emmery, and Yui, op. cit.
50. Cook, James, 1784. *A Voyage to the Pacific Ocean*, vol. 2, Lords Commissioners of the Admiralty, London, pp. 395–396. Later, River Turnagain was renamed Turnagain Arm, at the same time that Cook's River was renamed Cook Inlet, following the 1794 voyage of Captain George Vancouver, when it was determined that "Cook's River" was not a river.
51. This time difference between the moon's transit overhead and the following high water had many different names over the centuries, including *establishment of the port*, *vulgar establishment*, *high water full and change*, *tide hour*, and *lunitidal interval*.
52. Hughes, Paul, 2006. "The Revolution in Tidal Science," *Journal of Navigation*, vol. 59, pp. 445–459. The London tide table may have been the work of Abbott John Wallingford, who died in 1213. It had six columns and thirty rows. The first column listed the days after the new moon. The second and third columns gave the times, in hours and minutes, of "Flod at London Bridge," interpreted by most scientists today to have meant the times of high water, when the tidal current would still be a flood current, namely, flowing upstream. For the day of new moon at London Bridge this time was listed as 3:00 A.M., three hours after the moon's transit at midnight, and forty-eight minutes were added to it for each succeeding day. The fifth and sixth columns gave the amount of time that the moon was supposed to shine each night, which also increased by forty-eight minutes each day.
53. Howe, Derek, 1993. "Some Early Tidal Diagrams," *Mariner's Mirror*, vol. 79, no. 1, pp. 27–43. The Catalan Atlas for King Charles VI of France was produced by Abraham Cresques.
54. These tidal charts were produced by Guillaume Brouscon, who lived near Brest in Brittany, France. A 1543 version can be seen at the Digital Scriptorium of the Bancroft Library of the University of California at Berkeley, HM 46, fols. 3, 6. It was digitized from Dutschke, C.W., and R.H. Rouse, 1989. *Guide to Medieval and Renaissance Manuscripts in the Huntington Library*, Huntington Library Press, Henry E. Huntington Library and Art Gallery, San Marino, CA.
55. Cartwright, *Tides*, op. cit., pp. 20–22. The gear ratios in the tide clock were designed to simulate lunar time instead of solar time.

56. Ibid., pp. 28–35. Some of the stranger theories of tides from ancient times and the Middle Ages still found a few adherents.
57. The natural period of oscillation is also affected by depth, though it is a weaker effect. This period is directly proportional to the basin length and inversely proportional to the square root of the depth. Thus shallower basins have longer natural periods than deeper basins of the same length.
58. The story of Galileo's efforts to develop his tidal theory while avoiding being burned at the stake is told in Dava Sobel's *Galileo's Daughter: A Historical Memoir of Science, Faith, and Love*, Penguin, 2000.
59. The formula derived by Newton says that the tide-producing force is directly proportional to the mass of the moon (or sun), but inversely proportion to the *cube* of the distance to the moon (or sun). The sun is approximately 27 million times the mass of the moon, but because the sun is almost 400 times farther away (which when multiplied by itself three times becomes about 59 million), that means the sun's tide-producing force is about 46 percent of the moon's.
60. The primary tidal oscillations are referred to as *semidiurnal*, meaning twice a day. Tidal oscillations with one high water a day are referred to as *diurnal*. If the basin is the right size, hydrodynamics (the physics of how water oscillates in an ocean or a bay) can increase the amount of the diurnal energy in the tides, something we see, for example, in the South China Sea or along part of the U.S. Gulf Coast.
61. Halley, Edmund, 1697. "The true theory of the tides, extracted from that admired treatise of Mr. Isaac Newton, Philosophia Principia Mathematica; being a discourse presented with that book to the late King James," *Philosophical Transactions*, vol. 19, pp. 445–457.
62. Cartwright, *Tides*, op. cit., pp. 39–40.
63. *Resonance* is when energy is input at the same frequency as the natural period of oscillation, producing large oscillations. A simple pendulum's natural period of oscillation is the time it takes to swing back and forth through one complete cycle. If we hit the pendulum just as it reaches its highest point and is about to swing back, then we are inputting energy at precisely the right time to make the pendulum go higher. It is the same with the input of tidal energy in the same direction as the natural oscillation of the basin.
64. The size of an ocean basin can also increase the size of the tidal oscillation produced by the apogee-perigee effect (the moon's changing distance from the Earth due to the moon's elliptical orbit), as it does along the Atlantic Coast of the United States, where it is greater than the spring-neap effect. Hydrodynamics also makes Tahiti one of the few places in the world where high tides occur at almost the same time each day, around noon and around midnight. This is because at Tahiti the dimensions of the South Pacific reduce the tidal oscillation caused by the moon, but not the tidal oscillation caused by the sun. Since the period of the solar part of the tide is twelve hours (versus twelve hours and twenty-five minutes for the lunar part of the tide), high waters do not get progressively later in Tahiti; they always stay close to the time the sun is overhead (noon) or directly on the other side of the Earth (midnight). Tahiti is near what is called an *amphidromic point* for the lunar tidal oscillations, that is, a point with almost no tidal range. On opposite sides of the amphidromic point, the water surface slowly moves up and down, much like the water sloshing up and down at opposite ends of a bathtub. See: Marmer, H.A., 1927. "The Truant Tides of Tahiti," *Natural History*, vol. 27, no. 5, pp. 430–438.
65. When the tidal current is flowing up the bay, the Coriolis force pushes the water to the right in the Northern Hemisphere, raising the tide level on the right-hand coast (looking up the bay) and lowering it on the left-hand coast. Six hours later, when the tidal current

is flowing out of the bay, the opposite happens. So the right-hand coast has the higher high tides and the lower low tides, giving it a greater tidal range than the left-hand coast. The above directions are opposite in the Southern Hemisphere.

66. Harris, "Manual of Tides," chap. 7, pp. 422–437. Also see: Cartwright, *Tides*, op. cit., pp. 68–75. Both provide detailed explanations of Laplace's pioneering hydrodynamic work. Cartwright reproduces the first three pages (in French) of Laplace's groundbreaking 1776 paper, "Recherches sur plusieurs points du system du monde," *Mémoires de l'Académie Royale des Sciences*, vol. 89, pp. 177–264.
67. These equations of motion were based somewhat on equations that Euler had developed in 1755 for smaller-scale processes.
68. The conservation of momentum equations come from Newton's second law, which says that the force on an object is equal to its mass times its acceleration, or $F = ma$. *Momentum* is mass times velocity, so the *ma* in Newton's equation is the rate of change of momentum. To obtain a momentum equation useful in a tidal model, Newton's second law is put into a different form, in which the rate of change of momentum is balanced with the forces acting on the water, such as gravity and friction. Parker, op. cit., 2007.
69. For a clear explanation on exactly how the Coriolis force works, see: Parker, Bruce, 1998. "The Coriolis Effect: Motion on a Rotating Planet," *Mariners Weather Log*, vol. 42, no. 2, pp. 17–23.
70. We saw earlier in the chapter that *frequency* is the inverse of *period*. For example, the primary tide due to the sun has a period of 12 hours, which is equivalent to a frequency of 1 cycle per 12.00 hours, or 2.00 cycles per day. The primary tide due to the moon has a period of 12.42 hours, which is the equivalent of a frequency of 1.93 cycles per day. The portion of the tide due to the moon's elliptical orbit (namely, the effect of the slowly varying distance between the moon and the Earth) has a period of 12.66 hours, which is equivalent to a frequency of 1.90 cycles per day. And so on for other frequencies that come from the relative motions of the moon, Earth, and sun. The difference between the lunar frequency and the solar frequency may appear to be very small, but as they slowly go in and out of phase with each other, they vary in their combined effect over 14.8 days, the average time from a new moon to a full moon.
71. Hutchinson, William, 1787. *A Treatise on Practical Seamanship*, 2nd ed., 280 pages, p. 169. The first edition was in 1777.

CHAPTER 2

1. That additional tide-generating force due to the moon being at its closest to the Earth was even larger than the sun's tidal force, thanks to the hydrodynamic effect of the North Atlantic.
2. Olson, Donald W., and Russell L. Doescher, 1993. "Astronomical Computing: The Boston Tea Party," *Sky and Telescope*, vol. 86, no. 6, pp. 83–86.
3. Drake, Francis S., 1884. *Tea Leaves: Being a Collection of Letters and Documents Relating to the Shipment of Tea to the American Colonies in the Year 1773, By the East India Tea Company*, A.O. Crane, Boston, MA, 375 pages. This book includes many eyewitness accounts, as well as newspaper reports, letters, and excerpts from ships' log books. The quotation in the next paragraph is from page 77.
4. Carpenter, Daniel H., 1901. *History and Genealogy of the Carpenter Family in America*, The Marion Press, Jamaica, Queensborough, New York, 370 pages.
5. Allston, F.F.W., 1844. "On Planting and Preparing Rice," *Southern Agriculturist, Horticulturist, and Register of Rural Affairs*, vol. 4, no. 1, pp. 6–25. Rice-growing fields were irrigated by the flooding that occurred during high tides along the freshwater

portion of the tidal rivers. Before this technique was developed, rice growing in the Carolinas was never profitable.

6. In addition to tide predictions, Colonial almanacs provided a calendar with astronomical information such as sunrise, sunset, moonrise, and moonset, along with religious holidays, whose dates were often astronomically based. There were even weather predictions, which were almost never correct since the weather, unlike the tides, was not astronomically caused. Each almanac also had its own special added information. In the case of Franklin's it was his humorous writings. In the case of Benjamin Banneker's (see note 7) it was to lecture about the evils of slavery. Banneker was a free African American in Baltimore who was a self-educated scientist, astronomer, and inventor, and who did his own astronomical calculations for the almanac. He put his own picture on the cover to make it very clear to his readers that he was a black man. On the upper right side of the tide table for Chesapeake Bay he wrote, "The Slave-Trade, so disgraceful to humanity, began in the reign of Queen Elizabeth, about the year 1567." Below the tide table he gave three more historical facts, such as "Needles first made in London, by a Negro, from Spain, in the reign of Q. Mary; but he dying without teaching the art, it was lost till 1566, when it was taught by Elias Grorose, a German." Banneker sent a copy of his first almanac to Thomas Jefferson, then the U.S. Secretary of State. In the accompanying letter he took Jefferson to task for still owning slaves and not living up to his famous declaration that "all men are created equal." See: Bedini, Silvio A., 1999. *The Life of Benjamin Banneker*, 2nd ed., Maryland Historical Society, Baltimore, MD, 428 pages, pp. 156–165.
7. Banneker, Benjamin, 1792. *Benjamin Banneker's Pennsylvania, Delaware, Maryland, and Virginia Almanack, for the Year of Our Lord 1792*, printed and sold by William Goddard and James Angell, Baltimore, MD. Library of Congress Rare Books and Special Collections Division, Washington, DC, digital ID: rbcmisc ody0214. See note 6 for more information about Banneker and his almanac.
8. Staples, William R., 1845. *The Documentary History of the Destruction of the Gaspee, Compiled for the Providence Journal*, Knowles, Vose, and Anthony, Providence, RI, 56 pages. The man quoted was Ephraim Bowen, a participant in the *Gaspée* attack.
9. Parker, Bruce, 2009. "Tide Prediction in Colonial America," *Hydro International*, vol. 13, no. 4, pp. 28–31.
10. Ibid.
11. Fischer, David Hackett, 1995. *Paul Revere's Ride*, Oxford Univ. Press, 464 pages.
12. Ibid. See also: Olson, Donald W., and Russell L. Doescher, 1992. "Astronomical Computing: Paul Revere's Midnight Ride," *Sky and Telescope*, vol. 83, pp. 437–440.
13. Thomson, William, 1868. "Report of the Committee for the Purpose of Promoting the Extension, Improvement and Harmonic Analysis of Tidal Observations," in *Report of the 38th Meeting of the British Association for the Advancement of Science*, London, pp. 489–510.
14. Ferrel, William, 1874. *Tidal Researches*, U.S. Coast Survey Report, Washington, DC, 268 pages. Ferrel also went on to apply dynamic equations similar to the Laplace tidal equations to global atmospheric problems, greatly advancing the science of meteorology.
15. Darwin, George Howard, 1883. "The Harmonic Analysis of Tidal Observations," *British Association Report for 1883*, pp. 40–118. The primary U.S. scientist to improve the method was Paul Schureman. See: Schureman, Paul, 1924. *Manual of Harmonic Analysis and Prediction of Tides*, Special Publication 98, Coast and Geodetic Survey, U.S. Dept. of Commerce, Washington, DC, 317 pages.
16. Parker, Bruce B., 2004. "Tides," in *Encyclopedia of Coastal Science*, ed. M. Schwartz, Kluwer Academic, Encyclopedia of Earth Sciences Series, pp. 987–996. For a longer discussion, see: Parker, Bruce B., *Tidal Analysis and Prediction*, NOAA Special Publication NOS CO-OPS 3, U.S. Dept. of Commerce, Washington, DC, 378 pages.

17. A musical note represents acoustic energy at a particular frequency. The first overtone has twice the frequency of the primary note. Likewise, the first overtide (M_4) has twice the frequency (or half the period) of the primary tidal constituent (M_2). In shallow waterways one must also analyze the data for M_4 (with a period of 6.21 hours, roughly four cycles per day) and for M_6 (with a period of 4.14 hours, roughly six cycles per day). When these overtides are fairly large, the tide curve shows double high waters (or double low waters), and when they get very large, there is the very rapid rise to high water as seen in a tidal bore. The effects in shallow water are referred to as *nonlinear effects*, because in the hydrodynamic equations the mathematical terms that have more than one key variable (such as tidal height and current speed) are larger. The result is considerable energy exchange between certain frequencies.
18. Tide measurements have to be made relative to a fixed reference point, called a *datum*. The usual practice is to obtain a long set of tide data and determine a mean low water (MLW) value and use that as a reference point. But this *tidal datum* must be referenced to a *bench mark*, which is a brass disc hammered into a solid piece of bedrock on land (or some other solid object that will not move vertically unless the land as a whole moves). There are other types of tidal datums, as well as other types of vertical datums that are not tidal (see reference in note 65).
19. Palmer, Henry, 1831. "Description of a Graphical Register of Tides and Winds," *Philosophical Transactions of the Royal Society of London*, vol. 121, pp. 209–213. Palmer built the gauge for John William Lubbock, a tidal researcher and a producer of tide tables based on his own synthetic method, which consisted of correlations he had discovered after studying several years of tide data. Lubbock's tide tables were the most accurate in Britain in the nineteenth century, but his method was replaced by the more accurate harmonic method, which required only a month's worth of data.
20. Hunt, Edward Bissel, 1853. "Saxton's Self-Registering Tide Gauge," in Appendix of the *Report of the Superintendent, United States Coast Survey*, Washington, DC, pp. 94–96.
21. See also note 35 in Chapter 7.
22. Harris, Rollin A., 1898. "Manual of Tides," pt. 2, chap. 1, pp. 482–483, in Appendix 8 of the *Report of the Superintendent, U.S. Coast and Geodetic Survey, 1897*, Government Printing Office, Washington, DC. [This is the source, on page 483, of the picture in figure 2.1.] See also: "An Automatic Tidal Indicator," *New York Times*, Jan. 29, 1894.
23. Thomson, William, 1881. "The Tide-Gauge, Tidal Harmonic Analyzer and Tide Predictor," *Proceedings of the Institute of Civil Engineers*, vol. 65, pp. 4–74.
24. Ferrel, William, 1884. "Description of a Maxima and Minima Tide Predicting Machine," in Appendix 10 of the *Report of the Superintendent, United States Coast Survey, 1883*, Washington, DC, pp. 253–272. Ferrel's tide-predicting machine was a little different than Kelvin's. On the basis of slightly different mathematics, his machine directly calculated the times and heights of high waters and low waters, rather than drawing an entire tide curve and then picking the high and low waters off that curve. However, predicting an entire tide curve had many advantages, and the Coast Survey's second tide-predicting machine would use that procedure. See the next note.
25. Tierney, Samuel, 1912. "The Harris Tidal Machine," *Science*, n.s., vol. 35, no. 895 (Feb. 23), pp. 306–307. The tide-predicting machine is described in more detail in *Description of the U.S. Coast and Geodetic Survey Tide-Predicting Machine No. 2*, Special Publication 32, U.S. Coast and Geodetic Survey, Dept. of Commerce, Washington, DC, 1915, 35 pages. [This is the source of figure 2.2.] The U.S. Coast and Geodetic Survey was the new name for the U.S. Coast Survey as of 1878.
26. Zetler, Bernard D., 1991. "Military Tide Predictions in World War II," in *Tidal Hydrodynamics*, ed. Bruce B. Parker, Wiley, New York, pp. 791–797.

27. Ibid. Some of these harmonic constants had been published by the Japanese before the war and were available from the International Hydrographic Bureau in Monaco. At first there was suspicion that the Japanese might have deliberately published misleading data, but these harmonic constants compared well with constants from other sources.
28. Much of the material dealing with Tarawa comes from the following sources: McKiernan, Patrick, 1962. "Tarawa: The Tide That Failed," *U.S. Naval Institute Proceedings*, vol. 88, no. 2, pp. 38–49; Alexander, Joseph H., 1997. *Utmost Savagery: The Three Days of Tarawa*, Ivy Books, New York, 352 pages; Kernan, Michael, 1993. "Heavy fire...unable to land...issue in doubt," *Smithsonian Magazine*, Nov., pp. 118–131.
29. Olson, Donald, 1987. "The Tide at Tarawa," *Sky & Telescope*, vol. 74, no. 5, pp. 526–528.
30. U.S. Coast and Geodetic Survey, 1942. *Tide Tables, Pacific Ocean and Indian Ocean for the Year 1943*, U.S. Dept. of Commerce, U.S. Government Printing Office, Washington, DC, p. 278.
31. The Japanese tide table also included Tarawa as a subordinate station, but it used Auckland, New Zealand, as its reference station—even farther away. See: Hydrographic Department, Imperial Japanese Navy, 1940. *Tide Tables, Part II, 1941*, p. 145.
32. Alexander, Joseph H., 1995. "David Shoup, Rock of Tarawa," *Naval History*, vol. 9, no. 6, pp. 19–24. Years later Shoup became Commandant of the Marine Corps.
33. "Knox Upholds Plan of Tarawa Action," *New York Times*, Dec. 1, 1943.
34. Letter to the Hydrographic Office of the Navy Department, January 26, 1944. From archive at National Ocean Service, NOAA.
35. The naval portion of Operation Overlord, including the amphibious landings, was called Operation Neptune. See: Eisenhower, Dwight D., 1948. *Crusade in Europe*, Doubleday, Garden City, NY, 559 pages, p. 239n.
36. "When they come it will be at high water," Rommel had said to Second Lieutenant Arthur Jahnke, during an inspection of strongpoint W5, behind Utah Beach. From Carell, Paul, 1964. *Invasion. They're Coming*, transl. E. Osers, Bantam Books, New York, 288 pages, p. 29. Rommel repeated that opinion many times.
37. June 27, 1940, letter from Lieutenant Colonel E.I.C. Jacob to Captain R.V. Brockman. CAB 120/444, Public Record Office, Kew, London, UK.
38. June 30, 1940, letter from General H.L. Ismay to Captain R. Brockman, Admiralty. CAB 120/444, Public Record Office, Kew, London, UK.
39. July 1, 1940, letter from Captain R. Brockman to Lieutenant Colonel E.I.C. Jacob with accompanying report by Director of Navigation. CAB 120/444, Public Record Office, Kew, London, UK.
40. "Tide Tables Banned," *Liverpool Daily Post* (UK), Nov. 15, 1940. See also: "Notice by the Admiralty," *London Gazette*, Nov. 15, 1940.
41. Fergusson, Bernard, 1961. *The Watery Maze*, Holt, Rinehart, and Winston, New York, 445 pages, p. 300. The picture in figure 2.3 is from the Eisenhower Presidential Library in Abilene, Kansas.
42. Eisenhower, op. cit. See also: *Engineer Operations by the VII Corps in the European Theater*, vol. 2, section entitled "Breaching of Beach Obstacles," pp. 2–12, 1948.
43. Eisenhower, op. cit.
44. The H-Hours for the five Normandy beaches (Utah, Omaha, Gold, Juno, Sword) were 6:30, 6:30, 7:15, 7:15, and 7:15 A.M., respectively, according to a May 21, 1944, memorandum whose subject was "H Hour." The times for the last three beaches were revised to 7:25, 7:35/7:45, and 7:25 A.M. (where the two times for Juno were for "west brigade" and "east brigade"), in a May 25, 1944, memorandum. WO 205/530, Public Record Office, Kew, London, UK.

45. *Centenary Report and Annual Reports (1950–1945), Liverpool Observatory and Tidal Institute*, 1945, 24 pages. See also: "The Tides and Our Invasion. Liverpool's Part in Landings," *Liverpool Daily Post* (UK), July 25, 1944.
46. *Centenary Report*, op. cit.
47. "Tides of Time Change for the Eyes on the Hill. Observatory in Transfer to University," *Liverpool Daily Post* (UK), Apr. 1, 2000. Also from author's interviews of Liverpool Tidal Institute staff (see note 50).
48. *Centenary Report*, op. cit.
49. Morgan, Frederick, 1950. *Overture to Overlord*, Hodder and Stoughton, London, 296 pages. General Morgan was Chief of Staff to the Supreme Allied Commander.
50. Copies of the Doodson-Farquharson letters were obtained by the author from Thomas and Valerie Doodson (son and daughter-in-law of Arthur Doodson). Valerie joined the Liverpool Tidal Institute at the Bidston Observatory after the war, eventually becoming head of the Tidal Section. In 1987 Bidston became the Proudman Oceanographic Laboratory (POL). A few letters and newspaper articles were also obtained from the POL library and archive. Information was also obtained in March 2001 from a meeting at the Doodson home with former members of the Liverpool Tidal Institute, Joan Rossiter and Estelle Gilbert, who worked as assistants for Dr. Doodson during the war, and Joyce Scoffield, who joined the Institute after the war and who later wrote *Bidston Observatory: The Place and the People*, Countyvise, 2006, 344 pages.
51. April 22, 1944, handwritten letter from William Ian Farquharson to Arthur T. Doodson, on letterhead of the Hydrographic Department, Admiralty, Bath, 1 page. From private archive of Thomas and Valerie Doodson (son and daughter-in-law of Arthur Doodson).
52. Moitoret, Victor A., 1971. *Fifty Years of Progress, 1921–1971*, International Hydrographic Organization, Monaco, 21-V-1971, 30 pages. David Blackman of the Proudman Oceanographic Laboratory pointed out that these "code" numbers were actually IHB numbers.
53. Hydrographic Department, 1943. *The Admiralty Tide Tables*, sec. A. Home Waters (British Isles, Europe, and North-west Coast of Africa), pt. 2, 1944 ed., p. 41; station numbers 1592 (Merville-Ouistreham), 1593 (Courseulles), and 1594 (Port-en-Bessin) referenced to station number 1582 (Le Havre).
54. Ibid., pt. 1, p. 272.
55. Glen, N.C., 1985. "D-Day Bathymetric Survey," *Hydrographic Journal*, no. 35, pp. 5–7. Additional information in a letter from N.C. Glen to the author, August 22, 2001. Strutton, Bill, and Michael Pearson, 1959. *The Secret Invaders*, British Book Centre, New York, 286 pages.
56. October 9, 1943, handwritten letter from William Ian Farquharson to Arthur T. Doodson, on letterhead of the Hydrographic Department, Admiralty, Bath, 3 pages. From private archive of Thomas and Valerie Doodson (son and daughter-in-law of Arthur T. Doodson). See note 17 for a reminder of the characteristics of shallow-water constants.
57. Walters, Roy A., 1987. "A Model for Tides and Currents in the English Channel and Southern North Sea," *Advances in Water Resources*, vol. 10, no. 3, pp. 138–148. Le Provost, Christian, Gilles Rougier, and Alain Poncet, 1981. "Numerical Modeling of the Harmonic Constituents of the Tides, with Application to the English Channel," *Journal of Physical Oceanography*, vol. 11, no. 8, pp. 1123–1138.
58. "'Dr Tides' Retires," *Liverpool Daily Post* (UK), 1960. In this article Doodson is quoted as saying, "I must admit that a member of my staff and I guessed where the predictions were for." Copy of article from archive of the library at the Proudman Oceanographic Laboratory, Liverpool, UK.

59. Seiwell, H.R., 1947. "Military Oceanography in World War II," *The Military Engineer*, vol. 39, no. 259, pp. 202–210. [This is the source of figure 2.4.] See also: Seiwell, H.R., 1947. "Tidal Illumination Diagrams," *Science*, vol. 106, no. 2743, pp. 76–77.
60. The June 3, 1944, entry in Rommel's diary (probably written by Rommel's aide-de-camp) states, "5th–8th June 1944. Fears of an invasion during this period were rendered all the less by the fact that tides were very unfavourable for the days following," from *The Rommel Papers*, ed. B.H. Liddell-Hart, chap. 21 (written by Lieutenant General Fritz Bayerlain), p. 470. General Gerd von Rundstedt said (talking to B.H. Liddell-Hart): "The one real surprise was the time of the day at which the landing was made—because our Naval Staff had told us that the Allied forces would only land at high water." In Liddell-Hart, B.H., 1948. *The Generals Talk*, Quill, New York, p. 242.
61. Passmore, N.M., July 1944, "Omaha Beach—Tidal Observations." Former SECRET document signed by Lieutenant Commander N.M. Passlore, giving tidal observations made by the HMS *Gulnare*. These showed three days (July 5, 6, and 7) of predicted versus observed tides, with difference in heights less than or equal to 0.3 feet, and time differences less than or equal to twenty-two minutes. From archive of the UK Hydrographic Office, Taunton, UK.
62. *Report on Naval Combat Demolition Units in Operation "Neptune" as Part of Task Force 122*, submitted by H.L. Blackwell, Jr., D-V9G, USNR, July 5, 1944.
63. Ibid. At Utah Beach, Force U landed 1,500 yards south of its intended landing spot due to the tidal current, but because of this faced only light enemy fire; still, six demolition engineers were killed and eleven were wounded. By 8:00 A.M., the entire beach was cleared of enemy obstacles.
64. There was an in-between, and short-lived, reorganization before the establishment of NOAA in 1970, when ESSA (Environmental Science Services Administration) was created in 1965 by combining the then U.S. Weather Bureau and the U.S. Coast and Geodetic Survey, the latter temporarily known as the National Ocean Survey (conveniently also abbreviated as NOS).
65. Parker, B., K. Hess, D. Milbert, and S. Gill, 2003. "A National Vertical Datum Transformation Tool," *Sea Technology*, vol. 44, no. 9. pp. 1–15. The GPS measurement is the distance from the GPS receiver to three or more GPS satellites, allowing an exact position to be determined relative to a special reference datum, which itself has a known relationship to various tidal datums (see note 18), thus allowing the water-level height to be determined.
66. Cartwright, David E., 1991. "Detection of Tides from Artificial Satellites (Review)," in *Tidal Hydrodynamics*, ed. Bruce B. Parker, Wiley, New York, pp. 547–567. The radar determines this height by measuring the time it takes for a radio signal to travel to the ocean surface and back to the satellite. The sampling rate of satellite-measured water levels is very low, perhaps only a couple of measurements per day, but if there is a year or more of data, harmonic analysis can still calculate the tidal harmonic constants.
67. Parker, op cit., 2007, chap. 8, "The Use of Numerical Hydrodynamic Models for Prediction of Tides and Tidal Currents."
68. The accuracy of such predictions depends on how much the bay or river itself (namely, its depths and shoreline) has changed over the years, decades, or centuries since that historical event took place, because such changes modify the hydrodynamics of the waterway and thus modify the tides and especially the tidal currents (but the deeper the waterway, the less effect such changes have).
69. The PORTS system is operated by the part of NOS that runs the national tides and currents program (which had been established in the U.S. Coast Survey) the present organizational name being the Center for Operational Oceanographic Products and Services

(CO-OPS). The forecast numerical hydrodynamic models for PORTS were developed by the Coast Survey Development Laboratory. The author at different times headed up each of these organizations. The weather forecast models that provide the wind forcing to the ocean models are operated by the National Centers for Environmental Prediction in NOAA/NWS. The real-time current data from PORTS is also important for the efficient cleanup of oil spills when they occur in harbors. See: Parker, Bruce B., 1995. "P.O.R.T.S. Makes Navigation Safer," *Proceedings of the Marine Safety Council*, vol. 52, no. 5 (Sept.–Oct.), pp. 30–32. See also: Parker, Bruce B., 2002. "Integrated Real-Time/Forecast Information for Safe, Efficient, and Environmentally-Friendly Ports," *Singapore Maritime and Port Journal*, vol. 1, pp. 1–10.

70. Tyler, Patrick E., 2004. "19 Die as Tide Traps Chinese Shellfish Diggers in England," *New York Times International*, Feb. 8.
71. *BBC News*, March 24, 2006. This quotation and the preceding one are from Det. Supt. Mick Grad Gradwell of the Lancashire Police, in "Man Guilty of 21 Cockling Deaths," 18:39:20 GMT, http://news.bbc.co.uk/go/pr/fr/-/1/hi/england/lancashire/4832454.stm.
72. Lawrence, Felicity, et al., 2004. "Victims of the Sands and the Snakeheads," *Guardian* (Manchester, UK), Feb. 7.
73. *BBC News*, op. cit. The tragedy led to the passing of the Gangmaster Licensing Act, which went into effect October 2006, and the establishment of the Gangmaster Licensing Authority. It brought attention to the plight of illegal immigrants in the United Kingdom (most from the Fujian province of China), who had been taken advantage of at every turn, often being forced to live in overcrowded conditions.

CHAPTER 3

1. Eliot, John, 1900. *Hand-book of Cyclonic Storms in the Bay of Bengal*, 2nd ed., vol. 1, Meteorological Department of the Government of India, Calcutta, 310 pages, p. 210.
2. Ibid., pp. 211–213. The wind and weather comments quoted by Eliot came from ships' logs at locations around the Bay of Bengal.
3. Ibid., p. 212.
4. Ibid., p. 215.
5. Gastrell, J.E., and Henry F. Blanford, 1866. *Report on the Calcutta Cyclone of the 5th October 1864*, O.T. Cutter, Military Orphan Press, Calcutta, 175 pages, p. 33.
6. *A Brief History of the Cyclone at Calcutta and Vicinity, 5th October 1864*, 1865, O.T. Cutter, Military Orphan Press, Calcutta, 332 pages, p. 195.
7. Ibid., p. 195.
8. Ibid., p. 166.
9. Buckland, C.E., 1901. *Bengal Under the Lieutenant-Governors*, vol. 1, S.K. Lahiri & Co., Calcutta, 1,100 pages, pp. 298–299.
10. Graham, C., 1878. *Life in the Mofussil, Or, the Civilian in Lower Bengal*, vol. 2, Kegan Paul & Co., London, 284 pages, p. 275.
11. Eliot, op. cit., p. 216.
12. Eliot, op. cit., p. 217.
13. Graham, op. cit., p. 125.
14. Rao, S.R., 1973. *Lothal and the Indus Civilization*, Asia Publishing House, Bombay, 215 pages.
15. Frazer, James George, 1919. *Folk-Lore in the Old Testament*, vol. 1, Macmillan, London, 568 pages, pp. 183–185.
16. For example, Ryan, William, and Walter Pittman, 1998. *Noah's Flood: The New Scientific Discoveries about the Event That Changed the World*, Simon and Schuster, New York, 319

pages. Their theory was that glacial runoff built up in a closed-off Black Sea until it broke through at a weak point and caused a catastrophic flood.

17. Poebel, Arno, 1913. "Important Historical Documents Found in the Museum's Collection of Ancient Babylonian Clay Tablets," *Museum Journal*, vol. 4, no. 2, pp. 37–50, University Museum, Univ. of Pennsylvania. The Noah character in the Sumerian flood story was a man named Ziusudra.
18. Emerson, B.K., 1896. "Geological Myths," *Science*, vol. 4, pp. 328–344. The Babylonian flood story was in the *Epic of Gilgamesh*, written sometime between 1300 and 1000 BC, with a man named Ut-napishtim as the Noah character. The Greek version was written between 800 and 600 BC, with Deucalion as the man who built the ark. The Muslim version in the Koran (written much later, sometime in the late seventh century AD) uses many aspects of the biblical version, including Noah as the lead character. The Indian version was probably written about the same time as the Greek version.
19. Ibid.
20. Frazer, James George, 1916. "Ancient Stories of a Great Flood," *Journal of the Royal Anthropological Institute of Great Britain and Ireland*, vol. 46, pp. 231–283. The term *hurricane* was used in this translation, but hurricanes were not known at the time these clay tablets were written, nor was the concept of a rotating storm. The author must have meant a very large storm, and the translator thought that was conveyed by using *hurricane*, ignoring the technical meaning of the word. However, as will be seen, it actually could have been a hurricane in the Persian Gulf that produced a storm surge that flooded the lower Euphrates valley.
21. Suess, Eduard, 1885. *Das Antlitz der Erde* (*The Face of the Earth*), transl. B.C. Sollas, 1904, Clarendon Press, Oxford, UK, 482 pages, p. 71.
22. Parker, Bruce, 2000. "The *Perfect* Storm Surge," *Mariners Weather Log*, vol. 44, no. 2 (Aug.), pp. 4–12.
23. Louie, Kin-sheun, and Kam-biu Liu, 2003. "Earliest Historical Records of Typhoons in China," *Journal of Historical Geography*, vol. 29, no. 3, pp. 299–316. According to their survey of ancient literature, the Chinese character *ju jufeng* appeared for the first time in the book *Nanyue Zhi* (*Records of the South*), written sometime after AD 453 by Shen Huai-yuan and widely cited by later works.
24. Ibid. Louie and Liu provide this quotation from *Tai Ping Yu Lan* (*Tai-ping Reign-Period Imperial Encyclopedia*), written by Li Fang and others between AD 977 and 984.
25. In the Southern Hemisphere a cyclone rotates clockwise, so there these directions would be opposite.
26. Louie and Liu, op. cit.
27. Louie and Liu, op. cit. In *Jin Tang Shu* it states that on "day wu-shen of the eighth [lunar] month [of the eleventh year of the Yuan the reign-period], Rongzhou reported that jufeng occurred and seawater damaged the city wall." Another book published later confirmed this: "Strong rainstorm at Mizhou, sea surged, city wall damaged." Both attributed the damage to the sea surge rather than the wind. The first quotation is from Chapter 15 of *Jin Tang Shu* and is believed to have named the wrong city, Rongzhou, which is too far from the coast and has a Chinese character similar to Mizhou's. Louie and Liu verified this with a reference, *Sin Tang Shu*, published sometime between AD 1044 and 1066, in which they found the following statement. "6th lunar month [of the eleventh year of the Yuan reign-period], strong rainstorm at Mizhou, sea surged, city wall damaged." Mizhou is today called Gaomi City, in Shandong Province.
28. Needham, Joseph, 1954. *Science and Civilisation in China*, vol. 3, chap. 21, Cambridge Univ. Press. Needham quotes a book written by Meng Guan in the ninth century AD.
29. Louie and Liu, op. cit., p. 306. The key quotation from Liu Xun's *Ling Biao Lu Yi* (*Southern Ways of Men and Things*), 890 AD, is "When a jufeng arrives before the tide goes down,

the waves overflow onto the shore. Houses are thus flooded, crops washed away and vessels capsized. This is called *ta chao* [overlapping tides] in Nanzhong. . . . It is popularly called *hai chao* (sea surge) or *man tian* (seawater reaching the sky)." Louie and Liu have also found numerous poems from the eighth and ninth centuries AD that refer to *jufeng* and *jumu*, showing that these were well known by the Chinese of that era.

30. Marco Polo, who was with Kublai Khan during his invasions of Japan, describes one of them in his book. See: Polo, Marco, ca. 1299. *The Travels of Marco Polo the Venetian*, ed. John Masefield, 1908 edition, J.M. Dent, London, 461 pages, pp. 325–327. See also: Neumann, J., 1975. "Great Historical Events That Were Significantly Affected by the Weather: I. The Mongol Invasions of Japan," *Bulletin of the American Meteorological Society*, vol. 56, no. 11, pp. 1167–1171; Winters, Harold A., 1998. *Battling the Elements*, Johns Hopkins Univ. Press, Baltimore, MD, 318 pages, pp. 9–15.
31. Columbus's encounter with a hurricane was first reported in 1511 by Peter Martyr in his *de Orbo Novo* (*On the New World*), also referred to as *The Eight Decades of Peter Martyr d'Anghera* (see Book IV of the *First Decade*). Martyr wrote his book in Latin, and the exact spelling and meaning of some of the words he chose to use, including *furacane*, seem to be interpreted differently by different translators. There has been some disagreement on whether Columbus witnessed a hurricane or a waterspout (the over-ocean equivalent of a tornado), but the description that "the sea extended itself further in to the land, and rose higher than ever it did before in the memory of man, by the space of a cubit" indicates a hurricane. Part of the confusion was from the line "these tempests of the air (which Grecians call *Tiphones*, that is, whirl winds), they call *Furcanes*," because the Greek word *typhoon* was typically used for waterspouts (also see note 35 regarding the word *typhoon*). But this line goes on, "which they say do often times chance this Island: But that neither they nor their great grandfathers ever saw such violent and furious Furcanes, that plucked up great trees by the roots; Neither yet such surges and vehement motions of the sea, that so wasted the land." This translation is quoted in Ludlum (see note 34). Francis Augustus MacNutt (1912) in his translation added a note that says that Hurakán was the name of the god of winds in the mythology of Yucatán.
32. Emanuel, Kerry, 2005. *Divine Wind: The History and Science of Hurricanes*, Oxford Univ. Press, 285 pages, pp. 18–20.
33. It might be, however, that the spiral arms came from the spiral shape seen when a conch shell is cut open. The conch shell had been identified with the wind god Huracán. But, on the other hand, it is possible that the Caribbean Indians saw the hurricane as a much larger version of the narrow *waterspout* (or *whirlwind*), the over-ocean equivalent of a tornado, whose rotation was more easily observable.
34. Ludlum, David M., 1963. *Early American Hurricanes*, American Meteorological Society, 198 pages. See also: Mulcahy, Matthew, 2006. *Hurricanes and Society in the British Greater Caribbean, 1624–1783*, Johns Hopkins Univ. Press, 257 pages.
35. The origin of *typhoon* is still debated. There were many variations of the word as it was passed from culture to culture throughout the Indian Ocean and Far East region, although the direction of its passage is not always clear. See: Yule, Henry, and A.C. Burnell, 1903. *Hobson-Jobson: A Glossary of Colloquial Anglo-Indian Words and Phrases*, John Murray, London, 1,021 pages, pp. 947–950. "Typhoon" in Chinese is *taifeng*, but that word does not seem to have come into use until the eighteenth century. See: Louie and Liu, op. cit., p. 299. The first written documents using some form of the word *typhoon* were from European authors who had sailed in the Far East. In 1598 Richard Hakluyt wrote of *touffons*, one of which caused a "great wave of the Sea" that drove his ship beyond the shore, and when the "touffon" ended and the water receded, "wee found her a good mile from the Sea on drie land." See: Hakluyt, Richard, 1598–1600. *The Principal Navigations Voyages Traffiques & Discoveries of the English Nation*, vol. 3, J.M.

Dent, London, 370 pages, pp. 258–259. Captain William Dampier mentions *tuffoons* in his 1699 book based on his detailed journal written while sailing around the world three times. Dampier wrote that the wind direction veered around the compass and that there could be a period of calm. He also wrote of "the Sea... [being] driven off the shore by the violence of the Wind, so far, that some ships riding in the Harbour in 3 or 4 Fathoms Water, were a-ground; and lay so till the S.W. Gust came, and then the Sea came rowling in again with such prodigious fury, that it not only set them a-float, but dash'd many of them on the shore. One of them was carried up a great way into the Woods." See: Dampier, William, 1699. *Voyages and Descriptions*, vol. 2, pt. 3, chap. 6, pp. 279–293, ed. John Masefield, E. Grant Richards, London, 1906, 607 pages.

36. The English colony at Jamestown, Virginia, was hit on September 6, 1667, by "a very strange tempest which hath been seen in these parts (with us called a hurricane).... The sea swelled (by the violence of the wind) twelve feet above its usual height drowning the whole country before it, with many of the inhabitants, their cattle and goods." From: Anonymous, 1667. *Strange News from Virginia*, a pamphlet published in London.
37. Piddington, Henry, 1869. *The Sailor's Horn-Book for the Law of the Storms*, 5th ed., William and Norgate, London, 377 pages, pp. 10–11. In 1844 Piddington proposed "for all...circular or highly curved winds, the term '*Cyclone*' from the Greek Κυκλος (which signifies the coil of a snake) as neither affirming the circle to be a true one, though the circuit may be complete, yet expressing sufficiently the *tendency* to circular motion." This last comment on the "*tendency* to circular motion" was a diplomatic reference to the scientific debate that was then going on about the cause of tropical cyclones and just how circular their motion really was.
38. Some of that experience is mentioned in notes 35 and 36. There were a few isolated experiences earlier than the fifteenth century, when Europeans traveled to the Indian Ocean or to the Far East, for example, Marco Polo writing about the typhoon that crippled Kublai Khan's fleet and saved Japan. See note 30.
39. Tacitus, AD 117. *Annals*, bk. 1, chaps. 68–70, in *The Annals of Tacitus*, bks. 1–6, transl. George Gilbert Ramsay, 1904, John Murray, London, p. 83. He wrote, "The land was flooded; sea, shore, and fields, all presented one aspect." When the two Roman legions of Publius Vitellius (part of the army of Germanicus) were caught by a storm surge, "baggage, baggage-animals and corpses floated about and jostled against each other. All distinction of maniples [the units of each Roman legion] was lost. Some had their breasts above water, some their heads only; sometimes the ground would give way beneath them altogether; they would be thrown this way and that, and go under...brave men and cowards alike were swept along by the fury of the elements."
40. Pliny the Elder, AD 79. *Natural History*, bk. 16, in *Natural History: A Selection*, transl. John F. Healy, 1991, Penguin, p. 206. Thousands of these artificial dwelling mounds had dotted the coastal lowlands in all shapes and sizes. Later such a settlement mound became known as a *terp* (in the Dutch province of Friesland), a *torp* (in Denmark), or a *warft* (in Germany).
41. Jordan-Bychkov, T.G., and B.B. Jordan, 2002. *The European Culture Area: A Systematic Geography*, Roman and Littlefield, 437 pages, p. 72.
42. The term *polder* was apparently first used in 1138 in Flanders Plain. Ibid., p. 73. The drained land inside a polder would eventually compact and subside, so that its ground level dropped below sea level. Then draining the water at low tide via gravity would no longer work, and pumping became necessary, powered initially by windmills. Wind power drove a wheel with scoops that lifted the water; later the scoops would be replaced by an Archimedes screw. A village or town usually developed within each polder or group of joined polders. The political leader was often the *dike master* or *dike reeve*,

typically a rich landowner chosen by his neighbors, who, with the help of a *dike board* or a *water board*, kept the dikes strong.

43. Barry, Sally C., transl., 2005. "Flemish Lace in the Old Catholic Church on Nordstrand," Old Catholic Parish, Nordstrand, Uthlande-Verlag Nordstrand, p. 7.
44. Steensen, Thomas, 2008. "Das 'dänische Holland,' *Nordfriesland*," no. 162 (June), pp. 17–26. See also: Reinhardt, Andreas, ed., 1984. *Die erschreckliche Wasser-Fluth, 1634*, issue 9, North Local Association, Husum Printing, 204 pages.
45. Leeghwater's account is quoted in full in Rheinheimer, Martin, 2003. Mythos Sturmflut, *Demokratische Geschichte. Jahrbuch für*, Schleswig-Holstein, vol. 15, pp. 9–58.
46. The two largest remaining pieces would later be called the islands of Pellworm and Nordstrand, and many other small remaining pieces were called *Halligen*, described later as "mere shred of patches of land in the midst of the water, meadows with the greenest grass washed by every tide of the sea." From Ward, Adolphus W., 1921. *Collected Papers—Historical, Literary, Travel and Miscellaneous*, vol. 5, Cambridge Univ. Press, p. 83.
47. Barry, op. cit., p. 9; Reinhardt, op. cit. Some of the destruction of the 1634 storm surge is depicted in the etching in figure 3.1. It is titled "Die erschreckliche Wasser-Fluth" ("the terrible flood waters") and comes from the 1683 book by Eberhard Happel entitled *Die grőssten denkwürdigkeiten der Welt* (*Greatest Curiosities of the World*).
48. Those who survived the storm surge lost their houses, their windmills, and their livelihood, and so most Frisians left their home country. Some went to America, to the Dutch colony of New Netherlands and especially to New Amsterdam, which would later be called New York City when the English took it over. Jonas Bronck, after whom the Bronx is named, was a Frisian who first owned the land of that future New York City borough. The last governor of New Amsterdam, Peter Stuyvesant, may have called on his Frisian background when he built the protective wall that gave Wall Street its name.
49. Franklin, Benjamin, 1760. "On the northeast storms in North America," a letter to Alexander Small in London, May 12, 1760, from *The Works of Benjamin Franklin*, 1838, vol. 6, ed. Jared Sparks, Hilliard, Gray and Company, Boston, MA, 677 pages, pp. 219–222.
50. Franklin, Benjamin, 1750. A letter to Jared Eliot, February 13, 1750, from *The Works of Benjamin Franklin*, 1838, vol. 6, ed. Jared Sparks, Hilliard, Gray and Company, Boston, MA, 677 pages, pp. 105–108.
51. Franklin, Benjamin, 1747. A letter to Jared Eliot, July 16, 1747, from *The Works of Benjamin Franklin*, 1838, vol. 6, ed. Jared Sparks, Hilliard, Gray and Company, Boston, MA, 677 pages, pp. 79–80.
52. Franklin, op. cit., 1760.
53. Franklin, Benjamin, 1753. A letter to John Perkins, February 4, 1753, from *The Works of Benjamin Franklin*, 1838, vol. 6, ed. Jared Sparks, Hilliard, Gray and Company, Boston, MA, 677 pages, pp. 145–159.
54. Varenius, Bernhard, 1650. *A Compleat System of Geography, Explaining the Nature and Properties of the Earth*, transl. Mr. Dugdale in 1734, London, 521+ pages. In 1650 he wrote, "A Typhone is a strong swift Wind that blows from all Points, wandering about all quarters.... It rages not only at Sea, but on Land,... and carries great Ships a quarter of a Mile from the Sea.... The Cause of it, no doubt, is that the Wind rushing to a certain Point, is obstructed, and returns on itself, and is thus turned around, as we see in Water that turns about in a Vortex, when it meets with an Obstacle; or it may come from furious Winds meeting one another." Although his explanation was wrong, Varenius had correctly recognized the rotating air motion.
55. Langford, 1698. "Captain Langford's Observations of His Own Experience upon Huricanes and Their Prognosticks," *Philosophical Transactions*, vol. 20, pp. 407–416.

56. Capper, James, 1801. *Observations on Winds and Monsoons.* Quoted in Rosser, W.H., 1876. *The Law of Storms Considered Practically,* Charles Wilson, London, 108 pages, p. 7. Capper gave some guidance to ship captains: "If the changes are sudden and the wind violent, in all probability the ship must be near the centre or vortex of the whirlpool; whereas, if the wind blows a great length of time from the same point, and the changes are gradual, it may be reasonably supposed the ship is near the extremity of it."
57. Donnelly, J.P., S. Roll, M. Wengren, J. Bulter, R. Lederer, and T. Webb, 2001. "Sedimentary Evidence of Intense Hurricane Strikes from New Jersey," *Geology,* vol. 29, no. 7, pp. 615–618. This paleotempestology study found three significant overwash deposits, caused by very large storm surges. The middle sediment layer was from the storm surge caused by the 1821 hurricane. The study shows that another hurricane crossed the New York City location, but long before there was a New York City, sometime between 1278 and 1438. A third sediment layer, the uppermost and smallest layer, was from the storm surge caused by a northeaster in 1962.
58. Redfield, William C., 1831. "Remarks on the Prevailing Storms of the Atlantic Coast, of the North American States," *American Journal of Science and Arts,* vol. 20, pp. 17–51, pp. 21, 22.
59. Espy, James P., 1841. *The Philosophy of Storms.* Charles C. Little and James Brown, Boston, MA, 552 pages.
60. It is the Earth's rotation that is responsible for hurricanes having circular wind patterns, counterclockwise in the Northern Hemisphere and clockwise in the Southern Hemisphere. A hurricane has a low-pressure center area, and it is surrounded by higher-pressure air masses. At the low-pressure center the air is rising because of heat and moisture from the ocean, and air moves toward the center to replace it. The air currents coming from the north, south, east, and west are all pushed toward the right (in the Northern Hemisphere) by the Coriolis effect (due to the rotation of the Earth). These multiple pushes, however, drive the rotation around the low-pressure center in a counterclockwise direction, almost like small gears around one large gear in the middle, the large gear rotating in a direction opposite that of the small gears. See: Parker, Bruce, 1998. "The Coriolis Effect: Motion on a Rotating Planet," *Mariners Weather Log,* vol. 42, no. 2, pp. 17–23.
61. Ferrel, William, 1856. "An Essay on the Winds and the Currents of the Ocean," *Nashville Journal of Medicine and Surgery,* vol. 11, no. 4–5. An unusual journal to publish a landmark paper in meteorology. Ferrel was the brilliant but shy scientist who, as we saw in Chapter 1, invented the first U.S. tide-predicting machine. The journal's publisher was a friend.
62. Ibid., p. 191. See also: H.H.K., 1901. "What Is a Storm Wave?" *Monthly Weather Review,* Oct., pp. 460–463.
63. Piddington, op. cit., 1869, p. 195.
64. Piddington, Henry, 1853. "On the Cyclone Wave in the Sunderbunds. A Letter to the Most Noble the Governor-General of India, Calcutta." The letter is quoted in Prichard, 1869 (note 65), and cited in Piddington, op. cit., 1869. A copy of the letter is available at the Yale University Library.
65. Prichard, Iltudus T., 1869. *The Administration of India from 1859 to 1868,* vol. 1, Macmillan, pp. 227–230.
66. Buckland, op. cit., p. 407.
67. See note 38 in Chapter 6.
68. Norgate, Fred, 1874. "The Telegraph in Storm-Warnings," *Nature,* vol. 10, p. 125. A cyclone prediction using telegraphed wind data from locations where the cyclone had already been would be based on how those wind directions changed with time, as well as on studies of past cyclone tracks through the area.

69. Henry, Joseph, 1859. "On the Application of the Telegraph to the Premonition of Weather Changes," *Proceedings of the American Academy of Art and Sciences*, vol. 4, pp. 271–275.
70. Cox, John D., 2002. *Storm Watchers*, Wiley, 252 pages, pp. 51–56.
71. Ibid., pp. 75–83. See also: Anderson, Katherine, 1999. "The Weather Prophets: Science and Reputation in Victorian Meteorology," *Historical Society*, vol. 37, pp. 179–216; and Dry, Sarah, 2007. *Fishermen and Forecasts: How Barometers Helped Make the Meteorological Department Safer in Victorian Britain*, Discussion Paper 46, Centre for Analysis of Risk and Regulation, London School of Economics and Political Science, Oct., 30 pages. FitzRoy was earlier the captain of the HMS *Beagle* and the one who selected Charles Darwin as the ship's naturalist.
72. Those living by the coast often used "tide" to mean any change in water level, whether due to the astronomical tide or the effects of the wind. In this case, *Saxby Tide* implied the extra large high tide due to the effects of the wind from the Saxby Gale, that extra height being due to the storm surge.
73. *Moncton Times Daily Morning News* (Saint John, Canada), Friday, Oct. 8, 1969, p. 3.
74. Saxby, S.M., 1864. *Saxby's Weather System, or Lunar Influence on Weather*, 2nd ed., Longmans, London, 132 pages. (The first edition of this book in 1861 had been called *Foretelling Weather*, without his name in the title.) Saxby was attached to a guard ship of reserve in Sheerness called the *Cumberland*. He was involved in an odd assortment of other activities, for he was an agent to Lloyd's of London, an inventor of the "sphero-graph," and an inventor of improvements to winches, steam engine boiler plates, and window blinds. He wrote several books on marine steam engines, nautical astronomy, and most famously his own system of weather prediction. His weather prediction book included an attempted rebuttal to Robert FitzRoy's objections to Saxby's lunar methods for predicting the weather, which FitzRoy had included in his scientifically based *The Weather Book*. (Both books were published by the same publisher, probably to the great chagrin of FitzRoy.)
75. Saxby, S.M., 1868. Letter to *The Standard*, London, Friday, Dec. 25, no. 13,851, p. 5.
76. Saxby, S.M., 1869. Letter to *The Standard*, London, Thursday, Sept. 16, no. 14,078, p. 2.
77. Saxby also got lucky, in terms of hoped-for notoriety, when his letters in *The Standard* were noticed by Frederick Allison, an amateur meteorologist who lived near the Bay of Fundy, who published a monthly weather column in *The Evening Express* in Halifax. On October 1, in a letter to the editor of the same paper, he strongly expressed his belief in Saxby's prediction and added, "I believe that a heavy gale will be encountered here on Tuesday next, the 5th Oct." From Allison, Frederick, 1869. Letter to *Evening Express* (Halifax, Nova Scotia, Canada), Friday, Oct. 1, p. 2.
78. *Moncton Times Daily Morning News* (Saint John, Canada), Friday, Oct. 8, 1869, p. 2.
79. Udías, Agustín, 2003. *Searching the Heavens and the Earth: The History of Jesuit Observations*, Kluwer Academic, 369 pages.
80. Larson, Erik, 1999. *Isaac's Storm*, Crown, New York, 323 pages, pp. 71–72. See also: Cox, John D., 2002. *Storm Watchers*, Wiley, 252 pages, pp. 51–56.

CHAPTER 4

1. Vines, Benito, 1885. *Practical Hints in Regard to West Indian Hurricanes*, transl. Lieut. George L. Dyer, USN, U.S. Hydrographic Office, Bureau of Navigation, no. 77, Government Printing Office, Washington, DC, 15 pages.
2. Larson, Erik, 1999. *Isaac's Storm*, Crown, New York, 323 pages, pp. 71–72. See also: Cox, John D., 2002. *Storm Watchers*, Wiley, 252 pages, pp. 112–114.

3. In attachment to May 5, 1934, letter from the Forecast and Synoptic Reports Division of the Weather Bureau to Mrs. M.S. Douglas, showing the "Weather Conditions and General Forecast" for Sept. 4–8, 1900. Copy from the Galveston and Texas History Center, Rosenberg Library, Galveston, Texas.
4. Cox, op. cit., p. 118.
5. Another storm surge destroyed the Isle Derniere in Louisiana, 370 miles east of Galveston, killing more than two hundred people on August 10, 1856. See: Sallenger, Abby, 2009. *Island in a Storm*, Public Affairs, New York, 284 pages.
6. Although Cline had built up a good reputation for his forecasts, they were primarily forecasts for river-produced floods. His scientific understanding of hurricanes and especially storm surges was poor. As evidence that "it would be impossible for any cyclone to create a storm wave which could materially injure the city," Cline made two ridiculous statements in an article in the *Galveston Daily News*, also arrogantly describing an opposing opinion as "an absurd delusion." First, Cline said that the gently sloping shallow water in front of Galveston would keep storm waves from the city, when in fact the opposite is true, since this situation amplifies the storm wave. Second, he wrote, "The coast of Texas is according to the general laws of the motion of the atmosphere exempt from West India hurricanes and the two which have reached it followed an abnormal path which can only be attributed to causes known in meteorology as accidental." Cline was not alone in believing his first statement, although by 1900 there were many scientific papers explaining why this opinion was wrong, but the nonsense of his second statement is quite remarkable. See: Cox, op. cit., p. 118–120.
7. Cline, Isaac M., 1900. "Special Report on the Galveston Hurricane of September 8, 1900," in E.B. Garriott, "Forecasts and Warnings, West Indian Hurricane of September 1–12, 1900," *Monthly Weather Review*, vol. 28, no. 9, pp. 371–377.
8. And yet the signs of an approaching hurricane had been clearly laid out in a paper published by Wilfrid Stearns, the principal of the Rosenberg School in Galveston, whom Cline must have known. Stearns listed three important indications of an approaching storm: (1) an increase of surf intensity at the shore, (2) a great increase in the wavelength of the wind waves approaching shore (longer swell), and (3) "a marked and quite sudden rise in the tide [that] occurs without wind." Stearns, Wilfrid D., 1894. "Storms of the Gulf of Mexico and Their Prediction," *American Meteorological Journal*, vol. 10, pp. 497–504. Cline certainly should have been aware of the paper, since the same volume contains a paper by Cline (pp. 45–47), as well as a paper by his brother Joseph L. Cline (pp. 155–163). That volume also contained an announcement of an examination for a "professorship in the Weather Bureau" and named the competitors, which included Cline.
9. Larson, op. cit., p. 140. See also: Young, S.O., 1900. "Story of the Hurricane Which Swept Galveston," *Galveston Daily News*, Sept. 12, no. 172.
10. Larson, op. cit., p. 188.
11. Larson, op. cit. See also: Cline, Isaac Monroe, 1951. *Storms, Flood and Sunshine*, Pelican, 352 pages; Cline, Joseph Leander, 1946. *When the Heavens Frowned*, Mathis, Van Nort, 221 pages. The present summary is too brief to properly get across the drama of that night. For a much more detailed and dramatic account, see Erik Larson's *Isaac's Storm* (see note 2).
12. Probably many more died. A death toll between 8,000 and 12,000 is the number now most quoted. The picture in figure 4.4 of Galveston houses totally destroyed by the 1900 storm surge came from the NOAA Photo Library, but is also available at the Galveston and Texas History Center, Rosenberg Library, Galveston, Texas.
13. A later paper by Cline on storm surges was: Cline, Isaac M., 1920. "Relation of Changes in Storm Tides on the Coast of the Gulf of Mexico to the Center and Movement of

Hurricanes," *Monthly Weather Review*, vol. 48, no. 3, pp. 127–146. Many of the papers that he cites, however, were written before 1900. In Cline's time, a "storm surge" was still called a "storm wave," and "storm tide" referred to the storm surge added to the astronomical tide.

14. Cline, op. cit., 1900, p. 374.
15. Jarvinen, Brian R., 2006. *Storm Tides in Twelve Tropical Cyclones (Including Four Intense New England Hurricanes)*, Internal report for the National Hurricane Center, National Weather Service, NOAA, 99 pages. In this report a storm-surge model (the SLOSH model talked about later in this chapter) was used as a diagnostic tool to determine the size of storm surges produced by historical hurricanes in the United States, based on observed high-water marks that were representative of the storm tide, eyewitness accounts, and available meteorological data. According to this report, the 1915 hurricane actually produced a slightly higher storm surge than the 1900 hurricane.
16. Eliot, John, 1877. *Report of the Vizagapatam and Backergunge Cyclones*, Bengal Secretariat Press, Calcutta, 187 pages, p. 180. See also: Buckland, C.E., 1901. *Bengal Under the Lieutenant-Governors*, S.K. Lahiri, Calcutta, vol. 1, 1,100 pages, pp. 298–299, p. 575; "The Cyclone Wave in Bengal," *Leeds Mercury* (UK), Dec. 13, 1876, no. 12068; and "The Indian Cyclone and Famine, Terrible Effects of the Storm Wave," *New York Times*, Nov. 21, 1876. The first of three storm surge waves hit on the night of October 31–November 1, 1876, at midnight, near the time of a spring high tide, submerging populated islands to a depth of at least twenty feet.
17. IOC, 1998. *Project Proposal on Storm Surges for the Northern Part of the Indian Ocean.* IOC/EC-XXXI/7, Thirty-first Session of the IOC Executive Council, Paris, November 17–27, 1998, Intergovernmental Oceanographic Commission of UNESCO. Based on the table in app. 1.
18. Flierl, G.R., and A.R. Robinson, 1972. "Deadly Surges in the Bay of Bengal: Dynamics and Storm-Tide Tables," *Nature*, vol. 239, pp. 213–215.
19. This is much less important for the Gulf of Mexico, where the tidal range is very small.
20. Parker, Bruce, 1998. "The Coriolis Effect: Motion on a Rotating Planet," *Mariners Weather Log*, vol. 42, no. 2, pp. 17–23. See also: Parker, Bruce, 2000. "The *Perfect* Storm Surge," *Mariners Weather Log*, vol. 44, no. 2 (Aug.), pp. 4–12.
21. In the Northern Hemisphere the Coriolis effect pushes the wind-induced surface current slightly to the right (in the Southern Hemisphere, to the left). With increasing depth, each layer of water is further deflected to the right, the water speed decreasing with depth, producing an *Ekman spiral*. When averaged over the entire depth of the current, water transport is perpendicular (i.e., 90 degrees) to the right of the direction of the wind. If a coast in the Northern Hemisphere runs north and south, a wind blowing from the north will pile water up against the coast, raising the water level, while a wind from the south will push it away from the coast, lowering the water level.
22. To be as accurate as possible, oceanographers include the *inverted barometer effect*. For every atmospheric pressure drop of two and a half millibars, the water level rises an inch. Even at the center of a hurricane, with its very low pressure, this would account for a rise of only a couple of feet.
23. For example, the rules given in: Wheeler, W.H., 1906. *A Practical Manual of Tides and Waves*, Longmans, Green, London, 201 pages.
24. "Thames Floods, Investigation into Causes, a North Sea Surge," *Times* (London), Oct. 23, 1928, no. 45031. See also: Doodson, A.T., and J.S. Dines, 1929. "Meteorological Conditions Associated with High Tides in the Thames," *Geophysical Memoirs*, no. 47, U.K. Meteorological Office, London, 26 pages.
25. "Weather Forecast," *Times* (London), Jan. 30, 1953, no. 52533; "Weather Forecast," *Times* (London), Jan. 31, 1953, no. 52534.

26. Baxter, Peter J., 2005. "The East Coast Big Flood, 31 January–1 February 1953: A Summary of the Human Disaster," *Philosophical Transactions of the Royal Society*, A, vol. 363, pp. 1293–1312.
27. Grieve, Hilda, 1959. *The Great Tide*, County Council of Essex, Chelmsford, UK, Waterlow, 883 pages, p. 233.
28. Meadows, Dick, "1953 Floods Hero Returns to Remember," *BBC News*, Jan. 23, 2003, http://news.bbc.co.uk/2/hi/uk_news/england/2678215.stm.
29. Baxter, op. cit., p. 1302.
30. Ibid., p. 1308.
31. Gerritesen, Herman, 2005. "What Happened in 1953? The Big Flood in the Netherlands in Retrospect," *Philosophical Transactions of the Royal Society*, A, vol. 363, pp. 1271–1291, p. 1276.
32. Ibid. The storm surge warning service (Stormvloedwaarschuwingsdienst, or SVSD), which had been in operation since 1921, was run by the Royal Netherlands Meteorological Institute (Koninklijk Nederlands Meteorologisch Instituut, or KNMI).
33. The story originated in *Hans Brinker, or the Silver Skates*, written in 1865 by Mary Mapes Dodge.
34. Gerritesen, op. cit., p. 1281. To commemorate the saving of Holland in 1953, embedded in the dike is a statue of Captain Arie Evegroen and the *Twee Gebroders* (*Two Brothers*), the grain boat commandeered by the mayor of Nieuwerkerkaan de Ijssel, J.C. Vogelaar.
35. The warning system, now called the Storm Tide Forecasting Service, was later expanded to thirty-eight and then forty-five gauges around Great Britain. The Liverpool Tidal Institute is now part of the Proudman Oceanographic Laboratory.
36. Schneer, Jonathan, 2005. *The Thames*, Yale Univ. Press, New Haven, CT, 330 pages, pp. 249–262. A Thames barrier system had been called for many times in the past, but the case was apparently never made strongly enough to warrant the high costs. The 1953 storm surge finally pushed the political balance in favor of building the barrier, but still not without a political battle.
37. Gerritesen, op. cit., pp. 1285–1286.
38. Bijker, Wiebe E., 2002. "The Oosterschelde Storm Surge Barrier," *Technology and Culture*, vol. 43, no. 3, pp. 569–584.
39. Reid, Robert O., 1955. "Status of Storm-Tide Research," *Science*, vol. 122, no. 3178, pp. 1006–1008. See also: Defant, Albert, 1961. *Physical Oceanography*, vol. 2, Macmillan, New York, 590 pages, pp. 229–237.
40. Harris, D. Lee, 1962. "The Equivalence between Certain Statistical Prediction Methods and Linearized Dynamical Methods," *Monthly Weather Review*, vol. 90, pp. 331–340. See also: Pore, N.A., 1964. "The Relation of Wind and Pressure to Extratropical Storm Surges at Atlantic City," *Journal of Applied Meteorology*, vol. 3, no. 2, pp. 155–163.
41. Jelesnianski, C., 1967. "Numerical Computations of Storm Surges with Bottom Stress," *Monthly Weather Review*, vol. 95, no. 11, pp. 740–756.
42. Hansen, W., 1956. "Theorie zur Berechnung des Wasserstandes und der Stromungen in Randmeeren nebst Anwedungen," *Tellus*, vol. 8, no. 3, pp. 287–300. This was the first serious use of a numerical hydrodynamic model for storm surges (in this case in the North Sea).
43. The first successful weather satellite, TIROS-1, was launched on April 1, 1960, by the National Aeronautics and Space Administration (NASA) and was operated by the U.S. Environmental Science Services Administration (ESSA), the predecessor to NOAA. It was a polar-orbiting satellite that took visible and infrared (thermal) pictures of the Earth and could show cloud cover and map the thermal structure of hurricanes and other storms. This was followed by a series of TIROS and ESSA weather satellites, until by 1965 complete satellite coverage of the sun-illuminated side of the planet was achieved.

Because of satellite measurements, one no longer had to rely solely on airplane measurements or on the interpretation of pressure and wind measurements from ships that happened to be sailing in the right segment of the ocean, although that information was still important for estimating the strength of hurricanes. See: Burroughs, William J., 1991. *Watching the World's Weather*, Cambridge Univ. Press, 196 pages.

44. Dunn, Gordon E., 1962. "The Tropical Cyclone Problem in East Pakistan," *Monthly Weather Review*, vol. 90, no. 3, pp. 83–86. See also: Sullivan, Walter, 1970. Cyclone May Be the Worst Catastrophe of Century," *New York Times*, Nov. 22; Anderson, Jack, 1970. "Hurricane Victims, 'Died Needlessly in Pakistan,'" Washington Merry-Go-Round, *Washington Post*, Jan. 31.
45. Zeitlin, Arnold, 1970. "Story of the Pakistan Storm," Associated Press story in the *Oakland Tribune*, Dec. 6, p. 24C.
46. Ibid.
47. "The Survivors on Devastated Island Wait for Food, Water and Medicine," *Daily Telegraph* (London), Nov. 17, 1970, reprinted in *New York Times*, Nov. 18, 1970.
48. Ibid.
49. Moraes, Dom, 1971. *The Tempest Within*, Barnes & Noble, New York, 103 pages, p. 65.
50. Frank, Neil L., and S.A. Husain, 1971. "The Deadliest Tropical Cyclone in History?" *Bulletin of the American Meteorological Society*, vol. 52, no. 6, pp. 438–444.
51. Blood, Archer K., 2002. *The Cruel Birth of Bangladesh—Memoirs of an American Diplomat*, Univ. Press Limited, Dhaka, Bangladesh, 373 pages, p. 77. See also: Siddiqi, Abdul Rahman, 2004. *East Pakistan, The Endgame—An Onlooker's Journal, 1969–1971*, Oxford Univ. Press, 260 pages, p. 46.
52. Schanberg, Sydney H., 1970. "Foreign Relief Spurred," *New York Times*, Nov. 24 (written Nov. 22).
53. Schanberg, Sydney H., 1970. "Pakistanis Fear Cholera's Spread," *New York Times*, Nov. 23 (written Nov. 22).
54. Associated Press, 1970. "Disputes Snarl Cyclone Relief," *Charleston Daily Mail*, Nov. 23.
55. Siddiqi, op. cit., p. 48.
56. This very short summary of the events leading to the independence of Bangladesh, following the devastation of the 1970 storm surge, does not do justice to the drama; see the book by Archer K. Blood cited in note 51 for more details.
57. Haque, C. Emdad, and Danny Blair, 1992. "Vulnerability to Tropical Cyclones: Evidence from the April 1991 Cyclone in Coastal Bangladesh," *Disasters*, vol. 16, no. 3, pp. 217–229.
58. Back in the 1960s and early 1970s, when storm surge models were first used, computer power was inadequate. It could take a storm surge model longer to run than it took a real storm surge to move in and flood the shore. And even when computer power increased, real-time meteorological and oceanographic data for the models were insufficient. But that slowly changed over the next two decades.
59. Jelesnianski, C.P., Jye Chen, and Wilson A. Shaffer, 1992. *SLOSH: Sea, Lake, and Overland Surges from Hurricanes*, NOAA Technical Report NWS 48, National Weather Service, NOAA, 71 pages. Along with Jelesnianski's two coauthors above, Albion Taylor also contributed significantly to the development of the SLOSH model system. An earlier version of SLOSH was called SPLASH, which stood for "Special Program to List Amplitudes of Surges from Hurricanes."
60. Ibid., p. 64. Operational storm surge prediction at the National Hurricane Center does not use individual SLOSH model runs, but rather composites of many individual model runs, which are displayed graphically on a map. Each composite is called a *MEOW* (Maximum Envelope of Water), which shows maximum storm surge heights at every grid cell in the SLOSH model grid for a specific coastal region, produced from many

model runs for a given hurricane category, forward speed, and direction of motion, for a given tide level. Also produced is a higher-level product called a *MOM* (Maximum of MEOWs), it being a composite of the maximum storm surge heights for all simulated hurricanes of a given category (so there are typically five MOMs per coastal region). These products reduce variability and uncertainty and provide essentially worst-case scenarios, which are most useful to emergency evacuation officials.

61. Many universities as well as the U.S. Army Corps of Engineers also have storm surge models, such as the ADCIRC (Advanced Circulation) model. Newer models include the nonlinear interaction between the storm surge and the tide, as well as the effect of *wave setup*, namely, the increased water levels due to the large swells reaching the shore from the hurricane, even before landfall. For a recent review of the state of storm surge modeling, see: Resio, Donald T., and Joannes J. Westerink, 2008. "Modeling the Physics of Storm Surges," *Physics Today*, vol. 61, no. 9, pp. 33–38.
62. Parker, Bruce B., 1996. "Monitoring and Modeling of Coastal Waters in Support of Environmental Preservation," *Journal of Marine Science and Technology*, vol. 1, no. 2, pp. 75–84.
63. *Hurricane Katrina Advisory Number 19*, released by NHC at 10:00 P.M. on Saturday, August 27. *Hurricane Katrina Advisory Number 23* was released at 10:00 A.M. on Sunday, August 28. *Hurricane Katrina Advisory Number 24* was released at 4:00 P.M. on Sunday, August 28. The "Hurricane Katrina Forecast Timeline" from the NOAA National Hurricane Center, NOAA, Department of Commerce, summarizes events. This timeline was also attached to documents listed in notes 64, 65, and 66.
64. Glahn, B., A. Taylor, N. Kurlowski, and W.A. Shaffer, 2009. "The Role of the SLOSH Model in National Weather Service Storm Surge Forecasting," *National Weather Digest*, vol. 33, no. 1, pp. 3–14. See also: Committee of Homeland Security, 2006. *Hurricane Katrina: A Nation Still Unprepared*, Special Report of the Committee on Homeland Security and Governmental Affairs, U.S. Senate, S. Rept. 109–322, 109th Cong., 2nd session, 74 pages, pp. 42, 46.
65. Mayfield, Max, 2005. Written testimony for Oversight Hearing on NOAA Hurricane Forecasting before the Select Committee for Hurricane Katrina, U.S. House of Representatives, Sept. 22.
66. House Committee on Science, 2005. *Failing to Protect and Defend: The Federal Emergency Response to Hurricane Katrina*, Democratic Staff of the House Committee on Science, Oct. 20, 2004, ver. 1.0a.
67. Hurricane Katrina Forecast Timeline, op. cit.
68. Select Bipartisan Committee, op. cit.
69. Ibid.
70. The old record for highest U.S. storm surge had been twenty-four feet, set also at Pass Christian during Hurricane Camille in 1969. Katrina decreased to a Category 3 storm before hitting the Mississippi coast and was weaker than Camille, which was still a Category 5 storm when it hit the coast, but Katrina was spread over a larger geographical area than Camille, and it pounded the coast for a longer time. See: Knabb, Richard D., Jamie R. Rhome, and Daniel P. Brown, 2005. *Tropical Cyclone Report, Hurricane Katrina, 23–30 August 2005*. National Hurricane Center, National Weather Service, NOAA, Dec. 20 (updated Aug. 10, 2006). See also: Resio and Westerink, op. cit., Glahn et al., op. cit.
71. Mouawad, Jad, 2005. "No Quick Fix for Gulf Oil Operations," *New York Times*, Aug. 31.
72. Select Bipartisan Committee, 2006. *A Failure of Initiative*, Final Report of the Select Bipartisan Committee to Investigate the Preparation for and Response to Hurricane Katrina, Feb. 15, U.S. Government Printing Office, Washington, DC.

73. Shaw, A.M., 1929. "An Estimate of Storm-Tide Hazards to New Orleans," *Engineering News-Record*, vol. 102 (May 2), pp. 698–702. This was one of the earliest warnings about the vulnerability of New Orleans to storm surges from hurricanes.
74. Das, Saudamini, and Jeffrey R. Vincent, 2009. "Mangroves Protected Villages and Reduced Death Toll during Indian Super Cyclone," *Proceedings of National Academy of Sciences*, vol. 106, no. 18, pp. 7357–7360.
75. Although there may be additional uncounted deaths, no one has estimated more than 10,000 deaths for the 2007 storm surge in Bangladesh.
76. Some of the decrease in the number of deaths also might be due to the location of Sidr's landfall, its eye striking the southeastern part of the Sudarbans (and destroying almost a third of the mangrove forest), which is the least populated part of Bangladesh. But the strongest onshore winds and the largest storm surge heights were to the east of the eye and so would do the greatest damage to much of the same area as the 1970 Bhola cyclone and storm surge, which killed over 300,000. Cyclone Gorky in 1991 made landfall near Chittagong, which has a higher population density (but also better-built homes and higher ground to escape to). See: Paul, Bimal Kanti, 2009. "Why Relatively Fewer People Died? The Case of Bangladesh's Cyclone Sidr," *Natural Hazards*, vol. 50, no. 2, pp. 289–304. See also: Page, Jeremy, 2007. "Thousands Saved by Megaphone Messengers," *Times* (London), Nov. 24.
77. "A Year after Storm, Subtle Changes in Myanmar," *New York Times*, Apr. 29, 2009.
78. Webster, Peter J., 2008. "Myanmar's Deadly Daffodil," *Nature Geoscience*, vol. 1, pp. 488–490.
79. Resio and Westerink, op. cit. Even greater accuracy, however, would allow a better determination of which streets in a city will flood and need to be evacuated and which will not. Resio and Westerink's paper evaluates where storm surge models can still be improved.

CHAPTER 5

1. The *Queen Mary* set the speed record of 31 knots in 1938. See: Morison, Samuel Eliot, 1955. *The Battle of the Atlantic*, Little Brown, p. 306.
2. The two gray ghosts, as they were called because of their new paint jobs, made dozens of transits across the Atlantic, sometimes also carrying celebrities to USO shows, such as Bob Hope, Bing Crosby, Fred Astaire, and Mickey Rooney. Later Churchill himself took the *Queen Mary* west to meet with Roosevelt. After the war she made eleven war bride crossings to the United States and Canada. Satchell, Alister, 2001. *Running the Gauntlet: How Three Giant Liners Carried a Million Men to War, 1942–1945*. Naval Institute Press, 256 pages, p. 69.
3. *Gross tonnage* is the cargo volume of a ship, or the volume of the ship when not talking about cargo ships. It is not the weight of the ship. One *gross register ton* is equal to a volume of one hundred cubic feet.
4. Reynolds, David, 1995. *Rich Relations: The American Occupation of Britain, 1942–1945*, Random House, 555 pages, pp. 242–243.
5. Parrent, Erik, 1996. *47th Bomb Group (L)*, Turner Publishing, 104 pages, p. 11.
6. MacGregor, Mildred A., 2008. *World War II Front Line Nurse*, Univ. of Michigan Press, 456 pages, pp. 28–30. See also: Kelly, Carol. A., 2007. *Voices of My Comrades: America's Reserve Officers Remember World War II*, Fordham Univ. Press, 547 pages, pp. 78–79; Satchell, op. cit.
7. MacGregor, op. cit.
8. Parrent, op. cit.

9. Butler, Daniel A., 2002. *The Age of Cunard: A Transatlantic History, 1839–2003*, Lighthouse Press, 476 pages, pp. 316–317.
10. Satchell, op. cit. See also: Carter, Walter F., 2004. *No Greater Sacrifice. No Greater Love: A Son's Journey to Normandy*, Smithsonian Books, 224 pages, pp. 54–55.
11. Gleichauf, Justin F., 2002. *Unsung Sailors: The Naval Armed Guard in World War II*, Naval Institute Press, 456 pages, pp. 145–146. The author quotes Private First Class Raymond Roy, an army guard on the *Queen Mary*. See also: Butler, op. cit.
12. *Rutgers Oral History Archives of WW-II*, Interview with Charles Mickett, Jr., Oct. 24, 1997. See also: MacGregor, op. cit.
13. "Disaster Averted by the Queen Mary," *New York Times*, Oct. 1, 1943. The article quotes the naval correspondent for the *Daily Mail* in London. This near calamity has often been cited as the inspiration for the novel by Paul Gallico, *The Poseidon Adventure*, later made into two movies, but the quoted stories for Gallico's inspiration vary considerably. Most likely it was Gallico's 1937 voyage to Europe on the even-then often-rolling *Queen Mary* that was the original inspiration.
14. Ibid.
15. "Giant British Ships for Atlantic Only," *New York Times*, July 3, 1945.
16. Virgil's *Aeneid*, bk. 1, transl. Michael J. Oakley, 1995. Wordsworth Editions, 226 pages, pp. 6–7.
17. The exact wording differs with other translations, but they all capture the essential feel for the waves' great size. For example, "in a sheer cliff the water piled up" has also been translated as "a steep-climbing mountain of water." The word "billows" was sometimes used for "waves," and although not much in current use, its first meaning is "a wave, especially a great wave or surge of water."
18. Plutarch, ca. AD 100. *Lives of the Noble Greeks and Romans*, chap. on Themistocles, transl. John Dryden, Collier, 1909, 403 pages.
19. Neumann, J., 1973. "The Sea and Land Breezes in the Classical Greek Literature," *Bulletin of the American Meteorological Society*, vol. 54, no. 1, pp. 5–8. See also: Strauss, Barry, 2004. *The Battle of Salamis: The Naval Encounter That Saved Greece—and Western Civilization*, Simon and Schuster, New York, 294 pages.
20. This "channel effect of the narrow strait" means that the land on both sides of the strait forced the wind into the smaller volume over the waters of the strait and thus made the air move faster. The increased wind speed generated larger waves.
21. Plutarch, op. cit.
22. Strauss, op. cit.
23. Ibid.
24. The *sea breeze–land breeze*, a daily reversal in wind direction, is common in many warm coastal areas and in temperate coastal areas during the summer. During the day the land is warmer than the sea, so air rises over the land, with cooler air moving from the sea toward the land to replace it (the *sea breeze*). During the night the sea is warmer than the land, so air moves from the land toward the sea (the *land breeze*). The Greeks recognized this daily pattern, even though its intensity varied with cloudiness and other weather factors.
25. Aristotle, ca. 350 BC. *Problems*, bk. 23. "Problems Connected with Salt Water and the Sea," transl. W.S. Hett, William Heinemann, London, revised and reprinted 1957, 456 pages, pp. 13–41.
26. Plutarch, ca. AD 100. *Plutarch's Miscellanies and Essays Comprising All His Works Collected under the Title "Morals,"* ed. William W. Goodwin, 6th ed., 1898, vol. 3, *Natural Questions*, sec. 12, p. 503.
27. Ibid.
28. Pliny the Elder, ca. AD 77–79. *Natural History*, 37 vols., transl. Philemon Holland (in 1601), vol. 1, George Barclay, 1848, p. 141.

29. Bede, ca. 731. *The Complete Works of Venerable Bede*, transl. J.A. Giles, vol. 2, *Ecclesiastical History*, bks. 1–3, Whittaker, London, 1843, p. 315.
30. Leonardo da Vinci, ca. 1500. *Notebooks*, compiled by Irma A. Richter, Oxford Univ. Press, 2008, 392 pages, pp. 23–24.
31. Dampier, William, 1699. *Voyages and Descriptions*, vol. 2, ed. John Masefield, E. Grant Richards, London, 1906, 607 pages, p. 319. Captain William Dampier was a former pirate turned British naval captain. His book is based on his three voyages around the world.
32. Deacon, Margaret, 1997. *Scientists and the Sea, 1650–1900: A Study of Marine Science*, Ashgate, UK, 459 pages, p. 123.
33. Wind waves have short wavelengths and travel along the surface of deep water with a speed that is proportional to their period. The other waves we look at in this book (tides, storm surges, and tsunamis) have long wavelengths and move with a speed proportional to water depth. (In both cases when we say "short wavelength" or "long wavelength," we mean compared to the water depth.) Even short-wavelength wind waves become long waves when they move into very shallow water near the shore, at which point they are bent (refracted) by the variations in water depth.
34. Craik, Alex D.D., 2004. "The Origins of Water Wave Theory," *Annual Review of Fluid Mechanics*, vol. 36, pp. 1–29. See also: Darrigol, Olivier, 2003. "The Spirited Horse, the Engineer, and the Mathematician: Water Waves in Nineteenth-Century Hydrodynamics," *Archive for the History of Exact Sciences*, vol. 58, no. 1, pp. 21–95.
35. Snodgrass, F.E., G.W. Groves, K.F. Hasselmann, G.R. Miller, W.H. Munk, and W.H. Powers, 1966. "Propagation of Ocean Swell across the Pacific," *Philosophical Transactions of the Royal Society of London*, A, Mathematical and Physical Sciences, vol. 259, no. 1103, pp. 431–497. This study tracked swells produced by storms in the Indian Ocean using wave stations along a great-circle route from New Zealand to Alaska.
36. Although many references list wavelengths and wave speeds for particular wave periods, one set of examples is succinctly summarized in Bascom, William, 1964. *Waves and Beaches*, Anchor Books, Doubleday, 267 pages, pp. 62–63.
37. Cornish, Vaughan, 1910. *Waves of the Sea and Other Water Waves*, T. Fisher Unwin, London, 370 pages, p. 92.
38. The origin of the ninth wave was pondered by George Stokes, an oceanographer who developed much of classic water wave theory, including the concept of group velocity, around 1876 (although that concept was first mentioned in 1844 by another important oceanographer, Scott Russell). See: Craik, Alex D.D., 2005. "George Gabriel Stokes on Water Wave Theory," *Annual Reviews of Fluid Mechanics*, vol. 37, pp. 23–42.
39. Franklin, Benjamin, 1774. "Of the stilling of waves by means of oil. Extracted from sundry letters between Benjamin Franklin, LL.D.F.R.S., William Brownrigg, M.D.F.R.S. and the Reverend Mr. Farish," *Philosophical Transactions of the Royal Society*, vol. 64, pp. 445–460.
40. Ibid.
41. Ibid.
42. *Cat's paws* was the name given by sailors to the fleeting patches of ruffled water that moved forward and disappeared with the first breath of the wind, a hoped-for first sign of a freshening breeze to carry their ship forward.
43. Parker, Bruce, 1999. "How Does the Wind Generate Waves?" *Mariners Weather Log*, vol. 43, no. 1, pp. 17–22.
44. For a more complete history and explanation of oil calming waves, see: Tanford, Charles, 2004. *Ben Franklin Stilled the Waves*, Oxford Univ. Press, 267 pages. See also: Hühnerfuss, Heinrich, 2006. "Oil on Troubled Waters—A Historical Survey," in *Marine Surface Films*, ed. Martin Gade, Heinrich Hühnerfuss, and Gerald Korenowski, Springer, Berlin, 341 pages, pp. 3–12.

45. This incident was recorded in "Mémoires inédits de L'Abbé Morellet, deuxième édition, Paris, MDCCCXXII," Tome Premier, 204, and reproduced in *The Writings of Benjamin Franklin*, ed. Albert H. Smyth, vol. 1, Macmillan, London, 1905.
46. The speed of long-wave propagation is proportional to the square root of the water depth. There is also a frictional effect that reduces the wave speed and changes the wavelength in shallower water; this frictional effect is slightly different for different frequencies. We should make clear that we are talking about the speed with which the shape of the wave (the crest) moves, not the speed of individual water particles (the water particles move at much slower speeds).
47. Effects produced by shallow water, such as the steepening wave shape, breaking waves, and mass transport to the beach, are usually referred to as *nonlinear* because of the properties of terms in the hydrodynamic mathematical equations used to describe the waves. See note 20 in Chapter 6.
48. Another analogy with light is that water waves passing through the opening between two islands are *diffracted*—that is, the wave fronts leave the opening as concentric semicircles, similar to the patterns of light waves diffracted from a grating.
49. Longuet-Higgins, M.S., and R.W. Stewart, 1964. "Radiation Stress in Water Waves: A Physical Discussion, with Applications," *Deep-Sea Research*, vol. 11, pp. 529–562.
50. Panicked swimmers often try to swim against the rip currents, eventually wearing themselves out and drowning. To escape a rip current, a swimmer should swim parallel to the shore until he or she is out of the current and can then swim back to shore.
51. Lewis, David, 1975. *We, the Navigators—The Ancient Art of Landfinding in the Pacific*, University Press of Hawaii, Honolulu, 348 pages, pp. 84–96. Polynesians and Hawaiians could look at the pattern of waves and swells on the sea surface and recognize a swell coming from a specific direction. Some navigators could even recognize a swell by the feel of how their boats responded to the movement of the sea surface. Swells originated from the northeast trade winds (in the Northern Hemisphere), from the southeast trade winds (in the Southern Hemisphere), from the northwest monsoon, as well as from large storms.
52. Eddystone Light was rebuilt three more times. The third version was built in 1759 by John Smeaton. It was assaulted by a wave at least 50 feet high in 1840 and blew out a door. Figure 5.1 shows a perhaps exaggerated depiction of that event, taken from: *The Family Magazine or Monthly Abstract of General Knowledge*, 1843, vol. 1, Burgess and Zeiber, Philadelphia, p. 78. The fourth version of the lighthouse, a 167-foot tower made of solid rock, was built in 1882 and is still operating. Hardy, W.J., 1895. *Lighthouses, Their History and Romance*, The Religious Tract Society, 224 pages, pp. 120–139; Defoe, Daniel, 1704. *The Storm*, ed. Richard Hamblyn, Penguin, 2005, 228 pages, pp. 148–149. See also: Brayne, Martin, 2002. *The Greatest Storm: Britain's Night of Destruction, November 1703*, Sutton Publishing, 240 pages, pp. 77–87; and De Wire, Elinor, 1993. "Great Waves at Rock Lighthouses," *Mariners Weather Log*, Fall, pp. 38–40.
53. Bathurst, Bella, 1999. *The Lighthouse Stevensons*, Harper Collins, New York, 268 pages.
54. The dynamometer was a cast-iron cylinder with a strong steel spring inside, attached to a plate against which the wave pushed. The distance the plate was pushed by the wave indicated how much power the wave had. See: Stevenson, Thomas, 1886. *The Design and Construction of Harbours: A Treatise on Marine Engineering*, 3rd ed., Adam and Charles Black, Edinburgh, UK, 355 pages, pp. 52–53.
55. Ibid., p. 56. The 157-foot-high Skerryvore Lighthouse was on a tiny rock island that marked a dangerous rock reef next to the Hebrides off the west coast of Scotland.
56. De Wire, op. cit.

57. Stevenson, op. cit., pp. 27–30. Stevenson ignored wind speed in his formula and from his data came up with a coefficient that essentially represented a maximum possible wind speed, at least for the time period of his data.
58. Stevenson, op. cit., pp. 67–68. In the tidal race called the Merry Men of Mey, violent waves were produced when a westerly swell met the ebb tidal current. Though dangerous, the tidal race could also be a benefit because it formed a natural breakwater, since the waves' energy was dissipated in the tidal race and beyond it the water became calm. In the tidal race called the Bore of Duncansby, violent waves were produced when easterly swells met the flood tidal current.
59. Ibid., pp. 69–70.
60. An Army Corps of Engineers project to begin in 2010 will build an 840-foot-long, 40-foot-thick seawall to stop the erosion. The Corps hopes the new wall will do a better job than the seawall built in 1946, which was overrun by storm waves a few years later. The Surfrider Foundation objected to the project, because of fears the project would ruin the Alamo, a world-renowned surf break. See: Kilgannon, Corey, 2006. "For Montauk, It's Lighthouse vs. Surf's Up!" *New York Times*, Nov. 14.
61. Eisenhower, Dwight D., 1949. *Crusade in Europe*, Doubleday, New York, 559 pages, p. 70; Eisenhower quotes from Message 1406, from General Handy to General Marshall, Aug. 23, 1942, AGO.
62. Ibid.
63. Churchill, Winston S., 1950. *The Hinge of Fate*, vol. 4, *The Second World War*, Houghton Mifflin, Boston, MA, 917 pages, pp. 478–480. Churchill quotes a letter he sent Roosevelt Sept. 1, 1942, and the answer he received on Sept. 3.
64. Bates, Charles, 2005. *HYDRO to NAVOCEANO: 175 Years of Ocean Survey and Prediction by the U.S. Navy*, Corn Field Press, Rockton, IL, 329 pages, pp. 66–68. The first lieutenant (and shortly thereafter captain) was Harry Richard ("Dick") Seiwell, who had proposed and been given permission to establish the Oceanography Section the previous January, since at the time "there was not a single military or naval organization trained to evaluate information on the oceans and coast lines of the world and to transform it into the type of strategical and tactical intelligence required for military operations." See also: Seiwell, H.R., 1947. "Military Oceanography in World War II," *The Military Engineer*, vol. 39, no. 259, pp. 202–259.
65. In 1942 Sverdrup, along with his two hand-picked coauthors, finished the first comprehensive scientific book ever written on oceanography. The Navy immediately restricted its export (except to the British Admiralty), because "it would be a great aid to the enemy." See: Sverdrup, H.U., M.W. Johnson, and R.H. Fleming, 1945. *The Oceans: Their Physics, Chemistry and General Biology*, Prentice-Hall, New York, 1,087 pages.
66. Munk, Walter, and Deborah Day, 2002. "Harald U. Sverdrup and the War Years," *Oceanography*, vol. 15, no. 4, pp. 7–29.
67. Ibid. See also: Munk, Walter, 1980. "Affairs of the Sea," *Annual Reviews of Earth and Planetary Sciences*, vol. 8, pp. 1–17.
68. The theory used was later summarized in Sverdrup, H.U., and W.H. Munk, 1947. *Wind, Sea, and Swell: Theory of Relations for Forecasting*, pub. 601, Hydrographic Office, U.S. Navy Dept., 44 pages. Munk would later write modestly that his wave prediction scheme was "empiricism, pure and simple, with a few dispersion laws thrown in." Munk, op. cit., 1980.
69. Bates, op. cit., 2005. Munk was assisted by Mary Grier in finding data from ocean studies, wave tanks, lakes, or wherever they were buried. Grier had published the only oceanographic bibliography in existence, and usually worked for Lieutenant Mary Sears, who headed the Oceanographic Unit at the naval Hydrographic Office, which published

manuals based on Sverdrup and Munk's wave forecasting technique. See also: Williams, Kathleen B., 2001. *Improbable Warriors: Women Scientists and the U.S. Navy in World War II*, Naval Institute Press, Annapolis, MD, 280 pages.

70. No one, Munk said, was better than Sverdrup at "combining a noisy and disparate dataset into believable generalization." During most of this time both Munk and Sverdrup were under surveillance by the FBI and Naval Intelligence, because Munk had relatives in his native Austria and Sverdrup had relatives in Nazi-captured Norway. It did not seem to matter to the FBI and Navy that Munk had been in the United States since 1932 and had joined the Army in 1940, or that Sverdrup had applied for U.S. citizenship two months after Norway fell or that his sisters had been arrested and his younger brother killed by the Nazis. In spite of this, Captain Seiwell was able to get the Army Air Force to okay both men. After a concerted effort in the defense of both men by Roger Revelle, military intelligence finally concluded in January 1943 that Munk and Sverdrup were both loyal to the United States. Revelle had received his PhD from Scripps and was in the Navy with the responsibility for oceanography related to amphibious warfare. He would later become the Director of Scripps and one of the first to warn of the effects of greenhouse gases on global warming. See: Munk and Day, op. cit.
71. Analogous to *forecast*, which is a scientifically based prediction into the future, to "hindcast" is to make a scientifically based "prediction" into the past. Since one knows what happened in the past (from data taken at that time), making a hindcast is a good way to test a prediction model. In those hindcasts Munk and Sverdrup worried at first about occasional high wave conditions that did not seem to be properly predicted, until they finally realized that these wave spikes always occurred on Saturday nights. Munk and Day, op. cit.
72. Wave attenuation after traveling beyond the storm that generated them was estimated based on two effects—*wave dispersion* (waves of different periods spreading apart because they traveled at different speeds) and *geometric spreading* (wave energy spreading out in many directions so that waves heading toward a specific beach would have only a portion of the original energy).
73. Munk, op. cit., 1980.
74. Bates, Charles C., and John F. Fuller, 1986. *America's Weather Warriors, 1814–1985*, Texas A&M Press, College Station, 340 pages, pp. 70–71.
75. Ibid. Steere was supported by another meteorologist, six aerographer mates, and five radiomen.
76. Fuller, John F., 1990. *Thor's Legions—Weather Support to the U.S. Air Force and Army, 1937–1987*, American Meteorological Society, Boston, MA, 443 pages, p. 67.
77. Ibid.
78. Eisenhower, op. cit.
79. Fuller, op. cit., p. 70; Atkinson, Rick, 2002. *An Army at Dawn*, Owl Books, 682 pages, pp. 105–7. See also: Bates and Fuller, op. cit.
80. Patton, George S., 1995. *War as I Knew It*, Houghton Mifflin Harcourt, reissued (originally published 1947), 425 pages, pp. 8–9.
81. Fuller, op. cit.
82. Howe, George F., 1957. *Northwest Africa: Seizing the Initiative in the West*, Office of the Chief of Military History, Department of the Army, Government Printing Office, Washington, DC, 675 pages, pp. 69–70.
83. Boyle, Harold V., 1942. "French Fought On for Honor's Sake," *New York Times*, Nov. 16 (article written on Nov. 9).
84. Howe, op. cit.
85. Patton, op. cit.
86. Boyle, op. cit.

Chapter 6

1. Morgan, Frederick, 1950. *Overture to Overlord*, Hodder and Stoughton, London, 296 pages.
2. Marshall, George C., H.H. Arnold, and Ernest J. King, 1947. *The War Reports*, J.B. Lippincott Company, 801 pages, p. 183.
3. Munk, Walter, and Deborah Day, 2002. "Harald U. Sverdrup and the War Years," *Oceanography*, vol. 15, no. 4, pp. 7–29.
4. Munk, Walter, 1980. "Affairs of the Sea," *Annual Reviews of Earth and Planetary Sciences*, vol. 8, pp. 1–17.
5. The two Americans who had been trained at Scripps in the Sverdrup-Munk wave prediction method were Lieutenant Charles Bates and First Lieutenant John Cromwell. The British officer was Instructor Lieutenant H.W. Cauthery, who was trained in a technique developed in Britain by C.T. Suthons.
6. The Swell Forecast Section also included two American technicians and two WRNS (Women's Royal Naval Service) ratings from Britain. See: Bates, Charles C., 1949. "Utilization of Wave Forecasting in the Invasions of Normandy, Burma, and Japan," *Annals of the New York Academy of Science*, vol. 51, pp. 545–573. See also: Cromwell, J.C., and C.C. Bates, 1944. *Final Report of Swell Forecast Section, Admiralty*, vol. 1, Sept., Naval Meteorological Branch, Hydrographic Department, Admiralty, London, 17 pages.
7. Ibid, Bates.
8. Stagg, James M., 1971. *Forecast for Overlord*, Ian Allan, London, 128 pages. See also: Petterssen, Sverre, 2001. *Weathering the Storm*, ed. James R. Fleming, American Meteorological Society, Boston, MA, 329 pages. This was finished by Petterssen a short while before his death in 1974. An earlier version was tentatively titled *Of Storms and Men*, a copy of which is at the Eisenhower Presidential Library in Abilene, KS. See also: Shaw, R.H., and W. Innes, eds., 1984. *Some Meteorological Aspects of the D-Day Invasion of Europe, June 6 1944*, American Meteorological Society, Boston, MA, 170 pages.
9. Bates, op. cit., 1949; Bates, Charles C., and John F. Fuller, 1986. *America's Weather Warriors, 1814–1985*. Texas A&M Press, College Station, 340 pages, p. 91.
10. The second-choice date for the D-Day landing was two weeks later, when the tidal conditions would be similar but there would be no full moon, which would make the prelanding bombardments much more difficult. As it turned out, a violent storm hit Normandy during that period.
11. Eisenhower, Dwight D., 1949. *Crusade in Europe*, Doubleday, New York, 559 pages, p. 239.
12. Stagg, op. cit., pp. 104–114; Shaw and Innes, op. cit.
13. Stagg, op. cit., pp. 115–120; Eisenhower, op. cit., pp. 249–250.
14. At the 9:00 P.M. June 4 staff meeting, Admiral Bertram Ramsay, who was in charge of Operation Neptune and the amphibious landing, asked about the latest wave predictions and was told by Colonel Donald Yates, Stagg's deputy from the U.S. Army Air Force, that the waves would be above Ramsay's desired limits, at which point Eisenhower asked Ramsay if that would be a problem. Ramsay answered, "Hell no, I will get them on shore." See: Fuller, John F., 1990. *Thor's Legions—Weather Support to the U.S. Air Force and Army, 1937–1987*, American Meteorological Society, Boston, MA, 443 pages, p. 92. Fuller quotes from a postwar interview with Yates.
15. Chalmers, W.S., 1959. *Full Cycle: The Biography of Admiral Sir Bertram Home Ramsey*, Hodder and Stoughton, London, 288 pages, pp. 223–224; Vian, Philip, 1960. *Action This Day: A War Memoir*, Frederick Muller, London, 223 pages, p. 135; Edwards, Kenneth, 1946. *Operation Neptune*, Collins, London, 319 pages, pp. 116, 145–146; Lewis, Jon E.,

ed., 2004. *D-Day as They Saw It*, Carroll and Graf, New York, 314 pages, pp. 49, 62–63, 96, 127.

16. Vian, op. cit.; Chambers, op. cit.; Edwards, op. cit.
17. The date of this unpredicted storm, the worst to hit the English Channel in years, would have been the date of D-Day if the Normandy landings had been postponed because of bad weather on June 6. It destroyed one of the two Mulberries, portable harbors that the Allies had towed to Normandy from Britain.
18. The two Ocean Swell Section representatives on Omaha Beach were Donald Pritchard and Robert Reid, who years later would become two internationally known oceanographers in the United States. Bates and Fuller, op. cit., p. 97.
19. Seiwell, H.R., 1947. “Military Oceanography in World War II,” *The Military Engineer*, vol. 39, no. 259, pp. 202–259.
20. These hydrodynamic equations of motion are similar to the equations of Laplace mentioned in Chapter 1, but for different length and time scales, so that different forces are emphasized. The interaction between waves is represented in these equations by *nonlinear* terms (first mentioned in Chapter 2, note 17). These are mathematical terms that contain more than one key variable, such as wave height or orbital current speed, and thus affect each other. When an equation is *linear*, only one of these variables is in each term; their effects do not affect each other, and they can thus be simply added to each other. The simple wave model that was in use for decades was a linear model. The real world is nonlinear, but the strength of the nonlinear effects varies with location, usually being weak in the deep ocean. In shallow water or when strong ocean currents are involved, nonlinear effects are strong. But there must also be a nonlinear effect in deep water that generates rogue waves under the right conditions, although that is still being debated.
21. Phillips, Owen M., 1957. “On the Generation of Waves by Turbulent Wind,” *Journal of Fluid Mechanics*, vol. 2, pp. 417–445; Miles, John W., 1957. “On the Generation of Surface Waves by Shear Flow,” *Journal of Fluid Mechanics*, vol. 3, pp. 185–204. See also the classic paper by Jeffreys that first proposed the importance of the sheltering effect of the wave shape on wind wave growth: Jeffreys, Harold, 1925. “On the Formation of Water Waves by Wind,” *Proceedings of the Royal Society of London*, A, vol. 107, no. 742, pp. 189–206.
22. Mitsuyasu, Hisashi, 2002. “A Historical Note on the Study of Ocean Surface Waves,” *Journal of Oceanography*, vol. 58, pp. 109–120. See also: LeBlond, Paul H., 2002. “Ocean Waves: Half-a-Century of Discovery,” *Journal of Oceanography*, vol. 58, pp. 3–9.
23. Tolman, Hendrik L., 2009. *User Manual and System Documentation of WAVEWATCH III* TM, *version 3.14*, National Centers for Environmental Prediction, National Weather Service, NOAA, May 2009, 187 pages; WAMDI Group, 1988. “The WAM Model—A Third Generation Ocean Wave Prediction Model,” *Journal of Physical Oceanography*, vol. 18, pp. 1775–1810. WAVEWATCH III® (version 3.14) is a further development of the model WAVEWATCH developed at Delft University of Technology and WAVEWATCH II developed at NASA. The latest version of WAM is cycle 4. These models are based on the conservation of *wave action*, which is the wave energy at a particular frequency divided by that frequency.
24. Warwick, Ronald W., 1996. “Hurricane Luis, the *Queen Elizabeth 2* and a Rogue Wave,” *Marine Observer*, vol. 66, p. 134. See also: Warwick, Ronald, 1999. *QE2*, Norton, 224 pages, pp. 167–168. The rogue wave was calculated to be ninety-five feet high, since it was level with the bridge. A nearby Canadian weather buoy measured a wave ninety-eight feet high.
25. A few oceanographers did believe freak waves were real. See: Draper, L., 1964. “Freak Ocean Waves,” *Oceanus*, vol. 10, no. 4, pp. 13–15.
26. Churchill, D.D., S.H. Houston, and N.A. Bond, 1995. “The Daytona Beach Wave of 3–4 July 1992: A Shallow-Water Gravity Wave Forced by a Propagating Squall Line,” *Bulletin*

of the American Meteorological Society, vol. 76, no. 1, pp. 21–32. There is still some question about whether the wave that hit Daytona Beach should be considered a "wind wave" or a "mini storm surge."

27. The picture in figure 6.1 is an 1841 engraving by R.G. Reeve of a painting by W.J. Leatham. It shows the East India Company's iron war steamer *Nemesis* among huge waves in the Agulhas Current off the Cape of Good Hope. This copy of the picture is from the Fall 1993 issue of the *Mariners Weather Log*, p. 24.
28. Lavrenov, I.V., 1998. "The Wave Energy Concentration at the Agulhas Current Off South Africa." *Natural Hazards*, vol. 17, pp. 117–127. The Agulhas Current also focuses the wave energy along a line of fastest water flow, which is near the center of the current cross section.
29. Gaddis, Vincent H., 1964. "The Deadly Bermuda Triangle," *Argosy*, Feb., pp. 28–29, 116–119.
30. U.S. Coast Guard, 1964. *Summary of Findings, Marine Board of Investigation, March 17, 1964*, U.S. Coast Guard Headquarters, Washington, DC.
31. Cox, Mike, 2006. *Texas Disasters: True Stories of Tragedy and Survival*, Globe Pequot, 242 pages, pp. 167–176. See also: "Tragedies: Sulphur Queen's End," *Newsweek*, Mar. 4, 1963, pp. 29–30; "The Queen with the Weak Back," *Time*, Mar. 8, 1963.
32. Cameron, T. Wilson, 1993. "The Treachery of Freak Waves," *Mariners Weather Log*, Fall 1993, pp. 35–37; originally published in *The Marine Observer*, Oct. 1985. Of course, one cannot know for sure what sank the *Marine Sulphur Queen*. With her molten sulfur cargo, she was also susceptible to fires, although then there might have been time for an SOS. See also: U.S. Coast Guard, op. cit.
33. Whitemarsh, R.P., 1934. "Great Sea Waves," *U.S. Naval Institute Proceedings*, vol. 60, no. 8, pp. 1094–1103. The 112-foot height was actually a conservative calculation, so that no one could question the value. Previous scientists had made careful visual estimates of extreme wave heights, only to be dismissed by onshore skeptics. In 1826 a French scientist and naval officer, Captain Dumont d'Urville, reported a 108-foot wave and was ridiculed, even though he was backed up by three colleagues. Admiral Robert FitzRoy (who, as we learned in Chapter 3, began the British Meteorological Service) had once climbed to the top of a 60-foot mast on his ship and said the waves were much higher.
34. Petrow, Richard, 1969. "The Death of *World Glory*," *Popular Mechanics*, July, pp. 106–111, 198–199; Bamberger, Werner, 1968. "Tanker's Sinking Is Laid to Massive Wave," *New York Times*, July 28.
35. But there were a few exceptions. See: Draper, op. cit. In later years, as scientific data accumulated showing examples of very large waves, many oceanographers said they believed that there were very large waves. The only questions were, how frequent are they and what is the hydrodynamic mechanism that causes them.
36. Haver, Sverre, 2004. "A Possible Freak Wave Event Measured at the Draupner Jacket January 1, 1995," *Proceedings, Rogue Waves 2004*, Oct. 20–22, Brest, France.
37. Many of the rogue waves in the Goma data set were smaller than waves we have been talking about, but rogue waves are strictly defined as waves that are at least twice as high as the waves around them, or more precisely, at least twice as high as the significant wave height. It is these much larger waves that the wave models have been unable to produce, whether the prevailing sea state is great or not. See: Skourup, J., N.-E.O. Hansen, and K.K. Andreasen, 1997. "Non-gaussian Extreme Waves in the Central North Sea," *Transactions of the ASME*, vol. 119 (Aug.), pp. 146–150.
38. Wright, C.W., et al., 2001. "Hurricane Directional Wave Spectrum Spatial Variation in the Open Ocean," *Journal of Physical Oceanography*, vol. 31, pp. 2472–2488. The highest waves and longest wavelengths were observed in the right forward quadrant of the hurricane, as had been long ago stipulated by Henry Piddington in *The Sailor's Horn-Book*

for the Law of Storms, 1869, 5th ed., William and Norgate, London, 408 pages. (See Chapter 3 for more on Piddington.)

39. Waves are measured from satellites using two types of instruments: radar altimeters and synthetic aperture radar (SAR) systems. A radar altimeter measures the height of the ocean surface directly below it, to an accuracy of one or two inches, averaged over a square-mile or larger area (its *footprint*). The radar determines the distance from the satellite to the ocean surface by measuring the time it takes for a radio signal to travel to the ocean surface and back to the satellite. This is useful for sea level and tide measurements, but less so for wind wave measurement. However, the waves in that square-mile area scatter the radar signal pulse when it is reflected, distorting it in a way that can be analyzed to calculate the significant wave height. The heights measured by radar altimeters on satellites compare well with heights measured by wave buoys, which measure waves at a single location using an accelerometer, but there is no information on the direction the waves are moving, nor can individual waves be measured. SAR, on the other hand, *can* measure both wave direction and the size of individual waves, although it involves a much more complicated analysis process using mathematical algorithms that some scientists believe have serious problems.
40. Satellites equipped with radar altimeters, SARs, or both include in the United States NASA's JASON-1, JASON-2, and TOPEX/Poseidon (the latter with France) and the Navy's Geosat Follow-on; the European Space Agency's ERS-1, ERS-2, and Envisat; Canada's RADARSAT; and Germany's TerraSAR-X, which has the best resolution but was only recently launched.
41. They had found these ten rogue waves in three weeks of ERS-1 satellite data from August 1996. This had been part of the MaxWave project, which was followed by the more ambitious WaveAtlas program to find rogue waves in two years of raw ERS data. See: Rosenthal, Wolfgang, and Susanne Lehner, 2007. "Rogue Waves: Results of the MaxWave Project," *Journal of Offshore Mechanics and Arctic Engineering*, vol. 130, no. 2, pp. 21006–21013.
42. Wang, D.W., D.A. Mitchell, W.J. Teague, E. Jarosz, and M.S. Hulbert, 2005. "Extreme Waves under Hurricane Ivan," *Science*, vol. 309, no. 5736, p. 896. During this time a buoy deployed by NOAA's National Data Buoy Center measured the largest significant wave height they had ever measured, fifty-two feet. This buoy was slightly closer to the hurricane's eye than the NRL wave gauge, but it was not set up to measure individual wave heights, only significant wave height (which, as we saw at the beginning of this chapter, is the average of the highest one-third of the waves).
43. Liu, P.C., H.S. Chen, D.-J. Doong, C.C. Kao, and Y.-J.G. Hsu, 2009. "Freaque Waves during Typhoon Krosa," *Annales Geophysicae*, vol. 27, pp. 2633–2642.
44. Susman, Tina, 2005. "Nightmare Trip. Seven-Story Wave Slams into Norwegian Cruise Ship off Virginia," *Newsday*, Apr. 18.
45. Yamaguchi, Mari (Associated Press), 2008. "Four Die as Fishing Boat Capsizes off Japan Coast," *Seattle Times*, June 24. See also: *Japan Times*, June 24, 2008.
46. The physical mechanism that causes a rogue wave must involve a *nonlinear mechanism* (see note 20).
47. Some oceanographers have borrowed nonlinear equations from other oceanographic situations or even from other fields (such as the Schrödinger equation of quantum mechanics) and have shown that waves can grow very large, which is not surprising considering it is a nonlinear equation. But this does not tell us much, unless the nonlinear equation can be derived directly from the hydrodynamic equations that describe the physics of deep-ocean waves. Before scientists studied rogue waves, they had discovered that the physics allows a weak type of nonlinear interaction involving the interaction of three or four waves, but this did not seem strong enough to produce a rogue wave,

although some oceanographers are still looking into it. The only well-understood strong nonlinearities are those involved in a wave growing and becoming steeper in shallow water or when it is affected by a strong current.

48. Rosenthal and Lehner, op. cit.
49. Tamura, H., and T. Waseda, 2009. "Freakish Sea State and Swell-Windsea Coupling: Numerical Study of the *Suwa-Maru* Incident," *Geophysical Research Letters*, vol. 36, L01607. Their third-generation wave model was based on WaveWatch III with an improved nonlinear energy transfer scheme. Two significant swell systems were involved: a low-frequency (nine- to ten-second period) swell moving northwestward from the Pacific and a slightly higher-frequency (eight- to nine-second period), shorter swell moving east-northeastward from the southern coast of Japan. The shorter swell crossed at a 40-degree angle to the wind waves being produced by increasing local winds and as a result became much steeper and higher.
50. Sheffiff, Lucy, 2007. "Satellites Track Freak Waves to Réunion," *Register* (London), May 18. See also: "Huge Waves That Hit Reunion Island Tracked from Space," *ESA News*, European Space Agency, May 16, 2007.
51. Collard, F., F. Ardhuin, and B. Chapron, 2009. "Monitoring and Analysis of Ocean Swell Fields from Space: New Methods for Routine Observations," *Journal of Geophysical Research*, vol. 114, no. C7, C07023.

CHAPTER 7

1. The information in the first four paragraphs of this chapter comes primarily from the following documents: (1) Braddock, 1755. A letter from Mr. Braddock to Dr. Sanby (a chancellor of the Diocese of Norwich, England), written November 13, 1775, and later given by another person to Charles Davy, who in 1787 published it with other letters in his *Letters Addressed Chiefly to a Young Gentleman upon Subjects of Literature*, vol. 2, Bury St. Edmunds, pp. 12–60; with an introductory letter written October 24, 1780, by another, pp. 1–11; and another letter written November 12, 1780, which included eyewitness information from a pamphlet written by Antonio Pereira shortly after the earthquake and tsunami, pp. 114–127. (2) Various authors, 1755a. A collection of twenty-one eyewitness letters published in "An account of the Earthquake, November 1, 1755," *Philosophical Transactions of the Royal Society*, vol. 49 (1756), pp. 398–444. These authors were in Portugal, Spain, Morocco, and the islands of Madeira. (3) Chase, Thomas, 1775. "An account of what happened to Mr. Thomas Church, at Lisbon, in the great Earthquake: written by himself, in a letter to his Mother, dated the 31st of December, 1775," *The Gentleman's Magazine*, Feb. 1813, pp. 105–110, 201–206, 314–317.
2. Shocks felt by ships at sea, sometimes accompanied by agitation described as a "boiling sea," came to be known as *seaquakes* and were very different than tsunamis, the latter being water waves with very long wavelengths that travel great distances, while the water oscillations that accompany a seaquake have very short wavelengths and do not travel great distances or last long. See also the description of seaquakes near the 2004 epicenter in Chapter 9.
3. Shrady, Nicholas, 2008. *The Last Day*, Viking, pp. 20–23. Shrady quotes a letter from Queen Maria Ana Victoria to her mother found in: Caetano Beirão, 1947. Descrição Inédita do Terramoto de 1755 como o viu e viveu a Rainha D. Maria Victória, *Artes & Colecçoes*, vol. 1, no. 1, pp. 3–4.
4. Various authors, op. cit., 1755a. See also: Baptista, M.A., S. Hector, J.M. Miranda, P. Miranda, and L. Mendes Victor, 1998. "The 1755 Lisbon Tsunami: Evaluation of the Tsunami Parameters," *Journal of Geodynamics*, vol. 25, no. 2, pp. 143–157. The huge

tsunami wave swallowed the entire seaport of Setubal (20 miles south of Lisbon) and the city of Algarves (100 miles south), passed over a sixty-foot wall protecting the Spanish city of Cádiz (250 miles south), and broke against the walls of Tangier, Morocco. It even reached the city of Seville, Spain, 50 miles up the Guadalquivir River.

5. Various authors, 1755b. A collection of twenty-six eyewitness letters in "An account of the Earthquake, November 1, 1755," *Philosophical Transactions of the Royal Society*, vol. 49 (1756), pp. 351–398. Twenty-one authors were in Great Britain, two were in Ireland, one in Czechoslovakia, and two in the Netherlands. Additional eyewitness accounts are found in: Holdsworth, Henry, 1756. Extract of a letter from the Rev. Mr. Holdsworth, at Dartmouth, relating to the agitation of the waters observed there on the 1st of November, 1755, *Philosophical Transactions of the Royal Society*, vol. 49, pp. 643–644. At Portsmouth, England, his Majesty's forty-gun ship the *Gosport* suddenly moved backward a great distance for no apparent reason. On the Thames a barge suddenly heaved up three or four times from a rise of the water. Farther up the North Sea coast of England, at Yarmouth, there was "uncommon motion of ships," with water abruptly flowing six feet high. Near the western end of the English Channel, at Creston, the sea dropped five feet in two minutes, leaving a ferry on the mud bottom, and then returned to six-foot depth in less than eight minutes, floating the ferry again.
6. Affleck, 1755. "An Account of the Agitation of the Sea at Antigua, Nov. 1, 1755," by Captain Affleck of the *Advice Man of War*, written June 3, 1756, *Philosophical Transactions of the Royal Society*, vol. 49 (1756), pp. 668–670.
7. There was also a second kind of water oscillation felt all over Europe that day, much shorter in period, more like wind waves but with no wind blowing, described by many as looking like the "boiling of the sea in a pot." While the longer, larger water waves took a few hours to reach the shores of England, the shorter-wavelength, more-rapid water agitations tended to occur much earlier, very soon after the earthquake shocks were felt there. These rapid water oscillations even showed up in enclosed bodies of water that were not connected to the ocean, such as Loch Ness in Scotland. See the letter of William Borlase, 1756, in Various authors, 1755b. These rapid agitations were generated directly by local seismic vibrations, which as we will see travel through the solid earth much more quickly than the long water waves traveling on the ocean's surface.
8. Atwater, B.F., M.-R. Satoko, S. Kenji, T. Yoshinobu, U. Kazue, D.K. Yamaguchi, 2005. *The Orphan Tsunami of 1700*, Univ. of Washington Press, Seattle, 133 pages, p. 41. The term *tsunami* appears in the *Sumpuki*, an official diary written in January 1612 by an aide of the shogun Tokugawa Ieyasu, the first of fifteen shoguns to rule Japan from Edo. It was used to describe a tsunami in 1611 that killed at least 8,000 people on the north-east coast of Honshu. See also: Cartwright, J.H.E., and H. Nakamura, 2008. "Tsunami: A History of the Term and of Scientific Understanding of the Phenomenon in Japanese and Western Culture," *Notes and Records of the Royal Society*, vol. 62, pp. 151–166.
9. Like the tide and like wind waves, a tsunami wave is a gravity wave, which means that gravity is the restoring force that makes the water surface move up and down. When the water rises up to a crest above the equilibrium line (which is sea level without waves), gravity brings it back down. When the inertia of the falling water moves it past the equilibrium line to a low point the trough of the wave), the pressure of the water around it (pushed down by gravity) pushes it back up. This is more than a vertical oscillation, since the changing wave form is moving horizontally along the sea's surface. In terms of causes of a tsunami, by far the most common is a submarine earthquake, with landslides and volcanic eruptions fairly common, and an asteroid hitting the ocean extremely rare.
10. A caldera is a large basinlike depression in the center of a volcano created when the magma chamber is emptied by either an explosion or lava flows.

11. See note 35.
12. In the deep ocean the current of a tsunami still goes from the surface to the seafloor, but because the same volume of transported water is stretched over a much greater depth, these currents are extremely slow. And so, even though the tsunami feels the ocean bottom, the currents are too slow in deep water for there to be much frictional energy loss. This is why these long-wave tsunamis can travel thousands of miles before being worn down to nothing by bottom friction.
13. The estimation of eight tsunamis over twelve thousand years is based on examining sedimentological data from the seafloor of that region. Sediment cores taken near the bottom of the continental slope show eight layers (called *turbidites*), in each of which the sediment changes from coarse grain at the bottom to fine grain at the top. Radiocarbon dating techniques showed that the most recent turbidite was created around 1755. This was corroborated independently by data from a lagoon behind a sand spit near Cádiz, in which was found a coarse-grain deposit that was left by the 1755 tsunami when it swept up and over the sand spit. This evidence is summarized in the reference in note 14.
14. Gutscher, Marc-André, 2005. "Destruction of Atlantis by a Great Earthquake and Tsunami? A Geological Analysis of the Spartel Bank Hypothesis," *Geology*, vol. 33, no. 8, pp. 685–688. Most scholars do not believe that the story of Atlantis, as described in Plato's two dialogues the *Timaeus* and the *Critias*, is true. But it is possible that this tale could have been modeled after a real event in which a destructive earthquake and its very violent tsunami wiped out an entire island with a sizable population, somewhere to the west of the Pillars of Hercules, the name for the Strait of Gibraltar in Plato's time. The results of a 2003 high-resolution bathymetric survey west of the Strait of Gibraltar confirmed an earlier suggestion of an underwater feature called Spartel Bank that was above water twelve thousand years ago, not long after the end of the ice age, when sea level was much lower than it is today. (See: Collina-Girard, Jacques, 2001. "Atlantis off the Gibraltar Strait? Myth and Geology," *Sciences de la Terre et des planètes*, vol. 333, pp. 233–240, Comptes Rendus Académie des Sciences, Paris. The confirming bathymetric survey was described in Gutscher, 2005, op. cit.) In the *Timaeus*, Plato wrote, "Portentous earthquakes and floods, one grievous day and night befell them, when the whole body of your warriors was swallowed up by the earth, and the island of Atlantis in a like manner was swallowed up by the sea and vanished." (See: Plato, ca. 360 BC. *Timaeus*, 25c–d, transl. Robert G. Bury, 1929.) The one-day-and-night time frame could also have been used to describe the 1755 Lisbon event, since there were aftershocks and smaller tsunami waves throughout the day and night. Critias, the lead character in Plato's dialogue by the same name, claims that an Athenian lawmaker named Solon was told the Atlantis story by an old Egyptian priest in the city of Sais, and that Plato saw a half-finished poem that Solon was writing about it. Isaac Newton became interested in Plato's Atlantis and even suggested that it could have been the priests of Egypt who purposely exaggerated the story. (See: Newton, Isaac, 1728. *The Chronology of Ancient Kingdoms Amended*, Printed for J. Tonson in the Strand, and J. Osborn and T. Longman in Pater-noster Row.)
15. The Minoans had a written language before the Greeks developed one and numerous myths about King Minos, the Minotaur, and the labyrinth.
16. An analysis of radiocarbon-dated geological and archaeological data indicates that there are tsunami deposits from the time of the Thera eruption in the Minoan city of Palaikastro on the northeast coast. Palaikastro was apparently struck by tsunami waves estimated to have been thirty feet high. On the basis of these tsunamigenic deposits, a hydrodynamic model was used to estimate the largest tsunami wave generated at Thera. The model-produced tsunami that best fit the stratigraphic data had crests over a hundred feet high. See: Bruins, H.J., J.A. MacGillivray, C.E. Synolakis, C. Benjamini, J.

Keller, H.J. Kisch, A. Klügel, J. van der Plicht, 2008. "Geoarchaeological Tsunami Deposits at Palaikastro (Crete) and the Late Minoan IA Eruption of Santorini," *Journal of Archaeological Science*, vol. 35, pp. 191–212.

17. Thucydides, ca. 411 BC. *History of the Peloponnesian War*, bk. 3, par. 89. See also: Smid, T.C., 1970. *Tsunamis in Greek Literature. Greece and Rome*, 2nd ser., vol. 17, no. 1, pp. 100–104. The Peloponnesian War, between Sparta and Athens, lasted from 431 to 404 BC.
18. See: T.J.J., 1907. "On the Temperature, Secular Cooling and Contraction of the Earth, and on the Theory of Earthquakes Held by the Ancients," *Proceedings of the American Philosophical Society*, vol. 46, no. 186, pp. 191–299.
19. Ælianus (AD 175–235) in *De Natura Animalium* (*The Nature of Animals*).
20. One *shiro* is 15.13 acres.
21. *Nihongi, Chronicles of Japan from the Earliest Times to A.D. 697*, transl. W.G. Aston, vol. 2, Transactions and Proceedings, The Japan Society, London, suppl. 1, 1896. See also: Cartwright and Nakamura, op. cit., 2008.
22. Guidoboni, Emanuela, 1998. "Earthquakes, Theories from 1600 to 1800," in *Sciences of the Earth: An Encyclopedia of Events, People, and Phenomena*, vol. 1, ed. Gregory A. Good, Garland, Taylor & Francis, New York, pp. 205–214.
23. Kant proposed a mechanism for the propagation of seismic disturbances over great distances. He thought there were great air-filled caverns under the earth and even under the sea that ran parallel to mountains and rivers. Earthquake tremors were felt all over Europe, he thought, because hot compressed air, caused by chemical explosions, spread out through these subterranean tunnels over great distances. See: Guidoboni, op. cit. See also: Oeser, Ebhard, 1992. "Historical Earthquake Theories from Aristotle to Kant," in *Historical Earthquakes in Central Europe*, vol. 1, ed. R. Gutdeutsch, G. Grünthal, and R. Musson, Band 48, pp. 11–31. See also: Michell, op. cit., pp. 611–612.
24. The Marquis of Pombal was Sebastião José de Carvalho e Melo, who became prime minister after keeping the Portuguese government going after the Lisbon disaster.
25. The speed of a long water wave would years later be determined to be proportional to the square root of the water depth. See note 35. See: Michell, John, 1760. "Conjectures concerning the cause, and observations upon the phaenomena of earthquakes; particularly of that great earthquake of the first of November, 1755, which proved so fatal to the city of Lisbon, and whose effects were felt as far as Africa, and more or less throughout almost all Europe," *Philosophical Transactions of the Royal Society*, vol. 51, pp. 566–634.
26. There is still some debate on the epicenter's exact location, but most modern calculations and knowledge of the tectonic plates in the area put it about 175 miles southwest of Lisbon.
27. Michell, op. cit. Michell believed an earthquake was produced when water under the earth came in contact with subterranean fires, exploding and producing a vapor. His reasoning was based on noticing that earthquakes frequently occur in regions with volcanoes, also caused, he thought, by subterranean fires when they broke through the Earth's surface. His explanation of water hitting hot magma is, in fact, an eruption mechanism for volcanoes. For earthquakes he said that when subterranean water was heated by fires, it turned to steam and expanded, raising the earth above it and the seafloor above that. Then cold water from fissures supposedly cutting vertically through horizontal strata of earth would cool the steam, decreasing the vapor pressure and letting the earth above it fall, along with the seafloor. This, he said, created a depression toward which the ocean water would flow, causing the sea along the coast to retreat. When the steam was produced again, raising the earth and seafloor, ocean water would flow back to the shore as a great wave. His theory was wrong, but one can see his logic.
28. A few instruments have been invented at different times in history to detect the motion of the land caused by an earthquake, the first probably being the instrument invented in

China in AD 32 by Chang Hêng (or Zhang Heng), who used a pendulum inside a bronze jar that could move in one of eight directions. However, sometime after the sixth century that and similar instruments disappeared in China. A seismographic instrument, perhaps based on the Chinese instrument, apparently was used in thirteenth-century Persia. See: Needham, Joseph, 1959. *Mathematics and the Sciences of the Heavens and the Earth*, vol. 3, *Science and Civilisation in China*, Cambridge Univ. Press, 926 pages, pp. 626–635. Serious work began on seismometers in Europe in the mid-1700s, and although much progress had been made by 1854, they were not yet perfected. See note 49.

29. The newly established Tidal Division in the Coast Survey, headed by Count Louis F. De Pourtales, from Switzerland, also included Mary Thomas as one of the first *tide computers* ("computers" then were people). Thomas was the second woman professional in the U.S. federal government.
30. Bache, Alexander Dallas, 1856. "Notice of Earthquake Waves on the Western Coast of the United States, on the 23d and 25th December, 1854," app. 51, *Report of the Superintendent of the Coast Survey 1855*, A.O.P. Nicholson, Washington, DC, pp. 342–346.
31. The port of Simoda is now called Shimoda, and the island of Niphon is now called Honshu. The tsunami had also reached Oregon, but the sandbar at the entrance to the Columbia River probably kept it from being clearly detected upriver at the Astoria tide gauge.
32. Perry, Matthew C., 1856. *Narrative of the Expedition of an American Squadron to the China Seas and Japan*, compiled by Francis L. Hawks, U.S. Navy, Washington, DC, Appleton, New York, 624 pages, pp. 587–589. Perry wrote about the 1854 earthquake and tsunami in Japan, and sent information to Bache. Both mentioned the Russian frigate *Diana*, which rolled over on its side four times in the shallow water left by four different tsunami wave retreats, damaging its structure beyond repair. When the first wave rushed in, it spun the *Diana* around violently as though in a whirlpool, reportedly making forty-three complete revolutions in half an hour (one revolution every forty-two seconds), which made the crew dizzy.
33. A version of this story first appeared in *Atlantic Monthly* in 1896. Later it was turned into a children's book, which was used from 1937 to 1947 as a way to increase awareness of tsunamis in Japan and to teach what to do when a tsunami comes. More recently it was turned into a play. See: Hearn, Lafcadio, 1896. "A Living God," *Atlantic Monthly*, vol. 78, no. 470 (Dec.), pp. 833–841. See also: Atwater et al., op. cit., 2005; Ohta, H., T. Pipatpongsa, and T. Omori, 2005. "Public Education of Tsunami Disaster Mitigation and Rehabilitation Performed in Japanese Primary Schools," *Proceedings of the International Conference on Geotechnical Engineering for Disaster Mitigation and Rehabilitation*, World Scientific, Singapore, pp. 141–150.
34. The woodblock print in figure 7.2 comes from: Atwater et al., op. cit., 2005, p. 47. See also: Ohta et al., op. cit., 2005. The famous "Great Wave" woodblock print by the Japanese artist Katsushika Hokusai, created in 1832 (see figure 6.3), has often been used to represent a tsunami. However, it is not a tsunami but a large wind wave, which would be considered a rogue wave if it were twice as high as the other waves around it.
35. Bache knew that the speed of a tsunami, being a very long wave (about 200 miles from crest to crest in this case), was determined by the depth of the ocean. If he could determine the speed of the tsunami that crossed from Japan to California, he could then calculate the average depth of the Pacific Ocean along that route. He did just that using a simple formula: the speed of a long wave c is equal to the square root of the product of the depth h and the acceleration due to gravity g. The formula is $c = \sqrt{gh}$. Bache got the formula from George Airy's *Waves and Tides* paper in *Encyclopedia Metropolitana* (London, 1845, vol. 5, pp. 241–396), but the formula had been around since at least

Joseph-Louis Lagrange's work in 1781. Bache knew the time the submarine earthquake occurred (based on shocks felt in Japan) and the times when the waves arrived at San Diego (almost fourteen hours after the earthquake) and at San Francisco (about twelve and a half hours after). He also knew the distance from Simoda to San Diego (4,917 miles) and from Simoda to San Francisco (4,527 miles). He thus could calculate the average speed of the seismic sea wave to San Diego (355 miles per hour) and to San Francisco (slightly faster). He then used the formula from Airy to calculate the average depth of the Pacific Ocean between Simoda and San Diego and between Simoda and San Francisco, which he calculated as 12,600 feet and 14,000 feet, respectively. Both values were much closer to the actual depths of the Pacific than any previous estimates. Bache, op. cit., 1856. See also: Bache, A.D., 1855. "Note on Earthquake Waves on the West Coast of the USA, 23 and 25 December, 1854," *Proceedings, American Association for the Advancement of Science*, 9th meeting, pp. 153–160.

36. Hilgard, J.E., 1872. "On the Earthquake-Wave of August 14, 1868," *Report of the Superintendant of the U.S. Coast Survey during the Year 1869*, app. 13, pp. 233–234. See also: U.S. Navy, 1868. *Report of the Secretary of the Navy, With an Appendix Containing Bureau Reports*, Government Printing Office, Washington, DC, xix–xx; "Affairs in Peru, a Night of Horror-Rising Sea," *New York Times*, Sept. 7, 1868 (written and mailed Aug. 14, 1868). The U.S. Navy had a storeship and a gunboat at the port in Arica, both of which sat on a dry harbor bottom amid flopping fish after the sea withdrew. When the tsunami crest entered the harbor, it broke the storeship into pieces but carried the gunboat (the *Wateree*) inland more than a quarter mile.
37. U.S. Coast and Geodetic Survey, 1885. *Report of the Superintendent of the U.S. Coast and Geodetic Survey*, for the fiscal year ending June 1884, Government Printing Office, Washington, DC, 622 pages, pp. 63–64, 71. See also: "Washed by High Tides," *New York Times*, Aug. 31, 1883 (written Aug. 30 in Washington); "Volcanic Eruptions in Java," *Times* (London), Aug. 28, 1883 (written Aug. 27 in Batavia); "Volcanic Eruptions in Java," *Times* (London), Aug. 30, 1883 (written in Batavia Aug. 28–29). Similar tsunami oscillations were found on tide curves from C&GS gauges in Hawaii and on Kodiak Island, Alaska. "Krakatau" was the more common Javanese name, and it is still used in numerous reports and books instead of "Krakatoa." The name Krakatoa apparently came into common usage outside of Indonesia because it was misspelled in a telegram during the eruption in 1883. See note 39.
38. *Run-up height* is the greatest height the tsunami reaches when traveling inland, that height usually being (as the name implies) after running up the slope of a hill. This differs from the height of the tsunami wave just as it reaches the shoreline, which is usually a difficult number to obtain unless there are reliable eyewitness accounts. Such observations are usually guesses unless the wave height could be compared to known objects, such as trees or a building, or measured from visible marks on trees or buildings that survived the wave. In this case the 115-foot run-up height relates to a dramatic escape up a hill behind Anjer. The wall of water chased after people running up the hill and fell just short of hitting a house on the hilltop, 115 feet above sea level (told dramatically in the first reference of the next note).
39. Winchester, Simon, 2003. *Krakatoa: The Day the World Exploded: August 27, 1883*, Harper Perennial, London, 416 pages. This exceedingly well-researched and well-written book provides dramatic details of the horrors of that day. See also: Symons, G.J., ed., 1888. *The Eruption of Krakatoa and Subsequent Phenomena*, Report of the Krakatoa Committee of the Royal Society, London, 491 pages. See especially part 3 by Captain W.J.L. Wharton and Captain Sir R.J. Evans, "On Seismic Sea Waves Caused by the Eruption of Krakatoa, August 26th and 27th, 1883," pp. 89–151.
40. Winchester, op. cit., p. 241.

41. Simkin, Tom, and Richard S. Fiske, 1983. *Krakatoa 1883: The Volcanic Eruption and Its Effects*, Smithsonian Institution, Washington, DC, 464 pages. This is the most thorough and detailed academic study of Krakatoa, covering eyewitness accounts, scientific work written before 1983, and all previous reports, including the 1885 Dutch report by R.D.M. Verbeek; see note 46.
42. Clarke, Arthur C., 1957. *The Reefs of Taprobane: Underwater Adventures around Ceylon*, Harper, 205 pages. See also: Symons, op. cit.; Simkin and Fiske, op. cit. Elsewhere along the Ceylon coast there were other reports of the sea receding, and a gradual return that posed no serious threat. The rise was about eight feet at Batticaloa (on the east coast), at Trincomalee (on the northeast coast), and at Colombo (on the west coast) and twelve feet at Hambantota (on the south coast). Port Mathurin on the island of Rodrigues (3,200 miles from Krakatoa) saw six-foot heights; Port Elizabeth in South Africa (5,400 miles from Krakatoa) saw four-foot heights.
43. Tide curves in Europe displayed recognizable oscillations, though they were very, very small. For example, there were seven oscillations of five inches each at Socca, near Biarritz, France (in the Bay of Biscay, near Spain's border), 10,000 miles from Krakatoa. At Cherbourg, France, on the English Channel the oscillations were only two inches. See: Symons, op. cit.
44. Tambora was a much larger eruption than Krakatoa, but its tsunamis were not as large as the ones generated by Krakatoa, although there were still major casualties. Tambora had a much greater climatic effect than Krakatoa. In fact, 1816 was the "year without a summer": when snow fell in New England in June and the southern United States had frost on July 4.
45. Winchester, op. cit., pp. 205–206.
46. The Royal Society put together a large report in 1888. Symons, op. cit., 1888. The Dutch, who had ruled Indonesia as a colony since 1816, had produced their report in 1886. Verbeek, R.D.M., 1886. *Krakatau*, Government Printing Office, Batavia, Government of the Netherlands, 567 pages. One aspect that was not understood, and is still debated, was exactly how the tsunamis were produced. See: Nomanbhoy, N., and K. Satake, 1995. "Generation Mechanism of Tsunamis from the 1883 Krakatau Eruption," *Geophysical Research Letters*, vol. 22, no. 4, pp. 509–512.
47. Land movement in Batavia was sensed by changes in the magnetic field on a magnetograph. Propagation to much greater distances was deduced by the average periods of the tsunamis that were recorded around the world on tide gauges. See: Nagaoka, H., 1907. "The Eruption of Krakatoa and the Pulsation of the Earth," *Nature*, vol. 76, pp. 89–90. See also: Dewey, J., and P. Byerly, 1969. "The Early History of Seismometry (to 1900)," *Bulletin of the Seismological Society of America*, vol. 59, no. 1, pp. 183–227.
48. von Rebeur-Paschwitz, Ernst, 1889. "The Earthquake of Tokio, April 18, 1889," *Nature*, vol. 40, pp. 294–295. The earthquake was sensed on two horizontal-pendulum seismographs in Potsdam and Wilhelmshaven (Germany) about an hour after it occurred in Japan. This successful detection of an earthquake halfway around the world was a great impetus to setting up an international seismograph network.
49. Dewey and Byerly, op. cit., 1969. See also: Milne, John, 1900. *Seismological Investigations—Fifth Report of the Committee*, Report of the British Association for the Advancement of Science, pp. 59–108.
50. Of these, all were from submarine earthquakes except for one volcanic eruption—Kilauea Volcano in Hawaii in 1924. The twenty-eight tsunamis not generated locally came from fourteen different locations around the Pacific: Russia (four times), Japan (three), Chile (three), Alaska (three), Mexico (four, but three were from the same month, so two can be considered aftershocks), Solomon Islands (two), New Guinea (two), Ecuador, Indonesia, Philippines, Samoa, New Caledonia, Tonga, and California. These

data came from the Tsunami Database at NOAA's National Geophysical Data Center (NGDC), which is accessible online_(www.ngdc.noaa.gov/hazards/tsu_db.shtml). See also: Geist, E.L., and T. Parsons, 2008. "Distribution of Tsunami Interevent Times," *Geophysical Research Letters*, vol. 35, L02612.

51. Fryer, G.J., W. Philip, and L.F. Pratson, 2004. "Source of the Great Tsunami of April 1, 1946: A Landslide in the Upper Aleutian Forearc," *Marine Geology*, vol. 203, pp. 201–218.
52. "Hilo City Rolled Up by Crash of Water," *New York Times*, Apr. 2, 1946 (written Apr. 1). The story of the Laupāhoehoe school and many other stories of the survivors and victims of the 1946 tsunami in Hawaii are collected in Dudley, Walter C., and Min Lee, 1998. *Tsunami!* 2nd ed., Univ. of Hawaii Press, Honolulu, 362 pages. They can also be found at the Pacific Tsunami Museum in Hilo, Hawaii (www.tsunami.org). See also: Shepard, F.P., G.A. MacDonald, and D.C. Cox, 1950. "The Tsunami of April 1, 1946," *Bulletin of the Scripps Institution of Oceanography*, vol. 5, no. 6, pp. 391–528, Univ. of California Press, La Jolla.
53. "Hawaii Lacked First Wave Warning," *New York Times*, Apr. 2, 1945 (written Apr. 1). See also: Dudley and Lee, op. cit., 1998.
54. Zetler, Bernard D., 1965. "Tsunamis and the Seismic Sea Wave Warning System," *Mariners Weather Log*, vol. 9, no. 5, pp. 149–152. See also: Zetler, Bernard D., 1988. "Some Tsunami Memories," *Science of Tsunami Hazards: The International Journal of the Tsunami Society*, vol. 6, no. 1, pp. 57–61.
55. The seismic sea wave detector was a mechanical device with a pressure chamber. The whole device was put in the tide well of a tide gauge. To recognize a seismic sea wave (tsunami), the detector looked for rapid changes in water level taking place over a few minutes to an hour, after averaging out wind-wave effects that caused rapid changes over a few seconds. See: Zerbe, W.B., 1948. "A Seismic Sea Wave Detector," *Journal, Coast and Geodetic Survey*, Jan. 1, pp. 51–55.
56. Seismographs used a photographic technique to record data, and so the photographic record had to be developed before it was known that seismic waves had arrived. The photographic technique was used because it required no mechanical contact, thus allowing greater sensitivity for measuring very small seismic waves, the most usual type; that sensitivity was not needed for seismic waves from large earthquakes. By 1948, fifty-two *strong-motion seismographs* were in operation in the western United States and seven in South and Central America. These were automatically turned on only during strong earthquakes and consisted of accelerographs and displacement meters. Standard seismographs also had greater magnifications for detecting the numerous small earthquakes that occur daily; there were more than five hundred in 1948.
57. "CAA Stations to Help Detect Tidal Waves," *New York Times*, Aug. 15, 1949. See also: Zetler, op. cit., 1988. Even after the 1946 deaths at Hilo and the complaints about not being warned, C&GS could not get Congress to appropriate funding for the warning system. They managed to build it anyway (without congressional authorization or funding), and only in 1952, after the system successfully warned of a tsunami from Kamchatka, Russia, was a small amount of funding provided by Congress. Lack of funding would plague the warning system throughout its existence until 2005.
58. With reasonably fine resolution in representing the ocean depths, the calculated travel times matched the observed travel times quite well. See: Zetler, Bernard D., 1947. "Travel Times of Seismic Sea Wave to Honolulu," *Pacific Science*, vol. 10, no. 3, pp. 185–188. See also: U.S. Coast and Geodetic Survey, 1965. *Tsunami Travel Time Charts, for Use in the Seismic Sea Wave Warning System*, rev. ed., Coast and Geodetic Survey, U.S. Department of Commerce, Rockville, MD. [This is the source of figure 7.3.]
59. Zetler, op. cit., 1988.

60. This ranking of earthquakes is from the U.S. Geological Survey, which assigned the following magnitudes to these four earthquakes: 9.5 for 1960 (Chile); 9.2 for 1964 (Prince William Sound, Alaska); 9.0 for 1952 (Kamchatka, Russia); and 8.6 for 1957 (Aleutian Islands, Alaska). Others have given the 1957 earthquake a 9.1 magnitude.
61. Coast and Geodetic Survey (W.B. Zerbe), 1953. *The Tsunami of November 4, 1952 as Recorded at Tide Stations*, Special Publication 300, Coast and Geodetic Survey, U.S. Dept. of Commerce, Government Printing Office, Washington, DC, 62 pages. Elsewhere around the Pacific the tsunami had large heights at other locations, including nine and a half feet at Avila Beach, California, and twelve feet at Talcahuano, Chile, the latter more than 9,700 miles (and twenty-two hours in tsunami travel time) away. See also: Dudley and Lee, op. cit., 1998.
62. The Seismic Sea Wave Warning Center had passed the warning on to the Hawaiian Police, the 31st Air Weather Wing, CINCPAC, and even the Tokyo Central Meteorological Observatory. See: Salsman, Garrett C., 1959. *The Tsunami of March 9, 1957, as Recorded at Tide Stations*, Technical Bulletin 6, Coast and Geodetic Survey, U.S. Dept. of Commerce, Government Printing Office, Washington, DC, 18 pages. See also: Dudley and Lee, op. cit., 1998.
63. Dudley and Lee, op. cit., 1998. See also: Johnston, Jeanne B., 2003. *Personal Accounts from Survivors of the Hilo Tsunamis of 1946 and 1960: Toward a Disaster Communication Model*, Master's Thesis, Univ. of Hawaii, May 2003, 142 pages; Atwater, B.F., M. Cisternas, J. Bourgeois, W.C. Dudley, J.W. Hendley II, and P.H. Stauffer, 1999. *Surviving a Tsunami—Lessons from Chile, Hawaii, and Japan*, Circular 1187, U.S. Geological Survey, Dept. of the Interior, Government Printing Office, Washington, DC, 20 pages.
64. Only some shallow funnel-shaped bays saw larger heights, usually between twelve and seventeen feet. It might have been the funnel shape, which forced more water into a smaller cross section as the wave moved toward the narrower head of the inlet, or it might have been that the natural period of oscillation of the bay was close to the period of the tsunami waves, which added to the wave height.
65. In Chapter 5 we talked about *wave refraction*, namely, the bending of the wave direction toward shallow water because the deeper part of the wave travels faster than the shallower part, and will talk more about it in this chapter and in Chapter 9.
66. Zetler, op. cit., 1988. See also: Japan Meteorological Agency, 1963. *The Report on the Tsunami of the Chilean Earthquake, 1960*, Technical Report 26, Mar., Tokyo.
67. There were two factors that the Japanese did not consider. The first is the *geometric spreading* of a wave as it moves farther away from its source, by which the energy density must decrease along its ever-lengthening wave front. On a sphere (like the Earth), however, such spreading only continues until the wave travels one-quarter of the way around the globe, at which point it becomes *geometric compression* and the energy density increases again as the wave front shortens again. Japan was almost on the opposite side of the world from Chile, and so the waves heights would have begun to increase again. Second, the wave energy tends to be focused and travels along submarine ridges by a process of refraction, and is also affected by other variations in the ocean's bathymetry (its variations in depth). The submarine ridges in the Pacific (including those with tropical islands on them) tend to go from southeast to northwest, roughly from Chile toward Japan. See: Woods, M.T., and E.A. Okal, 1987. "Effect of Variable Bathymetry on the Amplitude of Teleseismic Tsunamis: A Ray-Tracing Experiment," *Geophysical Research Letters*, vol. 14, no. 7, pp. 765–768.
68. Two international groups were established: the International Tsunami Commission (in the International Union of Geodesy and Geophysics) and the Coordinating Group for Tsunami Warnings (inside the United Nations Intergovernmental Oceanographic Commission). The United States also offered to support the creation and maintenance

of an International Tsunami Information Center. See: Zetler, op. cit., 1965. See also: Sullivan, Walter, 1960. "U.S., Japan and Russia Set up Alarm System for Tidal Waves," *New York Times*, Aug. 5.

69. Dudley and Lee, op. cit., 1998. See also: Bryant, Edward, 2008. *Tsunami: The Underrated Hazard*, 2nd ed., Springer-Praxis, Chichester, UK, 330 pages. The title of this book was very appropriate when the first edition came out in 2001.
70. Whitmore, P., et al., 2008. "NOAA/West Coast and Alaska Tsunami Warning Center Pacific Ocean Response Criteria," *Science of Tsunami Hazards*, vol. 27, no. 2, pp. 1–21.
71. Following the 2004 tsunami, the area of responsibility of the West Coast/Alaska Tsunami Warning Center was expanded to include the U.S. Atlantic and Gulf of Mexico coasts, Puerto Rico, the Virgin Islands, and the Atlantic coast of Canada. An important organization that indirectly supports tsunami warning was established in 1966, the National Earthquake Information Center, which was created in the National Ocean Survey (the former Coast and Geodetic Survey) and later moved to the U.S. Geological Survey in 1973.
72. The first of the following references is a firsthand account of a father and his eight-year-old son, Howard and Sonny Ulrich, who survived the tsunami, but it also briefly describes what happened to Bill and Vivian Swanson, the husband and wife who were in the first boat that was lifted over trees and carried to the Atlantic Ocean. The Miller paper includes other eyewitness accounts. Ulrich, Howard, 1958. "Night of Terror (as told to Vi Haynes)," *The Alaska Sportsman*, Oct., pp. 11, 42–44; Miller, Don J., 1960. *Giant Waves in Lituya Bay Alaska*, Geological Survey Professional Paper 354-C, U.S. Geological Survey, Government Printing Office, Washington, DC, 86 pages. See also: Fritz, H.M., F. Mohammed, and J. Yoo, 2009. "Lituya Bay Landslide Impact Generated Mega-tsunami, 50th Anniversary," *Pure Applied Geophysics*, vol. 166, pp. 153–175.
73. Prior, D.B., et al., 1989. "Storm Wave Reactivation of a Submarine Landslide," *Nature*, vol. 341, pp. 47–50.
74. "Tidal Wave Sweeps East Coast of Burin," *Free Press* (St. John's, Newfoundland), Nov. 26, 1929, vol. 30, no. 18. See also: Ruffman, Alan, and Violet Hann, 2006. "The Newfoundland Tsunami of November 18, 1929: An Examination of the Twenty-eight Deaths of the 'South Coast Disaster,'" *Newfoundland and Labrador Studies*, vol. 21, no. 1, pp. 1719–1726.
75. Bondevik, S., S. Dawson, A. Dawson, and Ø. Lohne, 2003. "Record-Breaking Height for 8000-Year-Old Tsunami in the North Atlantic," *EOS, Transactions, American Geophysical Union*, vol. 84, no. 31, pp. 289, 293.
76. Ward, S.N., and S.J. Day, 2001. "Cumbre Vieja Volcano—Potential Collapse and Tsunami at La Palma, Canary Islands," *Geophysical Research Letters*, vol. 28, pp. 3397–3400.
77. The worst-case scenario of the study is not likely to happen, many believe, because sediment evidence found around La Palma Island indicates that, rather than each past landslide occurring as a single large block of rock rapidly moving down the volcano into the sea, there were several smaller landslides that produced smaller tsunamis. But even in the 2001 study's worst-case scenario, both the quantity of material dropped into the water and the time duration of the event are not great enough to generate the very long-wavelength tsunamis produced by earthquakes. With their shorter wavelength, they would break when they moved into shallow water, dissipating much of the energy. See: Masson, D.G., C.B. Harbitz, R.B. Wynn, G. Pedersen, and F. Løvholt, 2006. "Submarine Landslides: Processes, Triggers and Hazard Prediction," *Philosophical Transactions of the Royal Society*, A, vol. 364, pp. 2009–2039.
78. After studying the 1906 San Francisco earthquake, Harry Fielding Reid had proposed his "elastic rebound theory" for earthquakes, in which he suggested that an earthquake was the result of the release of built-up elastic strain in the ground, but he did not know what had caused that strain to build up (later explained by plate tectonics). Reid, H.F.,

1911. "The Elastic-Rebound Theory of Earthquakes." *Bulletin of the Department of Geology, University of California Publications*, vol. 6, no. 19, pp. 413–444.

79. For a detailed history of this story, see: Oreskes, Naomi, 1999. *The Rejection of Continental Drift: Theory and Method in American Earth Sciences*, Oxford Univ. Press, 420 pages. For a briefer description, see: Frankel, Henry, 1998. "Continental Drift and Plate Tectonics," in *Sciences of the Earth: An Encyclopedia of Events, People, and Phenomena*, ed. Gregory A. Good, Garland, New York, pp. 118–136.
80. Wegener, Alfred, 1924. *The Origin of Continents and Oceans* (English edition of 1922 German 3rd edition), Methuen, London, 212 pages.
81. Holmes, Arthur, 1944. *Principles of Physical Geology*, Thomas Nelson, London, 532 pages. He first proposed his theories on the mechanisms behind continental drift in several papers during the 1920s and 1930s.
82. Holmes also studied radioactivity, from the standpoint of heating the Earth's interior, and also its use in radioactive dating in geology. Holmes, Arthur, 1931. "Radioactivity and Earth Movements," *Transactions, Geological Society of Glasgow*, vol. 18, pp. 559–606.
83. Vine, F.J., and D. Matthews, 1963. "Magnetic Anomalies over Oceanic Ridges," *Nature*, vol. 199, pp. 947–949.
84. Technically, the tectonic plates are made up of crust plus some upper (and more rigid) mantle material, the combined layer being referred to as the *lithosphere*. The rest of the more plastic mantle in which the heat-driven and excruciatingly slow convection takes place is referred to as the *asthenosphere*.
85. This is how the strain built up that Reid had talked about in his "elastic rebound theory" for earthquakes. See note 79.
86. The tsunami data in this paragraph and those following came from the Tsunami Database at NOAA's National Geophysical Data Center (NGDC), which is accessible online (www.ngdc.noaa.gov/hazard/tsu_db.shtml). The earthquake data came from the U.S. Geological Survey earthquake database (http://neic.usgs.gov/neis/epic).
87. The Pacific Tsunami Warning Center, in Ewa Beach, Hawaii (near Honolulu), was originally in the tides and currents division of the U.S. Coast and Geodetic Survey (C&GS). In 1965 C&GS became part of the newly formed ESSA (Environmental Science Services Administration), and then in 1970 became part of the newly formed NOAA (National Oceanic and Atmospheric Administration), at which point C&GS's name was changed first to the National Ocean Survey and then to the National Ocean Service (NOS). Eventually the Warning Center was moved out of the tides and currents division in NOS to another part of NOAA. It now resides within NOAA's National Weather Service, while the research group that supports it is in the Pacific Marine Environmental Laboratory (PMEL) (see note 90).
88. The Next Generation Water Level Measurement System was developed and operated within NOS (see previous note), in what was the tides and currents division of the old U.S. Coast and Geodetic Survey, now called the Center for Operational Oceanographic Products and Services (CO-OPS) in NOS in NOAA.
89. Bernard, E.N., Mofjeld, H.O., Titov, V., C.E. Synolakis, and F.I. Gonzalez, 2006. "Tsunami: Scientific Frontiers, Mitigation, Forecasting and Policy Implications," *Philosophical Transactions of the Royal Society*, A, vol. 364, pp. 1989–2007.
90. The Pacific Marine Environmental Laboratory (PMEL), in Seattle, is part of NOAA's Office of Oceanic and Atmospheric Research (OAR). PMEL is the primary research group that supports the Pacific Tsunami Warning Center (PTWC), developing, in addition to DART, sophisticated hydrodynamic tsunami models and other forecast tools used by PTWC.
91. See Chapter 8.
92. The five tsunamis, in 1797 and the 1800s, were generated by earthquakes whose epicenters were west of Sumatra and various distances south of the epicenter for the

December 26, 2004, earthquake. See: Newcomb, K.R., and W.R. McCann, 1987. "Seismic History and Seismotectonics of the Sunda Arc," *Journal of Geophysical Research*, vol. 92, no. 81, pp. 421–439. See also: Natawidjaja, D.H., et al., 2006. "Source Parameters of the Great Sumatran Megathrust Earthquakes of 1797 and 1833 Inferred from Coral Microatolls," *Journal of Geophysical Research*, vol. 111, B06403, pp. 1–37.

CHAPTER 8

1. The Sunda Trench is 1,600 miles long, running from the eastern end of the island of Java, past Sumatra, to the northern end of India's Nicobar and Andaman island chains.
2. The Burma microplate is the Southeast Asian portion of the much larger Eurasian continental plate. This thousand-mile stretch of building pressure is called the Sunda megathrust fault. It is a subduction zone, where the Indian plate (moving north-northeast) is trying to push under the Burma microplate.
3. 7:59 A.M. local Indonesian time is 0:59 A.M. UTC (Universal Time Coordinated), also referred to as GMT (Greenwich Mean Time), since it is the time at the 0° longitude line that runs through Greenwich, England. The nations around the Indian Ocean have different time zones. We list here the start time of the earthquake in the local time for each of these nations (the number in parentheses is the number of hours between each time zone and UTC—as it was in 2004, because some of these time zones have since been changed): 8:59 A.M. in Malaysia (+8); 7:59 A.M. in Thailand and Indonesia (+7); 7:29 A.M. in Myanmar (+6½); 6:59 A.M. in Sri Lanka (+6); 6:29 A.M. in India, including the Nicobar and Andaman islands (+5½); 5:59 A.M. in the Maldives (+5); 4:59 A.M. in Oman, Mauritius, and the Seychelles (+4); 3:59 A.M. in Yemen, Somalia, Kenya, Tanzania, and Madagascar (+3); and 2:59 A.M. in Mozambique and South Africa (+2). At the Pacific Tsunami Warning Center in Hawaii, the time of the earthquake was 2:59 P.M. in the afternoon of the day before (–10). To minimize confusion throughout this chapter we will sometimes refer to times in terms of hours after the initial earthquake (thus avoiding mention of a time or time zone.)
4. The location below the earth where the earthquake begins is called the *focus* (or *hypocenter*), and the location on the earth's surface above the focus is called the *epicenter.*
5. Pietrzak, J.A., et al., 2007. "Defining the Source Region of the Indian Ocean Tsunami from GPS, Altimeters, Tide Gauges, and Tsunami Models," *Earth and Planetary Letters*, vol. 261, pp. 49–64. See also: Subarya, C., et al., 2006. "Plate-Boundary Deformation Associated with the Great Sumatra-Andaman Earthquake," *Nature*, vol. 440, pp. 46–51; Ammon, C.J., et al., 2005. "Rupture Process of the 2004 Sumatra-Andaman Earthquake," *Science*, vol. 308, pp. 1133–1139; Stein, S., and E.A. Okal, 2005. "Speed and Size of the Sumatra Earthquake," *Nature*, vol. 434, pp. 581–582; Kennett, B.L.N., and P.R. Cummins, 2005. "The Relationship of the Seismic Source and Subduction Zone Structure for the 2004 December 26 Sumatra-Andaman Earthquake," *Earth and Planetary Science Letters*, vol. 239, pp. 1–8.
6. The rupture was 90 to 125 miles wide for most of its length. The 900-mile-long rupture of the Sunda fault was the longest earthquake rupture in recorded history, in both geographic distance and duration. But it was not a smooth, continuous rupture. In some places the speed of the rupture slowed down, and in other places it speeded up, according to the geological characteristics of the crustal rock along the fault. There were probably three locations along the southern half of the rupture where the vertical motions of the seafloor were the largest and from where the largest tsunamis emanated; two other locations farther north generated smaller tsunamis. In a few places the primary rupture may have instigated other smaller ruptures (call *splay fault* ruptures) that added to the vertical motion of the seafloor and increased the size of the tsunamis emanating from those particular sections of the fault rupture.

7. The speed of a long wave is proportional to the square root of the water depth. See note 35 in Chapter 7.
8. Ando, M., M. Nakamura, Y. Hayashi, M. Ishida, and D. Sugiyanto, 2009. "Observed High Amplitude Tsunami 0.5–20 km away from the Northern Sumatra Coast during the 2004 Sumatra Earthquake," *Journal of Asian Earth Sciences*, vol. 36, pp. 98–109. This paper includes the personal accounts of ten fishermen who were at sea when the submarine earthquake and tsunami took place.
9. Mariners have called all this commotion at sea a *seaquake*.
10. Ando et al., op. cit.
11. Bearak, Barry, 2005. "The Day the Sea Came," *New York Times Magazine*, Nov. 27. This excellent and moving article provides one of the best accounts of the tsunami in Banda Aceh (Sumatra, Indonesia) from the perspective of six people who survived the tsunami but lost families and homes. See also: Ando et al., op. cit. Other information came from newspaper articles too numerous to list here. See also: Carnes, Tony, 2006. "Walking the Talk after Tsunami," *Christianity Today*, Mar. 1. The article is about Breueh villagers who were killed by the tsunami.
12. JTIC, 2009. *Smong: Local Knowledge of Tsunami among the Simeulue Community, Nangroe Aceh Darussalam*, Jakarta Tsunami Information Centre, 50 pages. See also: McAdoo, B., L. Dengler, M.G. Prasetya, and V. Titov, 2006. "Smong: How an Oral History Saved Thousands on Indonesia's Simeulue Island during the December 2004 and March 2005 Tsunamis," *Earthquake Spectra*, vol. 22, no. 53, pp. S661–S669.
13. Gaillard, J.-C., et al., 2008a. "Ethnic Groups' Response to the 26 December 2004 Earthquake and Tsunami in Aceh, Indonesia," *Natural Hazards*, vol. 47, pp. 17–38.
14. Park, J., K. Anderson, R. Aster, R. Butler, T. Lay, and D. Simpson, 2005. "Global Seismographic Network Records the Great Sumatra-Andaman Earthquake," *EOS, Transactions, American Geophysical Union*, vol. 86, no. 6, pp. 57, 60–61.
15. Those at the Pacific Tsunami Warning Center (PTWC) who reacted to the alarm, analyzed the data, and sent out bulletins included Charles McCreery (Director), Stuart Weinstein (Deputy Director), and Barry Hirshorn. At the same time at the Pacific Marine Environmental Laboratory (see note 90 in Chapter 7), the NOAA office that developed much of the tsunami forecast system for PTWC, Vasily Titov was the first to hydrodynamically model the 2004 tsunami. Information on the response and activities of PTWC came from several sources, including: Revkin, Andrew C., 2004. "How Scientists and Victims Watched Helplessly," *New York Times*, Dec. 31; *NOAA News*, 2005. "NOAA and the Indian Ocean Tsunami," *NOAA News*, story 2358, Jan. 28, www.noaanews.noaa.gov/stories2004/s2358.htm; Watson, P., B. Demick, and R. Fausset, 2005. "A Tremor, Then a Sigh of Relief, before the Cataclysm Rushed In," *Los Angeles Times*, Jan. 2; Kayal, Michele, and Matthew L. Wald, 2004. "At Warning Center, Alert for the Quake, None for a Tsunami," *New York Times*, Dec. 28; Kerr, Richard A., 2005. "Failure to Gauge the Quake Crippled the Warning Effort," *Science*, vol. 37 (Jan. 14), p. 201; Gower, J., and F. Gonzalez, 2006. "U.S. Warning System Detected the Sumatra Tsunami," *EOS, Transactions of the American Geophysical Union*, vol. 87, no. 10, pp. 105–112. See also: Thomas, Evan, and George Wehrfritz, 2005. "Tide of Grief," *Newsweek*, vol. 145, no. 2, pp. 22–45; Kelman, Ilan, 2006. "Warning for the 26 December 2004 Tsunamis," *Disaster Prevention and Management*, vol. 15, no. 1, pp. 178–189. *Tsunami Bulletins* put out by the Pacific Tsunami Warning Center were found in the PTWC message archive at www.prh.noaa.gov/ptwc/.
16. Other centers receiving seismic information included NOAA's West Coast/Alaska Tsunami Warning Center, USGS's National Earthquake Information Center (NEIC), the Japanese Meteorological Agency's tsunami warning center, Geoscience Australia, and Russia's Yuzhno-Sakhalilnsk Tsunami Warning Center.

17. 8:07 A.M. Indonesian time was 3:07 P.M. the day before in Hawaii and 1:07 UTC. See note 3.
18. Only a one-foot tsunami was detected at Jackson Bay on the northeast coast of New Zealand's South Island. The tsunami crossed the Tasman Sea to Australia but registered only six inches at Spring Bay, Tasmania. That the earthquake did not produce a significant tsunami was perhaps due to the fault slip being more horizontal than vertical. See: International Tsunami Information Centre (ITIC), 2005. "North of Macquarie Island," *Tsunami Newsletter*, vol. 37, no. 1, p. 6.
19. "False Alarms in Tsunami Warnings," Associated Press, Feb. 13, 2005.
20. See references in note 15.
21. Ando et al., op. cit., 2009. See also: Bearak, op. cit., 2005, specifically, the story of Jaloe the fisherman. See also: Yulianto, E., F. Kusmayanto, N. Supriyatna, and M. Dirhamsyah, 2009. *Surviving a Tsunami—Lessons from Aceh and Southern Java, Indonesia*, IOC Brochure 2009-1, Intergovernmental Oceanographic Commission (IOC), UNESCO, Jakarta Tsunami Information Centre; and Meek, James, 2005. "Life among the Ghosts of Banda Aceh," *Guardian* (UK), Dec. 23. One of the moving stories is about a fisherman who was at sea when the earthquake and tsunami occurred.
22. The descriptions in the following paragraphs are based on accounts from many newspaper articles and other sources too numerous to list, but a few representative ones are listed here: Lines, David, 2006. "An Eyewitness Account," in *Status of Coral Reefs in Tsunami Affected Countries: 2005*, ed. C. Willerson, D. Souter, and J. Goldberg, Global Coral Reef Monitoring Network, Australia Institute of Marine Science, 150 pages, p. 45; Chamim, Mardiyah, 2005. *History Grows in Our Kampong: Notes from Aceh, Tsunami Hot Zone, Chapter 12*, http://acehjourney.net/wp-content/uploads/2009/04/bab-12_1_eng.pdf. Other information about the destruction comes from scientific papers such as the following: Paris, R., et al., 2009. "Tsunamis as Geomorphic Crises: Lessons from the December 26, 2004 Tsunami in Lhok Nga, West Banda Aceh (Sumatra, Indonesia)," *Geomorphology*, vol. 104, pp. 59–72.
23. Paris, R., F. Lavigne, P. Wassmer, and J. Sartohadi, 2007. "Coastal Sedimentation Associated with the December 26, 2004 Tsunami in Lhok Nga, West Banda Aceh (Sumatra Island)," *Marine Geology*, vol. 238, pp. 93–106.
24. Meek, James, 2005. "From One End to Another, Leupeng Has Vanished As If It Never Existed," *Guardian* (UK), Jan. 1. See also: Lavigne, F., et al., 2009. "Reconstruction of Tsunami Inland Propagation on December 26, 2004 in Banda Aceh, through Field Investigations," *Pure and Applied Geophysics*, vol. 166, pp. 259–281; Kawata, Yoshiaki, et al., 2005. "Earthquake, Tsunami and Damage in Banda Aceh and Northern Sumatra," in *The 2004 Indian Ocean Disaster Report*, Kyoto Univ., Japan, chap. 2; and Borrero, J.C., C.E. Synolakis, and H. Fritz, 2006. "Northern Sumatra Field Survey after the December 2004 Great Sumatra Earthquake and Indian Ocean Tsunami," *Earthquake Spectra*, vol. 22, no. S3, pp. S93–S104.
25. Meek, op. cit., 2005; Lavigne et al., op. cit., 2009.
26. Webb, Sara, 2008. "Aceh's Former Fighters Guide 'Guerrilla Tourists,'" Reuters, Feb. 18. See also: Gaillard, J.-C., E. Clave, and I. Kelman, 2008b. "Wave of Peace? Tsunami Disaster Diplomacy in Aceh, Indonesia," *Geoforum*, vol. 39, pp. 511–526.
27. Meek, James, 2005. "In the Town of Ghosts, First Sign of Order," *Guardian* (UK), Jan. 3.
28. Gaillard et al., op. cit., 2008b; Aglionby, John, 2004. "Stench of Dead Bodies Is All Around," *Guardian* (UK), Dec. 31. See also: Suryana, A'an, 2005. "Tsunami Victims Tell of Their Fight for Survival," *Jakarta Post*, Jan. 3.
29. Bearak, op. cit., 2005; Gaillard et al., op. cit., 2008a; Lavigne et al., op. cit., 2009.
30. Bearak, op. cit., 2005.
31. Lavigne et al., op. cit., 2009. Many pictures showing the devastation in Sumatra caused by the 2004 tsunami can be found at the "Photo Gallery of Northwestern Sumatra" taken

by U.S. Geological Survey personnel during their January 20–29, 2005, survey of that area. See: http://walrus.wr.usgs.gov/tsunami/sumatra05/photos.html. [The picture in figure 8.1 comes from that photo gallery.]

32. Bearak, op. cit., 2005. See also: Meek, James, 2005. "Life among the Ghosts of Banda Aceh," *Guardian* (UK), Dec. 23.
33. "Woman Rescued after 5 Days Adrift," *CNN.com*, Jan. 3, 2005, http://cnn.com.
34. Bearak, op. cit., 2005.
35. When the tsunami arrived, Mayor Syarifuddin Latief, who stayed to help direct people rather than escape by car, was swept away by the wave. His body was not found until January 17. "Mayor of Banda Aceh's Body Found" (English title), Jan. 19, 2005, *Endonesia* (article is in Indonesian), www.endonesia.com/mod.php?mod=publisher&op=viewarticle&artid=308.
36. Bearak, op. cit., 2005. See also: Perlez, Jane, 2005. "For Many Tsunami Survivors, Battered Bodies, Grim Choices," *New York Times*, Jan. 6.
37. Elliott, Michael, 2005. "Sea of Sorrow," *Time*, vol. 165, no. 2, pp. 22–39.
38. The results of the special tsunami workshop held in Jakarta in 2003 are summarized in: Purwana, Ibnu, and Fauzi, 2003. "International Seminar/Workshop on Tsunami: 'In Memoriam 120 Years of Krakatau Eruption-Tsunami and Lesson Learned from Large Tsunami,' Jakarta-Anyer, Indonesia, 26–29 August 2003," *Tsunami Newsletter*, vol. 35, no. 5 (Aug.–Dec.), pp. 22–24.
39. Lander, J.F., L.S. Whiteside, and P.A. Lockridge, 2003. "Two Decades of Global Tsunamis, 1982–2002," *Science of Tsunami Hazards*, vol. 21, no. 1, 88 pages.
40. IOC, 2000. *International Co-ordination (IOC) Group for the Tsunami Warning System in the Pacific*, Seventeenth Session, Seoul, Republic of Korea, October 4–7, 1999, IOC/ITSU-XVII/3, Paris, Feb. 8, 65 pages.
41. There are also fifty-three uninhabited islands, islets, and rocky areas always above tidal low water.
42. Sundar, V., S.A. Sannasiraj, K. Murali, and R. Sundaravadivelu, 2007. "Runup and Inundation along the Indian Peninsula, Including the Andaman Islands, due to Great Indian Ocean Tsunami," *Journal of Waterway, Port, Coastal, and Ocean Engineering, ASCE*, vol. 133, no. 6, pp. 401–413. See also: Cho, Y.-S., C. Lakshumanan, B.-H. Choi, and T.-M. Ha, 2008. "Observations of Run-up and Inundation Levels from the Teletsunami in the Andaman and Nicobar Islands: A Field Report," *Journal of Coastal Research*, vol. 24, no. 1, pp. 216–223.
43. Glass, Nick, 2005. "Tsunami Disaster: 'Stone Age' Life of Island Tribes People Helped Them Survive 'Black Sunday,'" *Independent* (UK), Jan. 16. See also: Stone, Richard, 2006. "After the Tsunami: A Scientist's Dilemma," *Science*, vol. 313, pp. 32–35; Pandey, Geeta, 2005. "Eyewitness: Remote Tragedies," *BBC News*, Jan. 3; "Two Babies Find Life among Deadly Waters," *CNN.com*, Jan. 1, 2005, http://cnn.com; Bhaumik, Subir, 2005. "Tsunami Folklore 'Saved Islanders,'" *BBC News*, Jan. 20; Thangaraj, K., G. Chaubey, T. Kivisild, A.G. Reddy, V.K. Singh, A.A. Rasalkar, and L. Singh, 2005. "Reconstructing the Origin of Andaman Islands," *Science*, vol. 308, p. 996.
44. Misra, Neelesh, 2005. "Nat Geo TV Shows Help Tsunami Islanders Save 1,500," Associated Press, Jan. 7.
45. Gupta, Shishira, and Amitav Ranjan, 2004. "Govt Got Wind 1 Hr before Waves Hit Chennai," *Indian Express*, Dec. 30. See also: Agence France-Presse, 2004. "First Tsunami Alert Lost in Indian Bureaucracy," Dec. 30.
46. Earthquake magnitude scales (like the Richter scale, or the more recent *moment magnitude scale*) are logarithmic scales, so that an increase of one unit of magnitude is ten times greater.
47. "Strong Quake Hits Indonesia, At Least Nine Killed," Reuters, Dec. 26, 2004; "Powerful Earthquake Rocks Indonesia's Aceh," Associated Press, Dec. 26, 2004; "A Strong Quake

Jolts Sumatra Island," United Press International, Dec. 26, 2004; "Panic after Quake Rocks Indonesia's Sumatra Island," Agence France-Presse (AFP), Dec. 26, 2004.

48. Elliott, op. cit., 2005.
49. Thailand has the same time zone as Indonesia.
50. Kietpawpan, M., P. Visuthismajarn, C. Tanavud, and M.G. Robson, 2008. "Method of Calculating Tsunami Travel Times in the Andaman Sea Region," *Natural Hazards*, vol. 46, pp. 89–106.
51. Elliot, op. cit., 2005; Thomas and Wehrfritz, op. cit., 2005.
52. Ioualalen, M., J. Asavanant, N. Kaewbanjak, S.T. Grilli, J.T. Kirby, and P. Watts, 2007. "Modeling the 26 December 2004 Indian Ocean Tsunami: Case Study of Impact in Thailand," *Journal of Geophysical Research*, vol. 112, C07024.
53. Bell, R., H. Cowan, E. Dalziell, N. Evans, M. O'Leary, B. Rush, and L. Yule, 2005. "Survey of Impacts on the Andaman Coast, Southern Thailand Following the Great Sumatra-Andaman Earthquake and Tsunami of December 26, 2004," *Bulletin of the New Zealand Society for Earthquake Engineering*, vol. 38, no. 3, pp. 123–148. See also: Ghobarah, A., M. Saatcioglu, and I. Nistor, 2006. "The Impact of the 26 December 2004 Earthquake and Tsunami on Structures and Infrastructure," *Engineering Structures*, vol. 28, pp. 312–326.
54. Thanawood, C., C. Yongchalermchai, and O. Densrisereekul, 2006. "Effects of the December 2004 Tsunami and Disaster Management in Southern Thailand," *Science of Tsunami Hazards*, vol. 24, no. 3, pp. 206–217. See also: Kietpawpan et al., op. cit., 2008; Bell et al., op. cit., 2005; Ghobarah et al., op. cit., 2006; Ioualalen et al., op. cit., 2007.
55. Richburg, Keith B., 2005. "Hope Fades on Identifying Missing Foreigners," *Washington Post*, Jan. 2; Elliot, op. cit., 2005; Ghobarah et al., op. cit., 2006; Bell et al., op. cit., 2005.
56. Goodman, Peter S., 2004. "For Many, a Moment's Difference between Life and Death," *Washington Post*, Dec. 30. See also: Ioualalen et al., op. cit., 2007; Bell et al., op. cit., 2005; Ghobarah et al., op. cit., 2006.
57. On the northeast coast of Sumatra at Medan (about a third of the way down the Malacca Strait), the tsunami was already only five feet high, arriving there four hours after the quake. Before dying out, there were more frequent undulations as it went down the strait. See: Koh, H.L., S.Y. The, P.L.-F. Liu, A.I. Md. Ismail, H.L. Lee, 2009. "Simulation of Andaman 2004 Tsunami for Assessing Impact on Malaysia," *Journal of Asia Earth Sciences*, vol. 36, pp. 74–83. See also: Yalciner, A.C., D. Perincek, S. Ersoy, G.S. Presateya, R. Hidayat, B. McAdoo, 2005. *Report on December 26, 2004, Indian Ocean Tsunami, Field Survey on Jan 21–31 at North of Sumatra*, ITST of IOC, UNESCO, Mar. 8.
58. Satake, K., et al., 2006. "Tsunami Heights and Damage along the Myanmar Coast from the December 2004 Sumatra-Andaman Earthquake," *Earth, Planets and Space*, vol. 58, pp. 243–252.

CHAPTER 9

1. Goodnough, Abby, 2005. "Survivors of Tsunami Live on Close Terms with Sea," *New York Times*, Jan. 23. See also: "Elders' Knowledge of the Oceans Spares Thai 'Sea Gypsies' from Tsunami Disaster," AP Worldstream, Jan. 1, 2005; Arunotai, Narumon, 2008. "Saved by an Old Legend and a Keen Observation: The Case of Moken Sea Nomads in Thailand," in *Indigenous Knowledge for Disaster Risk Reduction: Good Practices and Lessons Learned from Experiences in the Asia-Pacific Region*, UN International Strategy for Disaster Reduction, Bangkok, Thailand, July, pp. 73–77.
2. Tilly Smith became a worldwide celebrity after her awareness of the signs of a tsunami saved many lives on Mai Khao Beach in Thailand. Her invited appearances at the United Nations and elsewhere have helped educate others on recognizing the natural signs of an

impending tsunami. See: "Girl, 10, Used Geography Lesson to Save Lives," *Daily Telegraph* (UK), Jan. 1, 2005.

3. John Chroston, the teacher from Scotland, was also a member of the Ochil Mountain Rescue Team. The Thai doctor who helped was Harpreet Grover. See: "Fast-Thinking Father Saved Family and Tourist Bus," *Times* (London), Jan. 5, 2005.
4. Richie Neustfisten and his boss, Bill O'Leary, were both Australians living in Thailand. See: Elliott, Michael, 2005. "Sea of Sorrow," *Time*, vol. 165, no. 2, pp. 22–39. See also: McMahon, Neil, 2005. "Survivors Thank Man Who Read the Waves," *Sydney Morning Herald*, Jan. 8.
5. Bendeich, Mark, 2005. "Elephants Saved Tourists from Tsunami," Reuters, Jan. 2.
6. Sabine, Charles, 2005. "Senses Helped Animals Survive the Tsunami," *NBC News*, Jan. 6.
7. Highfield, Roger, 2005. "Did They Sense the Tsunami?" *Telegraph* (UK), Jan. 8. See also: Reuter, T., and S. Nummela, 1998. "Elephant Hearing," *Journal of the Acoustical Society of America*, vol. 104, pt. 1, pp. 1122–1123; O'Connell-Rodwell, Caitlin E., 2007. "Keeping an 'Ear' to the Ground: Seismic Communication in Elephants," *Physiology*, vol. 22, pp. 287–294; Le Pichon, A., P. Herry, P. Mialle, J. Vergoz, N. Brachet, M. Garcés, D. Drob, and L. Ceranna, 2005. "Infrasound Associated with 2004–2005 Large Sumatra Earthquakes and Tsunami," *Geophysical Research Letters*, vol. 32, L19802.
8. Henley, Thom, 2005. "Tsunami Sense," *Greater Phuket Magazine*, July 7.
9. Sri Lankan Time (SLT) was six hours after UTC in 2004; in 2006 it was changed to five and a half hours after to match Indian Standard Time (IST).
10. These were the first satellite altimetric measurements ever made of a tsunami. Satellite altimetry uses radar to measure the distance from the satellite to the ocean surface by measuring the time it takes a radio signal to travel to the sea surface and back to the satellite. Four satellites with altimetric radars were in orbit when the tsunami occurred: Jason-1, TOPEX, Envisat, and GFO. Their measurements were, however, very limited in time and space, since they could be made only along a satellite track as each satellite moved across one very narrow section of the Indian Ocean, only a few times a day depending on the satellite. The Jason-1 measurement was made almost two hours after the initial earthquake, along a track that went roughly from north to south, crossing the lower end of the Bay of Bengal about halfway between Sri Lanka and the epicenter and capturing the first wave front when it was about 700 miles south of Sri Lanka. See: Gower, Jim, 2005. "Jason 1 Detects the 26 December 2004 Tsunami," *EOS, Transactions of the American Geophysical Union*, vol. 86, no. 4, pp. 37–38. See also: Smith, W.H.F., R. Scharroo, V.V. Titov, D. Arcas, and B.K. Arbic, 2005. "Satellite Altimeters Measure Tsunami," *Oceanography*, vol. 18, no. 2, pp. 11–13; Ablain, M., Jo. Dorandeu, P.-Y. Le Traon, and A. Sladen, 2006. "High Resolution Altimetry Reveals New Characteristics of the December 2004 Indian Ocean Tsunami," *Geophysical Research Letters*, vol. 33, L21602.
11. Wijetunge, J.J., 2009a. "Field Measurements and Numerical Simulations of the 2004 Tsunami Impact on the East Coast of Sri Lanka," *Pure and Applied Geophysics*, vol. 166, pp. 593–622. See also: Wijetunge, J.J., 2009b. "Field Measurements and Numerical Simulations of the 2004 Tsunami Impact on the South Coast of Sri Lanka," *Ocean Engineering*, vol. 36, pp. 960–973; Liu, P.L.-F., et al., 2005. "Observations by the International Tsunami Survey Team in Sri Lanka," *Science*, vol. 308, p. 1595; Elliot, op. cit., 2005.
12. Seismologists at Sri Lanka's only seismic monitoring station, at Pallekele, Kandy, knew of the earthquake off Sumatra within minutes of its initial shaking at the epicenter, but they thought it was too far away for a tsunami to reach Sri Lanka. They were not tasked to issue warnings (and did not), and they did not have the capability to interpret the seismic data in terms of the earthquake's size. The seismic monitoring station at Pallekele was

maintained for the U.S. Geological Survey and the international community by the Geological Survey and Mines Bureau of Sri Lanka. See: Bhattacharje, Yudhijite, 2005. "In Wake of Disaster, Scientists Seek Out Clues to Prevention," *Science*, vol. 307, pp. 22–23.

13. For this and the following two paragraphs, see: Wijetunge, op. cit., 2009a; Rohde, David, 2005. "In a Corner of Sri Lanka, Devastation and Divisions," *New York Times*, Dec. 31; Tucker, Neely, 2005. "Cloudy Future, as the Rains Fall, Tsunami Survivors Look toward a Harsh Tomorrow," *Washington Post*, Jan. 6; Lancaster, John, 2004. "For Rich and Poor, Waves Bring Ruin," *Washington Post*, Dec. 29.
14. Rohde, op. cit., 2005.
15. The story of Dayalan Sanders and the Sri Lankan orphans is told in: Lancaster, John, 2004. "Outracing the Sea, Orphans in His Care," *Washington Post*, Dec. 30.
16. At Patanangala Beach a tsunami wave over fifty feet high crashed over a beach restaurant and killed forty people, some sitting in jeeps waiting to tour the park. Many large trees were broken or uprooted and carried inland by the water almost a mile. The sea inundated the land at fifteen locations where there were no sand dunes or where there were river or lagoon outlets. At the edge of the park, at the Yala Safari Game Lodge, a large sand dune had been removed to improve the view toward the ocean. As a result the tsunami flattened the lodge, killing many guests. See: Fernando, P., E.D. Wikramanayake, and J. Pastorini, 2006. "Impact of Tsunami on Terrestrial Ecosystems of Yala National Park, Sri Lanka," *Current Science*, vol. 90, no. 11, pp. 1531–1534; "Life in Sri Lanka beyond the Beach," *Times* (London), Mar. 26, 2005; Highfield, op. cit., 2005.
17. The 10,000 people who died along the southern coast were out of a population that was 50 percent larger than the population of the east coast, where almost 15,000 died. See: Wijetunge, op. cit., 2009b.
18. Colombo suffered some damage but had relatively few casualties (seventy-nine) from waves that arrived later and lasted longer (from 9:52 A.M. local time through the rest of the day), but none of which compared in height to that second wave farther south in Peraliya. In fact, the largest wave at Colombo was the sixth wave, which came three and a half hours after the first wave. Here the situation might have been complicated by the tsunami reflecting when it hit the continental shelf off India, and some have suggested a reflection off the sudden depth change next to the Maldives, islands farther west. See: Pattiaratchi, C.B., and E.M.S. Wijeratne, 2009. "Tide Gauge Observations of 2004–2007 Indian Ocean Tsunamis from Sri Lanka and Western Australia," *Pure and Applied Geophysics*, vol. 166, pp. 233–258.
19. Wijetunge, J.J., 2005. "Field Measurements of the Extent of Inundation in Sri Lanka Due to the Indian Ocean Tsunami of 26 December 2004," *International Symposium Disaster Reduction on Coasts*, Scientific-Sustainable-Holistic-Accessible, Nov. 14–16, Monash Univ., Melbourne, Australia. See also: Steele, John, 2004. "One Train, More Than 1,700 Dead," *Guardian* (UK), Dec. 29; Tucker, Neely, 2005. "Village of Lost Souls, the Waters Recede along with Hope," *Washington Post*, Jan. 3.
20. See references in note 15. Two hours and eighteen minutes after the earthquake a small tsunami wave reached Cocos Island, a thousand miles south of the epicenter, but the data from the tide gauge did not reach the Pacific Tsunami Warning Center for another two hours. The island had no continental shelf, and only a little tsunami energy traveled south, most going west, so it saw a tsunami of only 1.1 feet. See: Merrifield, M.A., et al., 2005. "Tide Gauge Observations of the Indian Ocean Tsunami, December 26, 2004," *Geophysical Research Letters*, vol. 32, L09603.
21. Soetjipto, Tomi, 2004. "Strong Quake Hits Indonesia, at Least Nine Killed," Reuters, Dec. 26.
22. Gupta and Ranjan, op. cit., 2004; "First Tsunami Alert Lost in Indian Bureaucracy," Agence France-Presse, Dec. 30.

23. The Indian Meteorological Department (IMD) had its own seismographs, including one at Port Blair in the Andamans, 175 miles north of Car Nicobar, but that seismograph was an older analog model and gave faulty readings for an earthquake this large. IMD digital stations at Chennai, Vishakhapatnam, and Kolkata began receiving seismic signals minutes after the earthquake. IMD calculated the epicenter's location and the earthquake's magnitude. Since the earthquake's location was far from India, they felt they were under no obligation to notify the government's Crisis Management Group. No one was thinking of tsunamis. At 8:00 A.M. Indian time, an hour and a half after the quake, officials from the National Geophysical Research Institute in Hyderabad assured the media that the Sumatran earthquake posed no threat to India. At 8:25 A.M. IMD talked to its office in Port Blair, which apparently told them only of damage to their buildings from the earthquake, not mentioning a tsunami. At 8:54 A.M., nine minutes before the tsunami would hit Chennai, IMD finally decided to send a communiqué to the Crisis Management Group and to the residence of the Science and Technology minister (but they sent it to the wrong person). Members of IMD and the Science and Technology Department all learned of the tsunami the same way everyone else in India learned about it—on television. See: Gupta and Ranjan, op. cit., 2004; "First Tsunami Alert...," op. cit., 2004. See also: "Govt Not to Join World Tsunami Warning System, to Strengthen Its Own Network," *Tribune* (Chandigarh, India), Dec. 29, 2004; Bhattacharje, op. cit., 2005.
24. Chennai was hit at 9:05 A.M. local Indian Standard Time (IST), which was 9:35 A.M. Sri Lankan Time (SLT). Back in 2004 the time zone for Sri Lanka was six hours after UTC, while for India it was five and a half hours after UTC. In 2006 Sri Lanka changed its time zone to five and a half hours after UTC to match India.
25. The tsunami wave hitting Chennai had come almost 800 miles directly eastward from the ruptured fault west of North Andaman Island. The wave hitting Vishakapatnam had traveled northwestward over a somewhat shorter distance but at a slightly slower speed due to the shallower water. See: McFadden, Robert D., 2004. "Walls of Water Sweeping All in Their Path: Families, Communities, Livelihoods," *New York Times*, Dec. 27; Rajaraman, R., S.J. Winston, T.S. Murty, H. Achyuthan, and N. Nirupama, 2006. "Numerical Simulation of Tsunamis on the Tamil Nadu Coast of India," *Marine Geodesy*, vol. 29, pp. 167–178; Sundar, V., S.A. Sannasiraj, K. Murali, and R. Sundaravadivelu, 2007. "Runup and Inundation along the Indian Peninsula, Including the Andaman Islands, Due to Great Indian Ocean Tsunami," *Journal of Waterway, Port, Coastal, and Ocean Engineering, ASCE*, vol. 133, no. 6, pp. 401–413.
26. Tharangambadi is in the district of Nagapattinam in the state of Tamil Nadu, approximately forty miles south of Chennai. The wave arrived twelve minutes later than at Chennai, because of the wider continental shelf in front of Tharangambadi.
27. Waldman, Amy, 2004. "Motherless and Childless, an Indian Village's Toll," *New York Times*, Dec. 31. See also: Sheth, A., S. Sanyal, A. Jaiswal, and P. Gandhi, 2006. "Effects of the December 2004 Indian Ocean Tsunami on the Indian Mainland," *Earthquake Spectra*, vol. 22, no. S3, pp. S435–S473.
28. Lakshmi, Rama, 2004. "The Water Has Eaten My Child," *Washington Post*, Dec. 28. See also: "India: Mass Graves for Hundreds of Children," *Independent* (UK), Dec. 28, 2004; Waldman, Amy, 2004. "In Drowned Village, Grim Searches, Quick Burials," *New York Times*, Dec. 31.
29. Raman, Sunil, 2005. "Tsunami Villagers Give Thanks to Trees," *BBC News*, Feb. 16.
30. Thousands of worshipers had been there for Christmas, and the following morning many were on the holy beach after having attended Mass. The first tsunami wave (at least fifteen feet high) struck the shore at around 9:20 A.M., followed by four more waves over the next forty minutes. The waves rushed up the main road from the beach and also

up the river, which was even closer to the basilica, reaching more than a half mile inland. See: Sheth et al., op. cit., 2006.

31. There were two different hydrodynamic effects at work here, *refraction* and *diffraction*. We have already seen the effects of refraction. Because of the seafloor sloping up toward the island, the part of the tsunami in deeper water traveled faster than the part in shallower water, and the wave front bent around, the energy following the coast and eventually reaching India. Diffraction does not involve differences in water depth, but is simply a wave front bending around an object in the middle of the ocean (namely, not leaving a long shadow behind the island).
32. At Kanyakumari, the southernmost point of mainland India, the tsunami waves had refracted around Sri Lanka, crossed a deep section of ocean, partially reflected at the edge of the continental shelf, and then propagated over a very wide shelf. A number of large waves arrived there over a two-hour period. The first few large waves produced dramatic splashes at the famous Swami Vivekananada Rock, which that Sunday morning had a thousand tourists on it. They had come by ferry to see the temple on this island of solid rock. Next to the island was another large rock island on which was the ninety-five-foot-tall Tiruvalluvar statue. The larger tsunami waves topped the thirty-eight-foot-high pedestal on which the statue stood and sent rockets of spray up to the shoulders of the statue. See: Krishnakumar, R., 2005. "An Encounter at Kanyakumarki," *Frontline*, vol. 22, no. 2. See also: Kumar, P.S. Suresh, 2004. "400 Die, Thousands Homeless in Kanyakumari," *Hindu* (India), Dec. 27; Bapat, Arun, and Tad Murty, 2008. "Field Survey of the December 26, 2004 Tsunami at Kanyakumari, India," *Science of Tsunami Hazards*, vol. 27, no. 3, pp. 72–86.
33. Highfield, op. cit., 2005.
34. The holidays occurring on that weekend in countries around the Indian Ocean included: for Hindus and Buddhists it was a Full-Moon Day; for Christians it was the day after Christmas (Dec. 25); for Kenyans, Tanzanians, and Australians it was Boxing Day (Dec. 26); and for South Africans it was Day of Goodwill (Dec. 26). Some have referred to the December 26, 2004, tsunami as the "Boxing Day Tsunami," but this is a very inappropriate name, since none of the countries with large death tolls celebrate that holiday. Of the twelve countries around the Indian Ocean that suffered casualties from the tsunami, only Kenya and Tanzania have Boxing Day as a legal holiday.
35. See references in note 15 in Chapter 8.
36. Elliot, op. cit., 2005.
37. The revised estimate of 8.9 for the earthquake magnitude used Harvard's Centroid Moment Tensor solution. The calculation had been carried out automatically and was ready two hours after the earthquake, but no seismologist had been there on a Sunday to check the readout. See references in note 15 in Chapter 8. The 9.3 estimate came from measuring the Earth's free oscillations caused by the earthquake, from: Stein, S., and E.A. Okal, 2005. "Speed and Size of the Sumatra Earthquake," *Nature*, vol. 434, pp. 581–582. The scientific consensus now seems to be 9.2.
38. Titov, V., A.B. Rabinovich, H.O. Mofjeld, R.E. Thomson, and F.I. González, 2005. "The Global Reach of the 26 December 2004 Sumatra Tsunami," *Science*, vol. 309, pp. 2045–2048. Figure 9.1 shows the arrival times of the leading tsunami wave of December 26, 2004, at different locations around the Indian Ocean. This figure is a modified version of a map produced by a hydrodynamic tsunami model. The original map was obtained from the National Geophysical Data Center, in NOAA's National Environmental Satellite, Data, and Information Service. It can be found at: www.ngdc.noaa.gov/hazard/icons/2004_1226.jpg.
39. Fritz, H.M., C.E. Synolakis, and Brian G. McAdoo, 2005. "Maldives Field Survey after the December 2004 Indian Ocean Tsunami," *Earthquake Spectra*, vol. 22, no. S3, pp. S137–S154. See also: Nagarajan et al., op. cit., 2006; Merrifield et al., op. cit., 2005; Lay, T., et al., 2005. "The Great Sumatra-Andaman Earthquake of 26 December 2004," *Science*, vol. 308, pp. 1127–1133.

40. Diego Garcia was about 1,800 miles southeast of the epicenter. Oscillations due to the tsunami showed up on a water level gauge there with a maximum height of 1.8 feet but were not noticed by people by the shore. The Pacific Tsunami Warning Center had contacted the U.S. Pacific Command (PACOM) in Hawaii to advise them of the greater earthquake magnitude and the possibility of tsunamis in the western Indian Ocean. An hour and a half later PACOM indicated that they had not observed a destructive tsunami at Diego Garcia. Rodrigues is part of the island nation of Mauritius, whose main island is 370 miles farther west (and about 550 miles east of Madagascar). Probably because of the protective reef system that surrounds much of the island, there was only a ten-foot run-up at a few locations and some moderate flooding in Port Mathurin. See: Nagarajan, op. cit., 2006. See also: Okal, E.A., A. Sladen, and E.A.-S. Okal, 2006. "Rodrigues, Mauritius, and Réunion Islands Field Survey after the December 2004 Indian Ocean Tsunami," *Earthquake Spectra*, vol. 22, no. S3, pp. S241–S261.
41. The Seychelles are a nation of 115 islands about 900 miles southeast of the closest African coast. The tsunami arrived later than at Mauritius because of the shallow water on the extensive submarine plateau on which the Seychelles are located. The tsunami registered as a 3.6-foot oscillation on the water level gauge there, but 12-foot waves were witnessed by some. The Seychelles had a National Disaster Committee, which immediately set up a base of operation at the Police Command Centre in the capital of Victoria. The committee had the Seychelles Broadcasting Corporation send out a national alert, which the public received before the waves arrived. Hundreds of families had their homes damaged, but only two people died. See: "How Kenya, Seychelles Avoided Tsunami Disaster," Afrol News, Jan. 3, 2005. See also: Merrifield et al., op. cit., 2005.
42. Salalah on the coast of Oman is midway between the entrances to the Persian Gulf and the Red Sea. The tsunami wave was only a foot high on the water level gauge, but the third wave was observed elsewhere to be five feet high, and the highest run-up was ten feet. Its strong currents caused a 900-foot container ship to break its mooring in the Port of Salalah, after which it was caught in a system of eddies for several hours. See: Okal, E.A., H.M. Fritz, P.E. Raad, C. Synolakis, Y. Al-Shijbi, and M. Al-Saifi, 2006. "Oman Field Survey after the December 2004 Indian Ocean Tsunami," *Earthquake Spectra*, vol. 22, no. S3, pp. S203–S218.
43. Socotra Island is east of the entrance to the Red Sea and about 1,100 miles north of the Seychelles. See: Fritz, H.M., and E.A. Okal, 2008. "Socotra Island, Yemen: Field Survey of the 2004 Indian Ocean Tsunami," *Natural Hazards*, vol. 46, pp. 107–117.
44. About fifteen hundred buildings in Somalia were destroyed or damaged, forcing thousands to relocate. The tsunami also broke up some rusted containers of radioactive and hazardous waste that had been dumped offshore (a previous president having apparently received money from European firms for the dumping). Because of a long civil war, Somalia had had no real government since 1991. Thus, even if some "official" in Somalia received the same warning that Kenya did, the people certainly did not receive it. See: Clayton, Jonathan, 2005. "Somalia's Secret Dumps of Toxic Waste Washed Ashore by Tsunami," *Times* (London), Mar. 4. See also: Fritz and Okal, op. cit., 2008.
45. Kurian, N.P., M. Baba, K. Rajith, N. Nirupama, and T.S. Murty, 2006. "Analysis of the Tsunami of December 26, 2004, on the Kerala Coast of India (Parts I, II, and III)," *Marine Geodesy*, vol. 29, pp. 265–270. See also: Nagarajan, B., et al., 2006. "The Great Tsunami of 26 December 2004: A Description Based on Tide-Gauge Data from the Indian Subcontinent and Surrounding Areas," *Earth, Planets and Space*, vol. 58, pp. 211–215; Merrifield et al., op. cit., 2005.
46. See note 34 regarding Boxing Day.
47. See: Blomfield, Adrian, 2004. "Evacuation from Beaches Cut Deaths by Hundreds in Kenya," *Telegraph* (UK), Dec. 29; "How Kenya, Seychelles . . . ," op. cit., 2005; Revkin, op. cit., 2004.

48. Blomfield, op. cit., 2004. See also: Majtenyl, Cathy, 2004. "Waves Kill at Least 10 off Tanzania Coast," *VOA News*, Nairobi, Dec. 28; Merrifield et al., op. cit., 2005.
49. By this point, successive crests were forty-two minutes apart. See: Joseph, A., et al., 2006. "The 26 December 2004 Sumatra Tsunami Recorded on the Coast of West Africa," *African Journal of Marine Science*, vol. 28, pp. 705–712.
50. Titov et al., op. cit., 2005; Dragani, W.C., E.E. D'Onofrio, W. Grismeyer, and M.E. Fiore, 2006. "Tide Gauge Observations of the Indian Ocean Tsunami, December 26, 2004, in Buenos Aires Coastal Waters, Argentina," *Continental Shelf Research*, vol. 26, pp. 1543–1550. See also: Candella, R.N., A.B. Rabinovich, R.E. Thomson, 2008. "The 2004 Sumatra Tsunami as Recorded on the Atlantic Coast of South America," *Advances in Geoscience*, vol. 14, pp. 117–128.
51. On the other side of the North Atlantic, at Concarneau, France, forty-five miles south of Brest, tsunami amplitude reached 0.8 feet. See: Rabinovich, A.B., and R.E. Thomson, 2006. "The Sumatra Tsunami of 26 December 2004 as Observed in the North Pacific and North Atlantic Oceans," *Surveys in Geophysics*, vol. 27, pp. 647–677.
52. The 15,000-mile distance assumes the tsunami took the longer-distance route predicted by most hydrodynamic models, that is, following three successive submarine ridges, the Southeast Indian Ridge, the Pacific-Antarctic Ridge, and the East Pacific Rise. See: Titov et al., op. cit., 2005; Rabinovich and Thomson, op. cit., 2006.
53. Williamson, Lucy, 2009. "Tsunami Museum Opens in Indonesia," *BBC News*, Feb. 23; "Indonesia Opens Tsunami Museum," Associated Press, Feb. 23, 2009.
54. The peace talks were facilitated by former Finnish president Martti Ahtissari, who was awarded the Nobel Peace Prize in 2008 for this effort and for resolving other international conflicts. See: Billon, Philippe Le, and A. Waizennegger, 2007. "Peace in the Wake of Disaster? Secessionist Conflicts and the 2004 Indian Ocean Tsunami," *Transactions of the Institute of British Geographers*, vol. 32, no. 3, pp. 411–427. See also: Gaillard et al., op. cit., 2008b.
55. Silva, Kalinga T., 2009. "'Tsunami Third Wave' and the Politics of Disaster Management in Sri Lanka," *Norwegian Journal of Geography*, vol. 63, pp. 61–72. See also: Hyndman, Jennifer, 2009. "Siting Conflict and Peace in Post-Tsunami Sri Lanka and Aceh, Indonesia," *Norwegian Journal of Geography*, vol. 63, pp. 89–96. One difference between Aceh and Sri Lanka was that in Sri Lanka all three larger ethnic groups (Sinhalese, Tamils, and Muslims) were hurt by the tsunami, and although this initially created a sense of common tragedy, eventually it led to intense competition for aid, with the Tamils and Muslims on the harder-hit east coast feeling that they received less help than the majority Sinhalese on the less damaged west coast.
56. Some scientists considered the 2005 earthquake an aftershock of the December 2004 earthquake, but its 8.6 magnitude made it one of the strongest quakes in recent history, and it had eight significant aftershocks of its own with magnitudes up to 6.0. The tsunami was smaller than anticipated, apparently because the vertical movement of the seafloor due to the earthquake took place in shallower water than the earthquake back in December 2004. Only moderate damage was caused by the ten-foot wave that hit the seaport halfway down the east coast of Simeulue Island, by the six-and-a-half-foot wave that hit Nias Island, and by the eight-foot wave that hit the Aceh coast on Sumatra. Tsunami heights recorded farther from the epicenter were generally less than a foot. The Pacific Tsunami Warning Center bulletins for March 28, 2005, were found in the PTWC message archive, www.prh.noaa.gov/ptwc/. See also: "Tsunami Alert Shows Nations Better Prepared," Reuters, Mar. 29, 2005; Geist, E.L., S.L. Bilek, D. Arcas, and V.V. Titov, 2006. "Differences in Tsunami Generation between the December 26, 2004 and March 28, 2005 Sumatra Earthquakes," *Earth, Planets and Space*, vol. 58, pp. 185–193.

57. Aglionby, John, 2006. "Official Failed to Pass on Tsunami Warning," *Guardian* (UK), July 18. See also: Ammon, C.J., H. Kanamori, T. Lay, and A.A. Velasco, 2006. "The 17 July 2006 Tsunami Earthquake," *Geophysical Research Letters*, vol. 33, L24308.
58. The six bulletins put out by the Pacific Tsunami Warning Center for the Sumatra tsunami on September 16, 2007, were found in the PTWC message archive (www.prh.noaa.gov/ptwc/). See also: Borrero, J.C., R. Weiss, E.A. Okal, R. Hidayat, D. Suranto, D. Arcas, and V.V. Titov, 2009. "The Tsunami of 2007 September 12, Bengkulu Province, Sumatra, Indonesia: Post-tsunami Field Survey and Numerical Modeling," *Geophysical Journal International*, vol. 178, no. 1, pp. 180–194; Normile, Dennis, 2007. "Tsunami Warning System Shows Agility—And Gaps in Indian Ocean Network," *Science*, vol. 317 (Sept. 23), p. 1661.
59. "How the Indian Ocean Tsunami Warning System Works," Reuters, Oct. 28, 2009. See also: US IOTWS, *US Indian Ocean Tsunami Warning System (IOTWS) Program Final Progress Report, August 1, 2005 to March 31, 2008*, US Agency for International Development, US IOTWS Program Document no. 32-IOTWS-08, 61 pages.
60. NOAA, 2009. "Indian Ocean Tsunami Tests NOAA New Forecast System," Pacific Tsunami Warning Center media information, NWS, NOAA, August 2009, www.noaa.gov/features/03_protecting/tsunamiforecast.html.
61. About two hundred more seismographs have been installed since 2004. See: Yuan, X., R. Kind, and H.A. Pedersen, 2005. "Seismic Monitoring of the Indian Ocean Tsunami," *Geophysical Research Letters*, vol. 32, L15308.
62. Le Pichon, A., et al., 2005. "Infrasound Associated with 2004–2005 Large Sumatra Earthquakes and Tsunami," *Geophysical Research Letters*, vol. 32, L19802.
63. The tsunami produces internal gravity waves in the atmosphere just above the ocean. These moving oscillations of the density layers in the atmosphere eventually (in about fifteen minutes) reach the ionosphere (the uppermost layer of the atmosphere), where its electron density is caused to fluctuate, which affects radio wave propagation and is thus detectable with GPS. See: Occhipinti, G., P. Lognonné, E.A. Kherani, and H. Hébert, 2006. "Three-Dimensional Waveform Modeling of Ionospheric Signature Induced by 2004 Sumatra Tsunami," *Geophysical Research Letters*, vol. 33, L20104.
64. The 2004 tsunami was detected by hydrophones at Diego Garcia. These hydrophones were designed to detect high-frequency sound waves due to explosions (e.g., a nuclear bomb), but the hydrophones could be modified to better detect low-frequency waves. See: Okal, E.A., J. Talandier, and D. Reymond, 2007. "Quantification of Hydrophone Records of the 2004 Sumatra Tsunami," *Pure and Applied Geophysics*, vol. 164, pp. 309–323.
65. See note 10.
66. Personal communication from Stuart Weinstein, Deputy Director, Pacific Tsunami Warning Center (PTWC). See also: *Tsunami Bulletin Number 001*, May 9, 2010, at 0608Z, from the message archive at: www.prh.noaa.gov/ptwc/.
67. Kanamori, H., and L. Rivera, 2008. "Source Inversion of W Phase: Speeding up Seismic Tsunami Warning," *Geophysics Journal International*, vol. 175, pp. 222–238; Okal, E.A., 2008. "The Generation of T Waves by Earthquakes," *Advances in Geophysics*, vol. 49, pp. 1–65; Chew, S.-H., and K. Kuenza, 2009. "Detecting Tsunamigenesis from Undersea Earthquake Signals," *Journal of Asian Earth Sciences*, vol. 36, pp. 84–92; Weinstein, S.A., and R.R. Lundgren, 2008. "Finite Fault Modeling in a Tsunami Warning Center Context," *Pure and Applied Geophysics*, vol. 165, pp. 451–474; Salzberg, D.H., 2008. "A Hydroacoustic Solution to the Local Tsunami Warning Problem," American Geophysical Union, Fall Meeting, 2008, abstract #OS43D-1325. The work referenced above by Kanamori and Rivera and by Salzberg is being tested at the Pacific Tsunami Warning Center.
68. Blewitt, G., W.C. Hammond, C. Kreemer, H.-P. Plag, S. Stein, and E. Okal, 2009. "GPS for Real-Time Earthquake Source Determination and Tsunami Warning Systems,"

Journal of Geodesy, vol. 83, pp. 335–343. See also: Song, Y. Tony, 2007. "Detecting Tsunami Genesis and Scales Directly from Coastal GPS Stations," *Geophysical Research Letters*, vol. 34, L19602.

69. Arcas, Diego, and Vasily Titov, 2006. "Sumatra Tsunami: Lessons from Modeling," *Surveys in Geophysics*, vol. 27, pp. 679–705. See also: Gisler, Galen R., 2008. "Tsunami Simulations," *Annual Review of Fluid Mechanics*, vol. 40, pp. 71–90; Synolakis, Costas E., and Eddie N. Bernard, 2006. "Tsunami Science Before and Beyond Boxing Day 2004," *Philosophical Transactions of the Royal Society*, A, vol. 364, pp. 2231–2265.
70. A number of instruments can be used for above-water potential landslides such as tiltmeters, GPS receivers, extensometers, borehole inclinometers, and microseism monitors, with some versions of these usable on submarine slumps. See: Brune, Sascha, 2009. "Landslide Generated Tsunamis—Numerical Modeling and Real-time Prediction," PhD diss., Dept. of Mathematics and Natural Sciences, Univ. of Potsdam, 83 pages. See also: Prior, D.B., et al., 1989. "Storm Wave Reactivation of a Submarine Landslide," *Nature*, vol. 341, pp. 47–50.
71. Niu, F., P.G. Silver, T.M. Daley, X. Cheng, and E.L. Majer, 2008. "Preseismic Velocity Changes Observed from Active Source Monitoring at the Parkfield SAFOD Drill Site," *Nature*, vol. 454, pp. 204–208. See also: Cyranoski, David, 2007. "In the Zone," *Nature*, vol. 449, pp. 278–280; Campin, S., Y. Gao, and S. Peacock, 2008. "Stress-Forecasting (Not Predicting) Earthquakes: A Paradigm Shift?" *Geology*, vol. 36, no. 5, pp. 427–430; Uyeda, S., T. Nagao, and M. Kamogawa, 2009. "Short-Term Earthquake Prediction: Current Status of Seismo-electromagnetics," *Tectonophysics*, vol. 470, pp. 205–213.
72. "Villagers Damage Tsunami Siren after Snafu. False Alarm Triggers Panic in Region Hardest Hit by 2004 Killer Waves," Associated Press, June 7, 2007.
73. Perry, Michael, 2009. "Samoa Tsunami Toll May Exceed 100, Hundreds Injured," Reuters, Sept. 30. See also: interview on *AccuWeather* with Mase Akapo, head NWS meteorologist in Pago Pago, American Samoa, Oct. 2, 2009; Synolakis, Costas, 2009. "Being Ready for the Big Wave," *Wall Street Journal*, Oct. 4; U.S. Geological Survey, 2009. "Notes from the Field—American Samoa Run-up Heights (Measured October 5–6, 2009)," Oct. 23.
74. Martina Maturana had felt the tremors and seen the strange motion of the sea, so she rang the village bell to warn everyone, who then ran for higher ground just before their village was destroyed. See: Peachey, Paul, 2010. "How a 12-Year Old Girl Saved Her Chilean Island from Catastrophe," *Independent* (UK), Mar. 4.

CHAPTER 10

1. Kiladis, G.N., and H.F. Diaz, 1986. "An Analysis of the 1877–78 ENSO Episode and Comparisons with 1982–83," *Monthly Weather Review*, vol. 114, pp. 1035–1047.
2. Gootenberg, Paul, 1989. *Between Silver and Guano*, Princeton Univ. Press, Princeton, NJ, 234 pages.
3. The tragic human loss in India and China was reported to the world by John Russell Young, a reporter with the *New York Herald* who accompanied ex-President Ulysses S. Grant during his diplomatic journey across the Asian continent during the 1877–1879 El Niño. See: Young, John Russell, 1879. *Around the World with General Grant*, republished in 2002, ed. Michael Fellman, Johns Hopkins Univ. Press, Baltimore, MD, 449 pages.
4. Davis, Mike, 2001. *Late Victorian Holocausts—El Niño Famines and the Making of the Third World*, Vero, New York, 464 pages.
5. Jarvis, C.S., 1935. "Flood-Stage Records of the River Nile," *Transactions, American Society of Civil Engineers*, vol. 101, pp. 1012–1071.

6. This is from an overall average of winter temperatures for the entire United States; the East Coast had a somewhat warmer winter than normal.
7. Davis, op. cit. There appears to be evidence that in India the British government did not use their railroads or their stores of rice to help prevent the starvation of Indians, while at the same time the British public was trying to save lives through charitable donations.
8. The National Centers for Environmental Prediction (NCEP) is in NOAA's National Weather Service. See: Barnston, A.G., et al., 1999. "NCEP Forecasts of the El Niño of 1997–98 and Its U.S. Impacts," *Bulletin of the American Meteorological Society*, vol. 80, no. 9, pp. 1829–1852.
9. Changnon, Stanley A., ed., 2000. *El Niño, 1997–1998—The Climate Event of the Century*. Oxford Univ. Press, 215 pages. The 1997–1998 El Niño caused $1.1 billion in losses versus the 1982–1983 El Niño's $2.2 billion.
10. Chen, D., and M.A. Cane, 2008. "El Niño Prediction and Predictability," *Journal of Computational Physics*, vol. 227, pp. 3625–3640.
11. The warming of the globe due to greenhouse gases (water vapor, carbon dioxide, methane, and other gases) has always occurred and is in itself a good thing, being necessary to keep the Earth warm enough for living things to survive. Without the greenhouse effect the Earth would be too cold for life as we know it. The question, though, is whether there can be too much global warming, in this case due to too much carbon dioxide (and other greenhouse gases) in the atmosphere. The global air temperature of the Earth can change; for example, it has been rising since the last ice age. The concern about global warming is about the higher global temperatures predicted to occur due to an enhanced greenhouse effect as people add more carbon dioxide to the atmosphere and the effect of those higher temperatures on our climate and on life on our planet. It is a very complex and chaotic problem.
12. This is essentially the same thing as being able to predict when the Earth will move into an ice age or when an ice age will end, neither of which has been convincingly accomplished.
13. The gas produced by phytoplankton that affects cloud formation is dimethyl sulfide.
14. Walker, Gilbert, 1923. "Correlation in Seasonal Variations of Weather VIII: A Preliminary Study of World Weather," *Memoirs of the Indian Meteorological Department*, vol. 24, pt. 4, Calcutta, pp. 75–131. By the beginning of the twentieth century there existed a worldwide network of weather observation stations. It would take almost another century before oceanographers would make similar progress for oceanographic observations.
15. Bjerknes, Jacob, 1961. "'El Niño' Study Based on Analysis of Ocean Surface Temperatures 1935–57," *Bulletin*, Inter-American Tropical Tuna Commission, vol. 5, no. 3, pp. 219–303.
16. Wyrtki, Klaus, 1973. "Teleconnections in the Equatorial Pacific Ocean," *Science*, vol. 180, no. 4081, pp. 66–68.
17. This slow movement of warm water from the western Pacific to the eastern Pacific along the equator sometimes appears as a type of wave motion, the equatorial Kelvin wave, kept along the equator by the effect of the Earth's rotation (the Coriolis force). Since the Coriolis force deflects a current to the right in the Northern Hemisphere and to the left in the Southern Hemisphere, deflections on both sides of the equator are brought back toward the equator when the current flows toward the east.
18. Cushman, Gregory T., 2004. "Choosing between Centers of Action—Instrument Buoys, El Niño, and Scientific Internationalism in the Pacific, 1957–1982," in *The Machine in Neptune's Garden*, ed. H.M. Rozwadowski and D.K. van Keuren, pp. 133–182.
19. See note 17.
20. On a much longer time scale continental drift has affected the relative distributions of land and ocean, which affect ocean currents as well as the amount of near-polar land on which ice sheets can form, both important in very long-term climate change.

21. Global warming can cause regional differences in long-term sea level change due to differences in long-term changes in water temperature or in wind speed and direction for different regions of the globe. For an introduction to sea level's relationship to climate change, see: Parker, Bruce B., 1991. "Sea Level as an Indicator of Climate and Global Change," *Marine Technology Society Journal*, vol. 25, no. 4, pp. 13–24.
22. Bell, Robin E., 2008. "The Unquiet Ice," *Scientific American*, vol. 298, no. 2, pp. 60–67.
23. Pfeffer, W.T., J.T. Harper, and S.O. Neet, 2008. "Kinematic Constraints on Glacier Contributions to 21st-Century Sea-Level Rise," *Science*, vol. 321, pp. 1340–1343.
24. Velicogna, I., 2009. "Increasing Rates of Ice Mass Loss from the Greenland and Antarctic Ice Sheet Revealed by GRACE," *Geophysical Research Letters*, vol. 36, doi:10.1029/2009GL040222.
25. Overpeck, J.T., B.L. Otto-Bliesner, G.H. Miller, D.R. Muhs, R.B. Alley, and J.T. Kiehl, 2006. "Paleoclimatic Evidence for Future Ice-Sheet Instability and Rapid Sea-Level Rise," *Science*, vol. 311, pp. 1747–1750.
26. Barnett, T.P., J.C. Adam, and D.P. Lettenmaier, 2005. "Potential Impacts of a Warming Climate on Water Availability in Snow-Dominated Regions," *Nature*, vol. 438, pp. 303–309.
27. Ocean acidification, in which seawater pH is lowered, will, it is believed, reduce calcification and thus lead to the weakening of coral skeletons and shells of marine organisms such as clams and certain phytoplankton. This could have wide-ranging effects such as reduced food production in the world's oceans and reduced coastal protection by coral reefs. See: Cooley, S.R., H.L. Kite-Powell, and S.C. Doney, 2009. "Ocean Acidification's Potential to Alter Global Marine Ecosystem Services," *Oceanography*, vol. 22, no. 4, pp. 172–179.
28. Bullis, Kevin, 2010. "The Geoengineering Gambit," *Technology Review*, vol. 113, no. 1, pp. 50–56.
29. Hergerl, Gabriele C., and Susan Solomon, 2009. "Risks of Climate Engineering," *Science*, vol. 325, no. 5943, pp. 955–956.
30. Ross, Andrew, and H. Damon Mathews, 2009. "Climate Engineering and the Risk of Rapid Climate Change," *Environmental Research Letters*, vol. 4, pp. 1–6, doi:10.1088/1748—9326/4/4/045105.
31. The potential dangers of geoengineering techniques for cooling the Earth are not mentioned by Steven Levitt and Stephen Dubner in their best-selling *Super Freakonomics*, which otherwise provides an interesting and useful discussion of these proposed "solutions" to the global warming problem (2009, William Morrow, 270 pages).
32. Solomon, S., et al., 2007. *Fourth Assessment Report of the Intergovernmental Panel on Climate Change* (IPCC), Cambridge Univ. Press, Cambridge, UK.
33. In 1990 the Intergovernmental Oceanographic Commission (IOC) of UNESCO (United Nations Educational, Scientific and Cultural Organization) passed a resolution to establish GOOS. A year later IOC was joined by the World Meteorological Organization (WMO), the United Nations Environment Programme (UNEP), and the International Council for Science (ICSU) in setting up a GOOS support office and writing a development plan, which was implemented by the member nations of these four intergovernmental agencies.
34. Gould, W. John, 2003. "WOCE and TOGA—The Foundations of the Global Ocean Observing System," *Oceanography*, vol. 16, no. 4, pp. 24–30.
35. Carson, Rachel L., 1951. *The Sea around Us*, Oxford Univ. Press, New York, 230 pages, p. 216.

INDEX